Chemical Oceanography

VOLUME 6

2ND EDITION

Chemical Oceanography

Edited by

J. P. RILEY

and

R. CHESTER

Department of Oceanography,
The University of Liverpool, England

VOLUME 6

2ND EDITION

1976

ACADEMIC PRESS

LONDON NEW YORK SAN FRANCISCO

A Subsidiary of Harcourt Brace Jovanovich, Publishers

ACADEMIC PRESS INC. (LONDON) LTD.
24/28 Oval Road,
London NW1

United States Edition published by
ACADEMIC PRESS INC.
111 Fifth Avenue
New York, New York 10003

Library of Congress Catalog Card Number: 74–5679
ISBN: 0–12–588606–3

Printed in Great Britain by
PAGE BROS (NORWICH) LTD
NORWICH

Contributors to Volume 6

S. R. Aston, *Department of Environmental Sciences, The University, Lancaster, England*

S. E. Calvert, *Institute of Oceanographic Sciences, Wormley, Surrey, England*

R. Chester, *Department of Oceanography, The University, Liverpool, England*

Egon T. Degens, *Woods Hole Oceanographic Institution, Woods Hole, Massachusetts 02543, U.S.A.**

F. T. Manheim, *U.S. Geological Survey, Woods Hole, Massachusetts 02543, U.S.A.*†

Kenneth Mopper, *Woods Hole Oceanographic Institution, Woods Hole, Massachusetts 02543, U.S.A.*

N. Brian Price, *Woods Hole Oceanographic Institution, Woods Hole, Massachusetts 02543, U.S.A.*‡

* Present address: Geologisch-Paläontologisches Institut der Universität Hamburg, Von-Melle-Park 13, 2 Hamburg 13, Federal German Republic.

† Present address: Department of Marine Science, University of South Florida, St. Petersburg, Florida 33701, U.S.A.

‡ Present address: Grant Institute of Geology, University of Edinburgh, Edinburgh, Scotland.

Preface to the Second Edition

Rapid progress has occurred in all branches of Chemical Oceanography since the publication of the first edition of this book a decade ago. Particularly noteworthy has been the tendency to treat the subject in a much more quantitative fashion; this has become possible because of our much improved understanding of the physical chemistry of sea water systems in terms of ionic and molecular theories. For these reasons chapters dealing with sea water as an electrolyte system, with speciation and with aspects of colloid chemistry are now to be considered as essential in any up-to-date treatment of the subject. Fields of research which were little more than embryonic only ten years ago, for example sea surface chemistry, have now expanded so much that they merit separate consideration. Since the previous edition, there has arisen a general awareness of the potential threat to the sea caused by man's activities, in particular its use as a "rubbish bin" and a receptacle for toxic wastes. Although it was inevitable that there should be some over-reaction to this, there is real cause for concern. Clearly, it is desirable to have available reasoned discussions of this topic and also an examination of the role of the sea as a potential source of raw materials in view of the imminent exhaustion of many high grade ores; these subjects are treated in the second, third, fourth and seventh volumes.

Most branches of marine chemistry make use of analytical techniques; the number and range of these has increased dramatically over recent years. Consequently, it has been necessary to expand greatly and restructure the sections dealing with analytical methodology. These developments are extending increasingly into the very important and rapidly developing area of organic chemistry.

Many dramatic advances have taken place in all aspects of marine geochemistry during the last decade. Topics which have received increasing attention include the metalliferous sediments found at some active centres of sea floor spreading, the chemistry of interstitial waters, the formation of deep-sea carbonates and the chemistry and mineralogy of both atmospheric and sea water particulates. Many of the most important developments in the study of deep-sea sediments themselves have been linked to the Deep-Sea Drilling Project. This was initiated in 1968, and for the first time the entire length of the marine sedimentary column was sampled. The extent of these many advances has made it necessary to devote three volumes, i.e. the fifth, sixth and seventh, to various topics in marine geochemistry.

Both the range and accuracy of the physical constants available have increased since the first edition and a selection of tabulated values of these constants are to be found at the end of each of the first four volumes.

No attempt has been made to discuss Physical Oceanography except where a grasp of the physical concepts is necessary for a better understanding of the chemistry. For a treatment of the physical processes occurring in the sea the reader is referred to the numerous excellent texts now available on physical oceanography. Likewise, since the distribution of salinity in the sea is of greater relevance to the physical oceanographer and is well discussed in these texts, it will not be considered in the present volumes.

This series is not intended to serve as a practical handbook of Marine Chemistry, and if practical details are required the original references given in the text should be consulted. In passing, it should be mentioned that, although those practical aspects of sea water chemistry which are of interest to biologists are reasonably adequately covered in the "Manual of Sea Water Analysis" by Strickland and Parsons, there is an urgent need for a more general laboratory manual.

The editors are most grateful to the various authors for their helpful co-operation which has greatly facilitated the preparation of this book. They would particularly like to thank Messrs A. Dickson and M. Preston for their willing assistance with the arduous task of proof reading; without their aid many errors would have escaped detection. They would also like to acknowledge the courtesy of the various copyright holders, both authors and publishers, for permission to use tables, figures, and photographs. In conclusion, they wish to thank Academic Press, and in particular Mr. E. A. S. Cotton, for their efficiency and ready co-operation which has much lightened the task of preparing this book for publication.

Liverpool
January, 1976

J. P. Riley
R. Chester

CONTENTS

Chapter 32 *by* F. T. MANHEIM

Interstitial Waters of Marine Sediments

Chapter 33 *by* S. E. CALVERT

The Mineralogy and Geochemistry of Near-shore Sediments

Chapter 34 *by* R. CHESTER and S. R. ASTON

The Geochemistry of Deep-sea Sediments

Contents of Volume 1

Contents of Volume 2

Contents of Volume 3

Contents of Volume 4

Contents of Volume 5

Symbols and units used in the text

Concentration. There are several systems in common use for expressing concentration. The more important of these are the molarity scale (g molecules l^{-1} of solution = mol l^{-1}) usually designated by c_i, the molality scale (g molecules kg^{-1} of solvent* = mol kg^{-1}) designated by m_i and the mole fraction scale usually denoted by x_i, which is of more fundamental significance in physical chemistry. In each instance, the subscript i indicates the solute species; when i is an ion the charge is not included in the subscript unless confusion is likely to arise. Some other means of indicating the concentration are also to be found in the text, these include: g or mg kg^{-1} of solution (for major components), μg or ng l^{-1} or kg^{-1} of solution (for trace elements and nutrients) and μg-at l^{-1} of solution (for nutrients).

Activity. When an activity or activity coefficient is associated with a species, the symbols a_i and γ_i are used respectively regardless of the method of expressing concentration, where the subscript i has the significance indicated above. Further qualifying symbols may be added as superscripts and/or subscripts as circumstances demand. It is important to realize that the numerical values of the activity and activity coefficient depend on the standard state chosen. It should also be noted that since activity is a relative quantity it is dimensionless.

UNITS

Where practicable, SI units (and the associated notations) have been adopted in the text except where their usage goes contrary to established oceanographic practice.

LENGTH

Å	= Ångstrom	= 10^{-10} m
nm	= nanometre	= 10^{-9} m
μm	= micrometre	= 10^{-6} m
mm	= millimetre	= 10^{-3} m
cm	= centimetre	= 10^{-2} m
m	= metre	
km	= kilometre	= 10^{3} m
mi	= nautical mile (6080 ft)	= 1·85 km

* A common practice is to regard sea water as the solvent for minor elements.

WEIGHT

pg	= picogram	$= 10^{-12}$ g
ng	= nanogram	$= 10^{-9}$ g
μg	= microgram	$= 10^{-6}$ g
mg	= milligram	$= 10^{-3}$ g
g	= gram	
kg	= kilogram	$= 10^{3}$ g
ton	= metric ton	$= 10^{6}$ g

VOLUME

μl	= microlitre	$= 10^{-6}$ l
ml	= millilitre	$= 10^{-3}$ l
l	= litre	
dm^3	= litre	

CONCENTRATION

ppm	= parts per million (μg g^{-1} or mg l^{-1})
ppb	= parts per billion (ng g^{-1} or μg l^{-1})
μg-at l^{-1}	= μg atoms l^{-1} = (μg/atomic weight) l^{-1}

TIME

s	= second	h	= hour
ms	= millisecond	d	= day
min	= minute	yr	= year

ENERGY AND FORCE

J	= Joule	= 0·2390 cal
N	= Newton	$= 10^{5}$ dynes
W	= Watt	

General Symbols

a_x	activity of component x in solution
K	equilibrium constant
M_x	molarity of component x
m_x	molality of component x
P_G	partial pressure of gas G in solution
T	temperature in K
t	temperature in °C

Chapter 30

Chemical Diagenesis in Sediments

N. BRIAN PRICE

*Woods Hole Oceanographic Institution, Woods Hole, Massachusetts 02543, U.S.A.**

30.1. INTRODUCTION

Diagenesis may be defined as those processes which bring about alteration of a sediment. Within this definition sediments may be altered at, or close to, the sediment-sea water surface, or may suffer further alteration at greater depths during continued burial. The framing of any definition of diagenesis is difficult since the sediments are temporarily deposited while in transit from the source to the depositional sites, and during this time are susceptible to many changes including that of authigenic mineral formation. For the purpose of this chapter the changes taking place at the surface of, and within,

* Present Address: Grant Institute of Geology, University of Edinburgh, Edinburgh, Scotland.

sediments buried to a depth of one or two metres will be discussed. Only passing reference will be made to change taking place as a result of deep burial.

Chemical and biological transformations of substances contained in marine sediments which have been deposited under different environmental conditions are, to say the least, imperfectly known. This is particularly so in respect of the importance of biological systems within sediments. Observations on the differences in chemistry between sediments formed in anoxic (anaerobic) and in oxic (aerobic) environments have been known for some time. In addition, differences in the compositions of sediments forming in the deep-sea and those forming beneath coastal waters, or on continental shelves, are also well documented. However, less attention has been paid to environmental and chemical factors governing these changes. Geographical variations in the compositions of surface sediments in the oceans, and factors influencing changes occurring in them during burial, are in many ways better understood than those relating to sediments forming on continental shelves, fjords and semi-enclosed seas. Perhaps this is partly due to the relatively much greater area of the former. However, the significance of the latter is in all probability much greater, at least from a geological standpoint. The epicontinental seas that were so abundant during the Palaeozoic and other eras have left ample evidence of their importance in the deposition of neritic sediments.

Studies on diagenetic reactions taking place within sediments forming in deep and shallow waters are important not only from the geological point of view. It is perhaps more important to assess diagnetic reactions in the light of chemical fluxes between the sediments and the overlying sea water. The extent of the reactions between different elements contained in a sediment and those in the sediment interstitial water, as well as those contained in the overlying sea water, will provide much insight into understanding the importance of sediments as a control affecting the composition of seawater.

In many respects the shallow water marine sediments of continental shelves and platforms, which contain much organic matter and terrigenous detritus and thus may be inherently more susceptible to chemical and biological attack, could be a more useful medium for observing and understanding chemical changes than could deep-sea sediments. Hence, a greater stress will be laid on discussing chemical diagenesis of shallow water sediments. Nevertheless, many of the reactions occurring in these sediments can be translated to events happening in sediments from the deep-sea. However, certain distinctions between deep and shallow water sediments can be made which have some bearing on the degree and kinds of diagenesis within them. In the deep-sea much of the breakdown of particulate organic matter, which seems to be essential before many other diagenetic changes can proceed, takes place either at, or above, the surface of sediments. Thus, relative rates of diagenetic modification are likely to be much slower, and, consequently changes during the early period of burial are not easily discerned in deep-sea sediments.

Chemical diagenesis in sediments depends on a number of environmental factors all of which are directly or indirectly related, and it is difficult to attribute any change observed within sediments to any single factor. Total sediment accumulation rate, the amount and composition of organic matter within a sediment and its consumption by biological systems of different activity are predominant amongst these factors. Further, within anoxic systems the presence of dissolved sulphide, produced by bacterial activity on the dissolved sulphate in marine waters, not only influences metal sulphide formation, but to a large extent may control the equilibria of other mineral species.

The primary purpose of this work is not to review the published literature on chemical diagenesis, but to try to illustrate this process in the context of the above factors and others directly or indirectly associated with them. Without some prior knowledge of the probable reactions and the controls affecting them when a sediment is buried to shallow depths, it is difficult to place a meaningful interpretation on the changes occurring in sediments during deep burial.

30.2. Changes in Organic Constituents

30.2.1. Bacterial changes in sediments

Significant numbers of bacteria and allied organisms (for example: enzymes, actinomyces, yeasts, moulds and algae) are found in Recent sediments and seem capable of altering, perhaps biocatalytically, various substances, and of thus controlling the physico-chemical conditions within sediments. For this reason these organisms have a most important influence on chemical diagenesis in sediments.

There is considerable evidence of a correlation between the number and kinds of bacteria and the physico-chemical conditions within a sediment (Rittenberg, 1940; Zobell, 1942; Bordovskiy, 1965). Culture experiments have shown that by far the greatest bacterial population exists at, or near, the surface of both anoxic (Sorokin, 1964) and oxic (Zobell, 1964) sediments. For instance, Zobell (loc. cit.) measured far more (by several orders of magnitude) bacteria in the uppermost 5 cm of sediment, than in sediments buried at depths > 30 cm (Table 30.1). The marked decline in numbers with depth is most pronounced (Zobell, 1946; Kaplan and Rittenberg, 1963; Bordovskiy 1965), although occasionally certain buried sediment layers of unusually high content of organic matter can have large populations of bacteria (Zobell,

TABLE 30.1

Number of bacteria found per gram of wet marine sediment as a function of depth in

Sample position	32°26·4′N 117°41·3′W	33°03·3′N 117°25·5′W
Water depth (m)	950 m	431 m
Core depth (cm)	Bacteria per g	Bacteria per g
0–2·5	38 000 000	840 000
2·5–5·0	940 000	102 000
10–12·5	88 000	63 000
23–25·5	36 000	19 000
35·5–38·0	2 400	1 500
48–51	400	2 200
74–76	180	370
99–102	330	190
150–152	250	210
201–203	130	140
251–254	290	140

1964). The decrease in the rate of anaerobic bacterial sulphate reduction with depth has been shown by Sorokin (1962) to be due to the depletion of metabolizable organic compounds. Bacterial activity in oxic sediments is also likely to be dependent on the amount of metabolizable organic matter present, although pH, sediment particle size, temperature and the sequence of physical and chemical changes caused by bacteria themselves may also have some influence (Oppenheimer, 1960).

Although the energy requirements of heterotrophic bacteria and chemosynthetic autotrophs are different (the former are satisfied by the oxidation of organic matter, whereas the latter survive on the oxidation of such substances as NH_3, NO_2^-, CH_4, Fe^{2+}, H^+, molecular H, H_2S, S^0, $S_2O_3^{2-}$, SO_3^{2-} and CO), both fix CO_2 to form bacterial cell substances. Although both types of organisms are better known as decomposers or mineralizers of organic remains, bacteria–principally the chemilithotrophic bacteria–are significant synthesizers of organic compounds, which may be either metabolic products, ectocrines or microbial cell substances. In this connection, Zobell (1964) has claimed that 30–40% of the carbon content of the organic substrate in sediments is converted into new kinds of organic compounds. The other 60–70% is liberated, largely as CO_2, although under certain anaerobic conditions appreciable amounts of methane and traces of higher hydrocarbons may result from the fermentation of organic matter (Foree and McCarty, 1970).

Variations in the populations and activities of micro-organisms between shallow and deep-water sediments, and between those forming under oxic and anoxic conditions, are difficult to evaluate because different experimental procedures are employed by different workers. However, it seems that both the bacterial count, and the activity of the bacteria, are far higher in shallow water sediments than in those of the deep-sea. For example, both the aerobic and anaerobic microbial counts in the shallow oxygenated sediments of the Black Sea are more than an order of magnitude higher than those in the deep water anoxic sediments (Sorokin, 1964). Further, the activity of the organisms decreases many times more than would be suggested if only the difference in the total bacterial biomass in the two types of sediments were considered (Sorokin, 1964; cf. Skopintsev, 1964). This may be due, in some instances, to a lack of fresh organic matter reaching the sediment. It is known that dissolved organic substances in oceanic bottom waters resist microbial attack (Barber, 1968), and it is possible that the particulate organic matter occurring in deep-sea sediments is also resistant to breakdown (see Chapter 31).

Such changes in bacterial population and activity can at best only be regarded as an expression of bacterial behaviour in sediments of different character, as there is evidence that the rates of microbial decomposition of organic matter exposed to deep-sea conditions, and those found in controlled culture experiments are very different. Recently, Jannasch *et al.* (1971) have found that decomposition of organic matter is much slower (8–666 times slower) in the deep-sea than it is under refrigerated conditions in the laboratory. Such experiments suggest that there is a slowdown of life processes in the deep-sea which may be dependent on hydrostatic pressure (Jannasch *et al.*, 1971; cf. Zobell, 1968). Thus, much of the published data expressing the metabolic activity (in terms of oxygen consumption) of organisms in sediments may well have to be revised. For instance, the role of bacterial activity deduced from culture experiments is invariably far higher than that measured in *in situ* experiments (Smith and Teal, 1973).

In most sediments bacteria and other living organisms, e.g. benthic animals, are the most important agents responsible for the change in the physico-chemical character of the sediments. For instance, the pH and Eh at any position at depth within a sediment appear to be a function of the net balance between the types of bacteria, the amounts of metabolizable organic matter, the buffering and poising capacity of the sediments and the rate of diffusion of O_2 through the sediment. In this respect, much of the change in inorganic as well as organic constituents that occurs during burial is related directly, or indirectly, to the population and activity of organisms within a sediment.

30.2.2. DECOMPOSITION OF ORGANIC MATTER

Unlike most natural inorganic systems, for which the mineral and ionic components at equilibrium are determined by a few chemical parameters such as pH and Eh, organic substances are unstable and their geochemistry is dominated by non-equilibrium processes (Dayhoff *et al.*, 1964). During diagenesis, the less stable organic compounds are eliminated by reactions that lead to a randomizing of the ordered structures which were created by organisms. Prominent or typical among these are reactions based on the elimination of functional groups by deamination, decarboxylation and condensation, depolymerization, isomerization and certain intermolecular oxidation and reduction reactions, most of which are irreversible. The rates of such processes are usually very slow, and many substances that are thermodynamically unstable, such as higher normal alkanes, isoprenoids and many porphyrins, have survived in geologically ancient sediments (Blumer, 1967). Many of the organic groups which may show considerable modification in the initial stages of diagenesis may nonetheless survive unaltered for very long periods. This is attested by the existence of carbohydrates and amino acids in rocks of Palaeozoic and Precambrian age. It would seem therefore that transformations of biomolecules may be kinetically inhibited by the formation of meta-stable associations during the various stages of diagenesis.

Although the compositions and properties of the principal organic compounds in Recent sediments have been established, few detailed studies have been made. Moreover, there is little information on more than one group of organic compounds from the same sediment and this creates difficulty when attempts are made to assess the relative change of substances within one of these groups, or between different groups, during sediment burial. In addition, much of the uncertainty in the interpretation of the changes of organic substances within sediments often stems from an incomplete knowledge of the history of sedimentation within a sediment profile. For instance, knowledge of alternating changes in oxic and anoxic conditions within a sediment, or changes in the accumulation rate of a sediment, may be important when the composition and behaviour of organic compounds are being interpreted. Such variations are frequently overlooked by organic geochemists.

Analysis is prominent among the problems encountered in the interpretation of organic substances in sediments. Evidence on the distribution of certain active constituents in organic matter presupposes that sampling, extraction and quantitative evaluation are correct. Such assumptions are not always well founded. Estimations of a particular organic group by two or more extraction methods have, on numerous occasions, been shown to give different results. For instance, Farrington and Quinn (1971b) have shown that the extraction of fatty acids from sediments by saponification and organic solvent extraction methods produce very different results. Nevertheless, evaluations of published data on the organic substances contained in sediments can in many instances give some overall insight into the changes which they have undergone during early diagenesis.

Modification of organic matter in the surface sediments is to a large degree dependent on micro-organisms and the metazoan biomass. However, the relative importance of these two groups of organisms as agents of diagenesis is unknown. As the zone of vigorous biological activity is confined to the upper few decimetres in sediments because of oxygen availability and other factors, alteration of organic matter at depth is likely to be slower, and may be abiotic, taking place through such reactions as decarboxylation (Degens, 1965).

Although there is some loss of organic carbon during sediment burial, the decrease in concentration of specific organic substances may be considerable. This is particularly true of amino acids, carbohydrates and fatty acids which with humic and fulvic acids comprise the bulk of the active organic constituents in sediments. Concentrations of these substances are often highest at the sediment surface. After burial they show a marked loss which, relative to organic carbon, often represents a decrease of one or more orders of magnitude. For instance, Rittenberg *et al.* (1963) have observed a change in total amino acids from 350 to ~20 μg/g within the depth interval 0–170 m in oxidizing sediments off California. However, most of the change takes place in the uppermost one or two metres. Similar trends in sediments of the same lithology have also been shown by Degens (1967) and more recently in sediments from the Cariaco Trench, Venezuela (Hare, 1972) and the Argentine Basin (Degens, personal communication; see also Chapter 31). It is probable that the amino acid distribution in the uppermost layer of a marine sediment is directly correlatable with microbiological activity as in soils (Hare, 1969). For instance, the abundance of β-alanine in soils seems to be directly related to microbial activity (Degens, 1965). Where the content of total amino acids decreases relative to organic carbon, especially within the first few decimetres of a rapidly forming sediment, the change in concentration of one amino acid with respect to others is likely to be small (Jessup, personal communication). In contrast, sediments sampled from greater depths may show some relative change in the concentrations of different amino acids. Thus, a sediment core sampled to 3·5 m in Saanich Inlet, British Columbia (Brown *et al.*, 1972), showed some changes in the relative concentrations of the various acidic amino acids. The differences between the amino acid compositions in the uppermost two metres of a sediment and at depth have been used to study the relative stability of amino acids (Rittenberg *et al.*, 1963). It is probable that ornithine, serine, glutamic and aspartic acids, the leucines, threonine and glycine are more resistant to diagenetic change than are the other amino acids.

In order to gain some further insight into the possible mode of origin of amino acids work has recently been carried out on changes in optical configuration of isoleucines and other amino acids (Hare 1969; Hare and

Mitterer 1969; Kvenvolden *et al.*, 1970; Bada and Schroeder, 1972). These studies have centred on the change in optical configuration of L-isoleucine and D-alloisoleucine in sediments for which the racemisation reaction can be written as:

$$L\text{-isoleucine} \underset{k\ \text{rac. (allo)}}{\overset{k\ \text{rac. (iso)}}{\rightleftharpoons}} D\text{-alloisoleucine}$$

where k rac. (iso) and k rac. (allo) are the first order rate constants for the interconversion of isoleucine and alloisoleucine respectively. As isoleucine has one of the slowest racemisation rates, this reaction has been frequently used in interpreting reactions within deeply buried sediments. The examination of other amino acids and other substances with higher rates of interconversion promises to be a significant aid to understanding the origin and change of organic substances in sediments.

Both free and combined sugars, like amino acids, decrease in abundance with sediment depth. For amino acids the concentration decreases to zero, or almost zero, within the first few metres (Prashnowsky *et al.*, 1961; Degens, 1965). Their concentration within the uppermost 5 m of sediment offshore from California decreases by a factor of five or more (Rittenberg *et al.*, 1963) and their concentration at depth is similar to that reported for ancient sediments. Chemical changes which transform combined sugars under marine conditions are not understood. Taking all sources of information on the stability of sugars into consideration, Degens (1965) has suggested that in oxidizing environments the bulk of carbohydrates, which may constitute some 5% of the organic matter, is eliminated biochemically in the early stages of diagenesis. However, within anoxic environments the decline in carbohydrate concentration relative to that of organic carbon, with depth, which has been found in sediments from the Cariaco Trench (Mopper, 1973), is much less marked.

In a study of the straight chain saturated fatty acids in Recent and ancient sediments Cooper (1962) noted that the ratio $(C_{16} + C_{18})/2C_{17}$ changed from 14·9 to 7·1 respectively. However, the "even-odd" ratios for some Recent marine sediments are not always consistent. For example Kvenvolden (1967) and Parker (1969) have reported ratios as low as 2·3 for Recent sediments. It would seem that biological and/or geochemical processes that determine the fatty acid compositions in a given sediment are complex, and may differ significantly from one instance to another. However, the pattern of fatty acid distribution within Recent sediments is such that the total, and especially the undersaturated, fatty acids show a marked decrease with depth (Farrington and Quinn, 1971b). This supports findings by Parker and Leo (1965) for the change in fatty acids in a layered sequence of blue-green algae, and the qualitative data of Rosenfeld (1948) for the fatty acid distribution in rapidly accumulating anoxic sediments. Bacteria may be important in the production of certain fatty acids. For instance, some straight chain iso- and antiso- fatty acids in sediments could be derived from the lipids of bacteria and protozoa rather than have originated as a constituent of marine animals or plants (Cooper and Blumer, 1968).

The mechanism of fatty acid change in a sediment has been re-examined by Rhead *et al.* (1971) using *in situ* experiments involving oleic acid-9,10-^{3}H and oleic acid-1-^{14}C. An 11 % conversion of the labelled acids to $C_{12:0}$—$C_{20:0}$ was observed. The pattern of incorporation of the labels suggested that the saturated fatty acids, especially $C_{16:0}$, are formed by breakdown and resynthesis, in addition to the direct route of hydrogenation and chain shortening or lengthening.

Information on the relative loss, or change in behaviour, of specific groups or organic compounds in sediments is sparse. Comparison of data for amino acids and carbohydrates in sediment cores from the Californian borderlands (see Degens, 1965) indicates that the loss of amino acids with sediment depth is greater than that of sugars. Recent investigations by Jannasch *et al.* (1971) and Jannasch and Wirsen (1973) support this contention as a greater degree of decomposition of nitrogenous substances takes place, relative to the carbohydrates, during microbial attack on the organic matter in the deep-sea. Analyses of the organic matter in sediments from Saanich Inlet, British Columbia, an intermittently anoxic fjord containing sediments of varied lithology, have shown that chlorins, fulvic acids and amino acids decrease markedly with depth (Table 30.2) (Brown *et al.*, 1972). Chlorins seem to show the greatest decrease, followed by amino acids and fatty acids. Fulvic acids which are highly abundant at the surface also decrease markedly with depth,

TABLE 30.2

Change in organic matter as a percentage of the total organic carbon in the sediments of Saanich Inlet, British Columbia (after Brown *et al.*, 1972)

Core and Depth (cm)	Fatty acids	Chlorins	Amino acids	Humic acids	Fulvic acids	Remainder
4 0–15	0·25	0·14	6·24	39·9	28·3	25·1
4 200–210	0·15	0·11	3·61	31·0	30·7	34·4
3 790–820	0·18	0·09	3·55	36·8	26·0	33·3
3 1710–1740	0·05	0·04	2·45	21·3	18·7	57·3
3 2620–2650	0·09	0·01	1·28	32·0	8·9	57·6
3 3450–3480	0·04	0·01	0·78	31·0	8·7	59·1

Note: Data presented are for two cores which may not necessarily represent the same environment. Sediment rate $\sim$ 4 m/10^3 yrs.

unlike humic acids which surprisingly do not appear to show any significant change.

Some attempts have been made to assess the degradation of the constituents of organic matter in different environments. The total amino acid concentrations in sediments accumulating under oxidizing conditions off California show a more marked decrease with depth than in reducing sediments from the same area (Rittenberg *et al.*, 1963; Degens *et al.*, 1961, 1963). In contrast, an examination of oxygen-containing functional groups present in the humic and fulvic acid fractions from the upper part (<1 m) of sediments forming in oxidizing and reducing environments failed to show any significant change (Rashid and King, 1970). However, in the latter investigation, no attempt was made to examine possible changes at depth within sediments. Experimental work by Krause (1959) has shown that a major proportion of proteins and amino acids in a system can be decomposed by bacteria, both aerobic and anaerobic, within a period of a few months. Therefore, the differences in the decomposition of sediment amino acids under oxidizing and reducing conditions, found by Rittenberg *et al.* (1963) and Degens *et al.* (1961, 1963) may be related more to the presence or absence of active metazoan life in the surface layers of the sediments than to the abundance and activity of bacteria alone.

30.2.3. DIAGENESIS OF CARBON, NITROGEN, PHOSPHORUS AND SULPHUR

Of the major elements in sediment pore waters some of the greatest concentration changes that occur during shallow burial are exhibited by carbon, nitrogen, phosphorus and sulphur, each of which is directly or indirectly due to the decomposition of organic matter by micro-and metazoan organisms.

Oxidized carbon is added to sediment interstitial waters by the decomposition of organic matter, and is normally present as CO_2 aq and H_2CO_3 aq (Berner, 1971). Where bacterial sulphate reduction is evident it will also be added as HCO_3^- aq. Within oxidized sediments both aerobic bacteria and higher organisms are responsible for the evolved CO_2 which will tend to show a net build up in the subsurface environment, and hence may lead to the development of acid conditions. Within anaerobic systems CO_2 may form through fermentation processes.

Berner (1971) has calculated that the aerobic oxidation of organic matter has a higher energy yield than that developed by reductive fermentation, and as a result organic matter is largely destroyed in environments where constant aeration takes place. In constrast, in anaerobic (anoxic) sediments much of the organic carbon may escape bacterial destruction. This view, which appears to be generally held, is only in part supported by data on the change in abundance of active organic substances in sediments. This explanation is often used to account for the differences between the organic carbon contents of oxic and anoxic sediments. However, this assumption has been questioned by Stumm and Morgan (1970) who favour the concept that organically enriched environments become by necessity anoxic.

The change in pH which occurs during anaerobic processes depends upon the net balance of CO_2 (and organic acids) evolved during fermentation, sulphate reducing processes and the generation of nitrogenous bases. During sulphate reduction by bacteria of the genus *Desulfovibrio*, bicarbonate formed adds to that (CO_2 gas) produced by fermentation reactions. This is because the bacteria reducing the sulphate ion use organic carbon as a reducing agent. Measurements of the amounts of dissolved sulphate and dissolved HCO_3^- for pore water from Somes Sound, Maine (Berner, 1971) clearly show this relationship (Fig. 30.1).

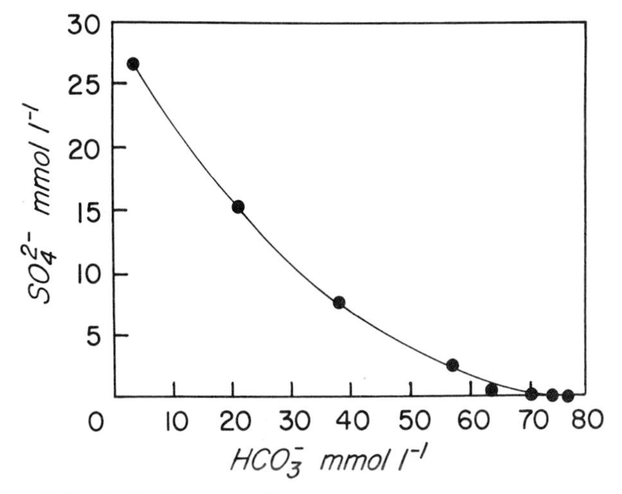

FIG. 30.1. Plot of dissolved sulphate vs. dissolved HCO_3^- for the interstitial waters in anoxic sediments from Somes Sound, Maine. (after Berner, 1971). The relative increase of HCO_3^- when SO_4^{2-} is low or exhausted relates to HCO_3^- developed by decomposition of organic matter by non-sulphate reducing bacteria.

The geochemistry of nitrogen during diagenesis is dominated by the proteinaceous matter in the sediments rather than by the NO_3^- content of the pore water within the sediment. In fact, the latter almost invariably plays a very small part in influencing the distribution of nitrogen-bearing substances in sediment interstitial waters (see Emery, 1960).

The decomposition of protein and the high loss of amino acids and amino sugars relative to other organic constituents from sediments during their burial may be reflected in the change in C/N ratios with depth. Changes in

C/N ratios with sediment depth, which seem to correspond to a decrease in the "protein" content of the organic matter have been observed by Bordovskiy (1965), Rittenberg *et al.* (1963), Emery *et al.* (1964), Stevenson and Tilo (1970), and Stevenson and Cheng (1972). The changes observed for C/N ratios and the amino acid content of the organic matter below the zone of expected biological activity in sediments of the Argentine Basin (Stevenson and Cheng, 1972; see Table 30.3) suggest that the main mechanism for C/N

TABLE 30.3

Variations of organic carbon, total and organic nitrogen, fixed NH_4^+-N and amino acid-N in a profile of Recent sediment from core V-15-142 from the Argentine Basin. Continued but more fluctuating change is observed at greater depths which appear contiguous to glacial advances during late Pleistocene times (after Stevenson and Cheng, 1972)

Depth (cm)	Org. C (%)	Total N (%)	Org. N (%)	NH_4^+-N (%N)	Amino acid-N (ppm)	Org. C / Total N	Org. C / Org. N
2	0·92	0·120	0·110	8·3	305	7·6	8·3
10	0·68	0·080	0·070	12·5	230	8·5	9·7
20	0·58	0·072	0·055	23·6	220	8·1	10·5
30	0·59	0·067	0·050	25·4	185	8·8	11·8
33	0·63	0·064	0·048	25·0	165	9·8	13·1
36	0·65	0·066	0·051	22·7	155	11·2	12·7
38	0·75	0·067	0·056	16·4	155	11·9	13·4
40	0·81	0·068	0·055	19·1	160	11·8	14·7
43	0·79	0·067	0·054	19·4	180	10·5	14·6
50	0·68	0·065	0·054	16·9	185	8·7	12·6

change at such depths is chemical, rather than biological, and may occur through reactions of amino compounds with phenolic constituents and reducing sugars (Stevenson and Tilo, 1970). The presence of large amounts of fixed ammonium in certain clay minerals in sediments at depth (Stevenson and Cheng, 1972), may account for spuriously low C/N ratios of $<5{\cdot}0$ (compared with the more normal values of $>7{\cdot}0$) that have been observed in some sediments (Mohamed, 1949; Arrhenius, 1950; Bader, 1955; Bordovskiy, 1965). The hypothesis of Arrhenius (1950) that a low C/N ratio in sediments is characteristic of those poor in organic matter has recently been explained by Stevenson and Cheng (1972) on the basis that the percentage of nitrogen as fixed ammonium will be highest where organic matter has been highly depleted through biological activity.

Deamination of nitrogenous organic matter, which occurs largely through the agency of biological processes in the presence or absence of dissolved oxygen, leads to the development of ammonia. Within aerobic environments the ammonia may be subsequently oxidized through microbial mediation to nitrite and eventually nitrate. In contrast, in anaerobic environments the ammonia may build up to considerable levels and may sometimes exceed concentrations of 10^{-3}M (Rittenberg *et al.*, 1955), and hence may raise the pH value of the subsurface environment.

The most stable and predominant form of dissolved phosphorus in marine sediment pore waters is orthophosphate. Most of the phosphorus developed in the pore waters of marine sediments is produced by decomposition of organic matter (Redfield *et al.*, 1963; Richards, 1965). Unlike carbon and nitrogen, phosphorus is not involved directly in redox reactions in sediments. However, its possible association with Fe^{2+}, Fe^{3+}, Ca^{2+}, and to a lesser extent Al^{3+}, could produce differences in the concentration of phosphorus in both the interstitial water and the sediments themselves.

Although interstitial waters of oxic sediments often differ little in PO_4^{3-} concentration from the overlying sea water, anoxic sediment interstitial waters may contain very high concentrations of phosphate (Rittenberg *et al.*, 1955; see Baturin, 1969). These differences in the PO_4^{3-} contents of interstitial waters of sediments from different environments are only partly explained by differences in the composition and abundance of organic matter. The lack of phosphorus in the pore water of some oxygenated sediments (Fanning and Pilson, 1971) may be caused by the adsorption of phosphorus on clay minerals (Weaver and Wampler, 1972), and the precipitation of phosphates. For instance, the formation of iron phosphates, or the adsorption of phosphorus onto hydrous ferric oxides, may also be an important control on the geochemistry of phosphorus, especially in the upper levels of oxic sediments. The changes in the relative uptake of phosphorus into iron and calcium compounds during diagenesis is well illustrated in estuarine sediments of the Rappahannock River (Nelson, 1967). Here, there is an increasing tendency for calcium phosphate to precipitate in preference to iron phosphate within those sediments forming in marine conditions where the pH and the activity of Ca^{2+} ions are presumably higher.

It is well known that bacterial reduction of SO_4^{2-} to HS^- or H_2S occurs only in anoxic environments (Kaplan *et al.*, 1963; Berner, 1964; Richards, 1965). The role of organic sulphur in the marine formation of sulphide ions is thought by most authorities to be insignificant (e.g. Kaplan *et al.*, 1963), although in some lakes, which contain very little SO_4^{2-}, the H_2S may be derived from sulphur-containing proteins and amino acids. Although sulphate is activated by various carriers, the precise mechanism of biological sulphate reduction is by no means clear (Nicholas, 1967). The generalized reaction exemplifying the overall stoichiometry of the reduction can be written:

$$2CH_2O + SO_4^{2-} \rightarrow H_2S + 2HCO_3^-,$$

where two atoms of organic carbon are used to reduce one atom of sulphate. It seems likely that the bacteria that mediate the reduction tend to metabolize small molecules produced by micro-organisms rather than long chain molecules (Sorokin, 1962). Thimann (1963) has used hydrocarbons, fatty acids, carbohydrates and amino acids as substrates for sulphate reducing bacteria in the above reaction.

The interrelationships of the various dissolved nutrient elements in sediment pore waters are not as well understood as they are in sea water (Redfield *et al.*, 1963; Richards, 1965) as the problem is complicated by the greater likelihood of their precipitation within sediments. The amounts of ammonia and phosphate released by the decomposition of marine organic matter in sediments are often linearly related. This is especially true in reducing sediments in which ammonia oxidation is unlikely. The relationship is illustrated (Fig. 31.2a) for two short (<1 m) sediment cores formed under slightly different environmental conditions. A change in the NH_3/PO_4^{3-} relationship in the uppermost sediments, i.e. where dissolved ammonia and phosphate are low, implies some that in one of the cores ammonia has undergone some oxidation.

Redfield *et al.* (1963) and Richards (1965) have also observed a linear relationship between the amount of sulphate reduction and the concentration of ammonia in anoxic waters. However, within sediments linearity only prevails where SO_4^{2-} is present (Fig. 30.2b). A similar trend may be seen between HCO_3^- and SO_4^{2-} (Berner, 1971; see also Fig. 30.1) in the interstitial waters of Somes Sound, Maine. At low concentrations of SO_4^{2-} there is a considerable change in the relationship between dissolved SO_4^{2-} and other constituents. Such trends imply that ammonia (and PO_4^{3-} and HCO_3^-) is produced in sediments, not only by micro-organisms which reduce sulphate, but also by a group, or groups, of micro-organisms which are active after interstitial sulphate has been exhausted.

The pattern of sulphide formation with depth in both shallow water marine and deep-sea sediments has been studied by noting the change in either the SO_4^{2-} concentration, the SO_4^{2-}/Cl^- ratio, or the content of dissolved sulphide in interstitial water (Bischoff and Ku, 1971; Nissenbaum *et al.*, 1972; Sayles *et al.*, 1973). The change in SO_4^{2-} concentration in the interstitial waters of the anoxic sediments from Saanich Inlet shows that there has been a very rapid loss of SO_4^{2-} within the uppermost 50 cm of sediment. Below this depth SO_4^{2-}, and dissolved sulphide, are usually absent (Nissenbaum *et al.*, 1972). Similar losses of sulphate from the interstitial water have been observed in the anoxic sediments of the Cariaco Trench (Sayles *et al.*, 1973), in which the sulphate concentration, or SO_4^{2-}/Cl^- ratios assume an upward exponen-

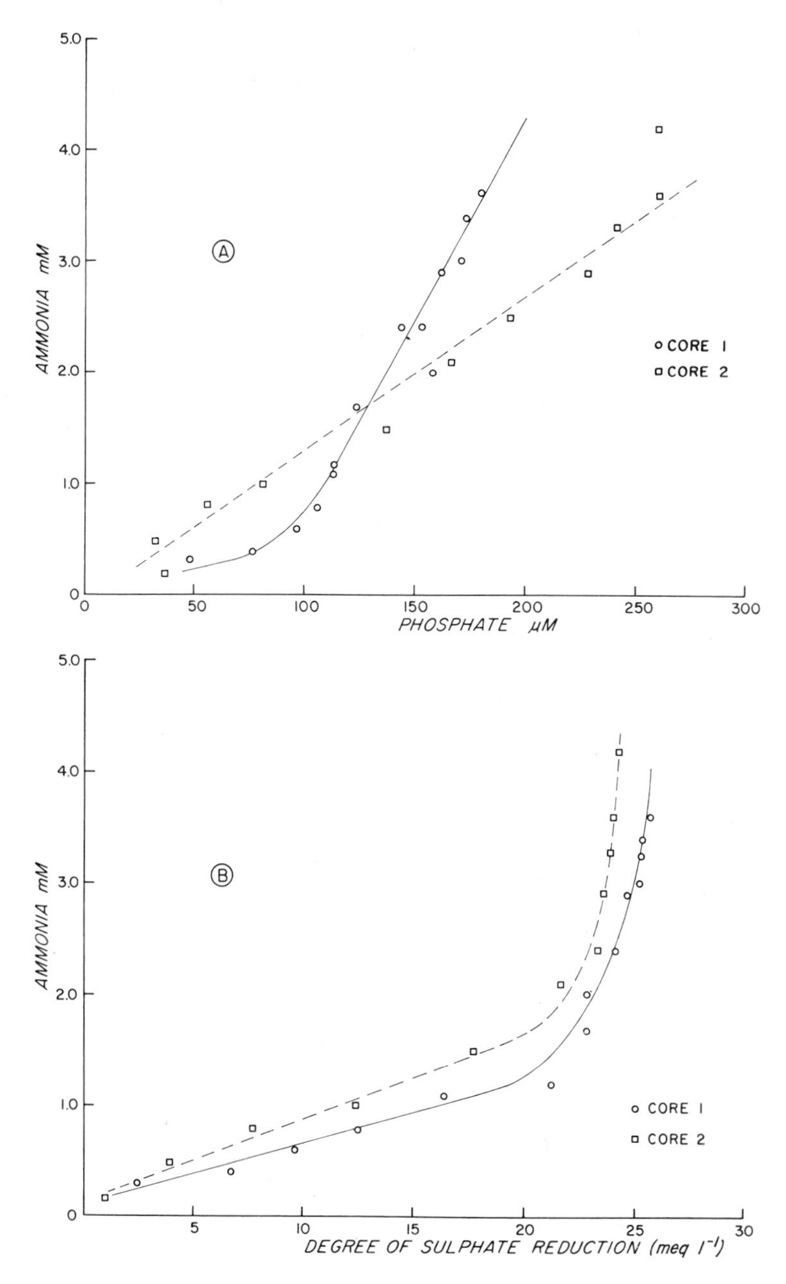

FIG. 30.2. Relationship of ammonia and (a) phosphate and (b) the extent of sulphate reduction in the interstitial waters of two shallow (<1m) sediments cores from Loch Duich, Scotland. (Ian Davies, unpublished data).

tial trend. Likewise, interstitial waters from deeply buried sediments can also show a progressive decrease in their SO_4^{2-}/Cl^- ratios with depth (Sayles *et al.*, 1973). Such trends are not easy to explain as at depth (e.g. >100 m) since bacteria, essential as mediators of sulphate reduction, are commonly assumed to be absent or inactive.

30.3. Redox and pH Controls in Sediments

Few elements are not influenced to some degree by organic matter. The predominant participating elements, besides iron and manganese, governing redox processes are relatively few in number; these are C, N, O, H, and S which are all closely linked to various biological processes. Respiration and photosynthesis and the decay products of organisms cause changes in the carbon dioxide and ammonia content of sediment pore waters, and both of the latter determine to a large extent the pH of the interstitial waters. Variations in the populations of organisms and the abundance of organic matter will thus considerably influence the acidity of the interstitial environment within sediments.

pH and Eh measurements of Recent marine sediment pore waters by Zobell (1946), Bass Becking *et al.* (1960) and others provide only a qualitative expression of their limits in different sedimentary environments. Although these cannot be used for thermodynamic purposes, they do perhaps give some indication of environmental changes in sediments. Such data seem to show that each sediment type has its own characteristic Eh and pH limits. Eh values ranging from $+0{\cdot}350$ to $-0{\cdot}5$ V have been observed in a variety of samples of Recent sediments in which the pH ranged from 6·4 to 9·5. However, most sediments fall within the pH range 7·5–9·0. As a very general rule the pH tends to increase, and the Eh to decrease, with depth in a sediment; that is conditions become more alkaline and more reducing at depth. However, the gradient of the change of pH in sediments of oxic and anoxic origin, may differ markedly with depth from one to another. This is due mainly to the high degree of oxidation of ammonia in the former environment. Within anoxic sediments the build up of ammonia may reach such proportions (Rittenberg *et al.*, 1955) that precipitation of calcium and other elements may occur (see Berner, 1971).

The practical difficulties involved in the electrochemical measurement of redox potentials are considerable (Stumm and Morgan, 1970). The principal dissolved species influencing the redox potential of anoxic sediments at normal pH (7–8) values are HCO_3^-, NH_4^+, HS^-, H_2S, CH_4 and SO_4^{2-}. Measurements of Eh in such sediments often do not reflect the Eh expected for any of the half-cell couples using combinations of these species since SO_4^{2-}, N_2, CH_4 and HCO_3^- are not electroactive, and other half-cell couples produce very low currents at the platinum surface (Stumm, 1966). However, in sediments where there are more electroactive species, Eh measurements may be more meaningful (Berner, 1963). In a sediment of heterogenous character, particularly one which is rich in organic life, oxidation levels different from those of the sediment in general may occur within a localised biotic environment. Diffusion and dispersal of reaction products from the micro-environment to the macro-environment, and even the loss of such materials from the sediment to the overlying water may give a false impression of redox conditions in such sediments.

To test for internal equilibrium in reducing sediments from coastal areas Thorstenson (1970) has used data from coastal sediments off California, and from Harrington Sound, Bermuda, to calculate the Eh values in a sediment from the measured concentrations of the principal dissolved species of the system S–N–C–H–O, and has compared these for each redox couple. Calculation of pE* or Eh for each of the pairs SO_4^{2-}–HS^-, HCO_3^-–CH_4 and N_2–NH_4^+ should give the same result if complete internal equilibrium exists (Table 30.4). The calculated Eh values for SO_4^{2-}–HS^- are in much better agreement with those of N_2–NH_4^+ than with the values calculated for HCO_3^-–CH_4 (also see Berner, 1971, p. 120). In spite of certain limitations such calculations are useful in predicting pE, and the exchange of interstitial water with overlying sea water. The data from Harrington Sound suggests that although element exchange between sea water and sediment interstitial water may take place in the uppermost 30–40 cm of sediment, below this, the sediments and their interstitial waters seem to act as a closed chemical system.

Sillén (1966) has suggested a thermodynamic model for the sequence of redox reactions involving elements, such as Mn, Fe, S, N and C, which may be applied to oxidation/reduction reactions during sediment burial. In this idealized model the oxidation of organic matter will supply electrons firstly to the lowest unoccupied electron level, i.e. dissolved O_2, and then successively to the next highest level, e.g. NO_3^-, NO_2^-, Mn^{4+}, and so on, until the redox potential is sufficiently low for fermentation and reduction of SO_4^{2-} and CO_2 to take place. If it is assumed that the pH of the subsurface environment is constant (pH 8·1), and that thermodynamic equilibrium is reached for every reduction reaction, or alternatively that the kinetics of each reaction is fast, the redox potential in a sediment would change with depth, in a step-like manner between the values of pE = 12·51 (corresponding to air-saturated

* pE is defined (Jørgensen, 1945) as the negative logarithm of the activity of electrons in solution

i.e. $pE = -\log_{10} [e^-]$

or $pE = \dfrac{Eh\, F}{2{\cdot}303\, RT}$

where Eh is the reversible redox potential. (RTF^{-1} *ln* 10) is 59·16 mV at 25°C.

TABLE 30.4

Distribution of molality concentrations of HS^-, NH_4^+ and HCO_3^- in interstitial waters of sediments from Devil's Hole, Harrington Sound, Bermuda and the Santa Barbara Basin, California. Calculated values of Eh_s, pE_s; Eh_N, pE_N and Eh_c, pE_c for the couples SO_4^{2-}; HS^-; N_2: NH_4^+ and HCO_3^- : CH_4 are also shown (after Thorstenson, 1970). *Analytical data for the Santa Barbara Basin interstitial waters* (adapted from Emery and Hoggan, 1958).

Depth cm	pH	$M_{HS^-} \times 10^4$	$M_{NH_4^+} \times 10^4$	$M_{HCO_3^-} \times 10^4$	pE_s	Eh_s	pE_N	Eh_N	pE_c	Eh_c
Devils Hole, Harrington Sound, Bermuda										
10–26	7·28	1·8	3·0	—	−3·76	−0·222	−3·84	−0·227	—	—
26–38	7·20	3·9	5·9	—	−3·71	−0·219	−3·83	−0·226	—	—
38–51	7·05	4·2	8·1	—	−3·54	−0·209	−3·68	−0·218	—	—
51–63	7·01	4·5	8·8	—	−3·50	−0·207	−3·63	−0·215	—	—
63–76	7·00	3·9	11·0	—	−3·49	−0·207	−3·65	−0·216	—	—
76–90	7·00	3·8	12·0	—	−3·49	−0·207	−3·65	−0·216	—	—
Santa Barbara Basin, California										
4–28	7·9	—	15·2	1·21	—	—	−4·88	−0·289	−5·15	−0·305
28–52	7·7	—	30·6	2·70	—	—	−4·71	−0·279	−5·06	−0·299
52–76	8·0	—	41·0	4·33	—	—	−5·16	−0·305	−5·42	−0·321
76–100	8·0	—	52·5	5·90	—	—	−5·19	−0·307	−5·44	−0·322
100–124	8·2	—	65·8	9·70	—	—	−5·50	−0·325	−5·62	−0·332
124–148	8·1	—	107·0	8·70	—	—	−5·44	−0·322	−5·53	−0·327

water) and $-4{\cdot}13$, at which CH_4 is formed by reduction of CO_2. The poising of the redox potential at each reduction step would depend largely on the nature and amount of the participant being reduced.

However, many oxidation and reduction reaction rates within a natural environment, (e.g. a rapidly forming sediment), are very slow. These slow reactions include certain metal-ion oxidations, oxidation of sulphides, sulphate reduction and aging of oxides or hydroxide precipitates (Stumm and Morgan 1970). For equilibrium to be approached or attained in many of these reactions, especially those involving organic matter, the presence of an enzyme system associated with bacteria is required. These enzymes act only as redox catalysts for reactions that are thermodynamically feasible. However, kinetic rather than thermodynamic factors often control the oxidation and reduction reactions occurring in sediments, and it is probable that several redox reactions take place simultaneously. This means that gradual changes in composition will occur as the redox potential of the system varies, rather than the stepwise ones which would be expected if thermodynamics was the sole controlling factor.

30.4. Diagenesis of Major and Minor Elements

30.4.1. Terrigenous Constituents

The chemical alteration of detrital minerals, especially clay minerals, in recently deposited marine sediments has received renewed attention in recent years. Two lines of research are being used to show the extent of this alteration: (1) studies involving the change or modification of clay minerals within sediments; (2) investigations of changes in the composition of sediment interstitial waters.

Diagenetic reactions of particular terrigenous minerals within sediments are very difficult to observe. Part of the problem has been the identification and characterization of the composition of the different components reaching a sediment. Besides silicates most sediments may contain varying amounts of oxides, carbonates, phosphates, sulphides and organic matter, each of which contributes to the total chemical make-up of the sediment. The complexity of the assemblage invariably complicates our understanding of the chemical changes associated with a particular mineral fraction, e.g. clay minerals, in a sediment during its burial. Despite this many workers have attempted to correlate mineralogical changes in the clay minerals with diagenesis.

There is no doubt that detrital silicates in deeply buried sediments undergo diagenetic changes as a result of the appreciable temperature and pressure variations to which they are subjected (Velde and Hower, 1963; Perry and Hower, 1970; Weaver *et al.*, 1971). However, the time taken to form the uppermost few metres of a sediment is comparatively small, taking at most only a million years or so. It is very questionable whether changes in the clay mineral structure and composition in this surface metre or so of sediments should be attributed to diagenesis. With such short times and low temperatures any recrystallization of minerals, which might thermodynamically be favoured, may not be detectable. Many of the gross changes in mineralogy and chemistry that are sometimes seen in the upper parts of a sediment and reported as being due to diagenesis, more often than not, can be better related to changes in sediment supply, or to factors influencing sedimentation.

The distribution of different clay minerals at the surface of oceanic sediments appears largely to be the result of a combination of river run-off and aeolian transport processes (Biscaye, 1965; Griffin *et al.*, 1968; Rateev *et al.*, 1969), and Chapter 26. Further, their distribution in near shore sediments appears to be due to an interplay between weathering of the parent rocks, stream drainage and shallow water current patterns (Griffin, 1962). The clay mineral analyses used in defining such distributions, are at best semi-quantitative. Although these distributions can be interpreted as implying a detrital origin for most of the clays they cannot by themselves imply that these minerals are not forming, or that existing clay minerals remain unmodified at the surface or within sediments. For instance, Mackenzie and Garrels (1966) have estimated on theoretical grounds that in order to maintain steady state conditions in the sea only about 7% of the clay minerals in marine sediments need to be formed authigenically. Using the same assumptions, and revised data, Berner (1971) has recalculated this figure to be about 4%. Mackenzie and Garrels (1966) considered that in a steady state ocean system (Sillén, 1961; Holland, 1965) authigenic clay minerals originate in the vicinity of rivers by the interaction of the dissolved constituents of stream discharge. In such a system HCO_3^-, in excess of that which precipitates as $CaCO_3$, reacts with stream derived x-ray amorphous cation-free clay and Na^+, K^+ and Mg^{2+} ions to form authigenic clays, e.g.

$$2K^+(aq) + 2HCO_3^-(aq) + 3Al_2Si_2O_5(OH)_4(amorph) \rightarrow 2KAl_3Si_3O_{10}(OH)_2 + H_2O + 2CO_2.$$

In essence, reactions of this type are the reverse of those which occur during weathering. Subsequent work has provided little evidence for this mechanism or others, such as clay mineral transformations in the oceans (Sillén, 1961) or the precipitation of different clay minerals from solutions that approximate to sea water (Harder, 1971). On the contrary, the distribution studies outlined above together with isotope studies on clay minerals (Hurley, 1966; Dasch, 1969; Savin and Epstein, 1970) have tended to show that clay minerals are resistant to gross alteration in the marine environment and have apparently undergone little change since their formation in the zone of weathering.

Several workers have found that ion exchange reactions occur when fresh water clays are brought into contact with sea water or waters simulating sea water (Carroll and Starkey, 1960; see Keller, 1963; Holland, 1965). For instance, Whitehouse and McCarter (1958), have shown by laboratory experiments that fresh water illites undergo rapid ion exchange when brought into contact with artificial sea water. More recently, Russell (1970) and Drever (1971a, b) have used different approaches to study the changes occurring in river-borne clays of the Rio Ameca, W. Mexico, as they enter the sea and are buried. Russell (1970) showed that surface clays near the mouth of the river, and those in the river, are mineralogically identical and have the same number of equivalent cations. For this reason it appears that the clay minerals react with sea water only to the extent that some exchange between the Ca^{2+} of the river-borne clay and K^+, Na^+ and Mg^{2+} may take place. It also seems likely that the total cation exchange capacity of montmorillonite-rich clay remains unaffected when it is brought into contact with sea water. There is evidence that clay minerals in highly reduced sediments may have higher levels of non-exchangeable magnesium, and lower iron contents, than clays deposited in less reducing environments (see Fig. 30.3 and Drever, 1971a, b).

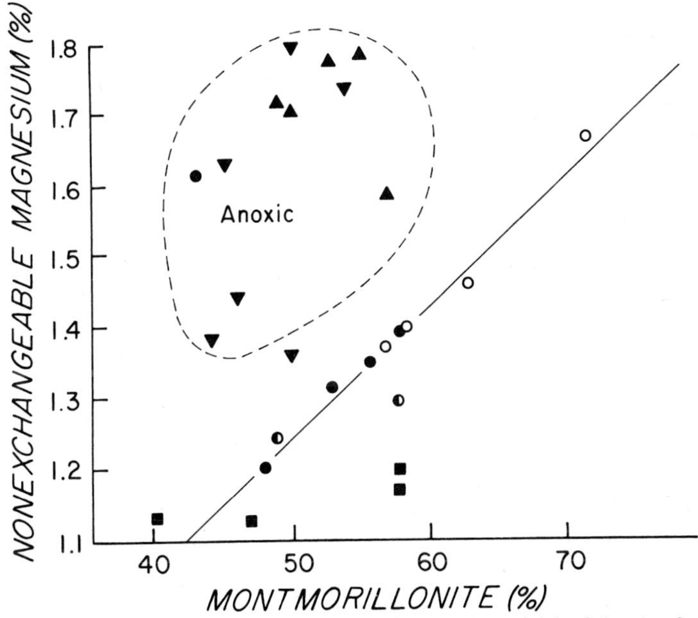

FIG. 30.3. Non-exchangeable magnesium content (percent by weight) of the clay fraction of recent sediments from Banderas Bay, Mexico plotted against montmorillonite. Each symbol represents a different core. The straight line is the predicted value on the assumption that all the variation in magnesium content is due to variation in montmorillonite content. The higher concentrations of exchangeable magnesium in clays from H_2S areas are due to replacement of Fe by Mg when Fe is extracted from the clay mineral to form a sulphide (after Drever, 1971a, 1971b).

Drever has suggested that this may be due to a release from montmorillonite, of iron, which is subsequently converted into an iron sulphide. Mg then enters the structural site in the clay left vacant by the iron and so impoverishes the interstitial water with respect to Mg^{2+}.

Field data on the reactions of ions with clay minerals have also been provided by Powers (1959) and Nelson (1963) who noted some uptake of K^+ and Mg^{2+} on degraded illite and chlorite respectively in estuarine sediments. There seemed to be a greater uptake of Mg^{2+} ions relative to K^+ in the near surface environment; at depth the reverse was true (Powers, 1959). It is uncertain whether these reactions are solely restricted to ion exchange, as opposed to changes in the structure of the clay minerals. It should also be pointed out that provenance changes with time, could also have produced similar effects.

There is a lack of evidence showing that changes in the structure of clay minerals do occur. Only Whitehouse and McCarter (1958) have observed, under laboratory conditions, a substantial alteration of montmorillonite to chlorite and illite. Experimental data given by Russell (1970) indicated that upon prolonged soaking, montmorillonite will take up Mg^{2+} from sea water especially when pH values are >8. This results in the formation of small amounts of a chlorite-like clay mineral and a decrease in the alkalinity of the system. Mackenzie and Garrels (1965) have also noted that clay minerals are unstable in sea water since they observed, again under laboratory conditions, some release of silicon from clay minerals, notably montmorillonite.

The inhibiting action of dissolved organic matter in interstitial waters may in part explain the difference between reactions involved in the chemical and mineralogical reconstruction of clay minerals in the laboratory and the sea. For instance, while Whitehouse and McCarter (1958) showed that appreciable clay mineral alteration could take place in the presence of artificial sea water, the addition of raffinose and fucodin to the system severely restricted the alteration of montmorillonite to chlorite and illite. Perhaps the main cause underlying the uncertainties in establishing whether the theoretically predicted structural changes in clay minerals occur in the marine environment, is one of identification. At present both chemical and mineralogical techniques are incapable of observing with any degree of certainty whether such a change is occurring, since it affects only a small part of the total clays in a sediment.

Volcanic products are usually much more subject to alteration than are terrigenous clay minerals. However, factors determining their rate of decomposition are still obscure. For instance, although glass shards in Mezozoic sediments may be unaltered, in some Quaternary deposits they may be altered to montmorillonite minerals or to phillipsite (Arrhenius, 1963). In fact, it appears that much of the montmorillonite and Fe-rich montmorillonite (nontronite) clays in areas of the bottom of the deep Pacific are derived from the devitrification of glass and other volcanic products (Bonatti, 1963; Nayudu, 1964), and are not due to the alteration of other clay minerals (Peterson and Griffin, 1964). It is also possible that some of the nontronite in deep-sea sediments may have been formed during the interaction between hot submarine volcanic lavas and sea water, and could be unstable with respect to a Fe-poor montmorillonite at the temperature of deep ocean water (Mackenzie and Garrels, 1966). Further, although kaolinite can form by the alteration of basaltic glass under acid weathering conditions (Sigvaldson, 1959); it has not been found to any extent as a product of marine pyroclastics, and its absence is perhaps accounted for by the more alkaline conditions and the higher cation activity in the sea. In addition to the clays, other minerals are also produced by devitrification. These include palagonite and zeolites of the phillipsite-harmotome series, the latter may comprise more than 50% of the total sediment in some Pacific Ocean areas (Arrhenius, 1963).

Recent analyses of deep-sea occurrences of tholeiitic basalts exposed to sea water for different periods show systematic chemical trends with their age and/or exposure to weathering (Hart, 1970). Upon hydration the basalts commonly show increases in their potassium contents suggesting that this element and possibly others have been taken up from sea water. Using a more sophisticated approach, Melson and Thompson (1973) have examined the *in situ* products of submarine weathering by analysing smectites formed on basalt. The composition and abundance of these clay minerals, which are potassium-rich, strongly indicate that much of the potassium and magnesium needed for their formation comes from sea water.

30.4.2. MAJOR ELEMENT COMPOSITION OF SEDIMENT INTERSTITIAL WATERS

Many studies have shown that the interstitial waters of marine sediments differ significantly in composition from the overlying sea water (see for example Bruevich, 1965; Siever *et al.*, 1965; Shishkina, 1966; Shishkina and Bykova, 1962; Manheim and Sayles, 1971). Such differences are thought to be due to slow reactions between the solids of the sediments and the entrapped interstial water. The concentration gradients that exist between interstitial water and the overlying sea water lead to diffusive fluxes of constituents across the sea water-sediment interface. A knowledge of these gradients will greatly help our understanding of the geochemical mass balance of elements in the marine system. For instance, average gradients of $<0{\cdot}1\%$ per cm depth would produce in global terms fluxes of many constituents (Mg, K, Na, SO_4^{2-} etc.) that could approach in amount the total river input of these elements to the sea (Sayles *et al.*,1973b).

In terms of sediment diagenesis the concentrations of major ions in interstitial waters do not necessarily relate to the total amount of modification of sedimentary particles, but represent solely the net effect of such a change. Hence, from a study of the composition of interstitial waters alone, interpretations on the diagenesis of silicates in sediments are difficult. These difficulties are increased when an element is held in sediment particles other than silicates. For instance, changes in the calcium and magnesium concentration of interstitial waters may be due as much to carbonate dissolution or precipitation as to silicate alteration; changes in the iron and manganese contents of interstitial waters more often than not reflect the diagenesis of non-silicates. Much of the silicon in interstitial waters is more likely to arise from dissolution of diatom frustules and similar organic remains than from alteration of inorganic silicates. Because of these factors, sensible interpretation of the significance of complex mixtures of silicates and biogenic detritus is extremely difficult. In less complex sediments, e.g. biogenic oozes, some diagenetic reactions can be identified with considerable confidence.

The results of early studies of the composition of sediment interstitial waters exhibit few coherent trends. In instances where changes were found, as with K/Cl or Mg/Cl ratios, these were interpreted as arising from feldspar hydrolysis or incipient dolomite formation, as well as being due to uptake by degraded clay minerals. However, interpretation of such data has been hindered by the quality of the analyses. Inadequacies in both the storage of core material and in the procedure used for extraction of interstitial water can produce misleading results. For instance, it has been shown (Mangelsdorf *et al.*, 1969; Bischoff *et al.*, 1970) that univalent cations in interstitial waters tend to be enriched when their extraction from the sediment is carried out under warm conditions; in contrast, divalent cations show a depletion under these conditions. On a quantitative basis potassium is affected far more strongly by temperature than any of the other major cations, and increases of as much as 28% over the *in situ* temperature (4°C) concentrations are often observed (Sayles *et al.*, 1973). Considerable changes in silicon concentrations (~50%) have also been observed in waters extracted at different temperatures (Fanning and Pilson, 1971). The effect of such experiments has been to invalidate much of the early work on the behaviour of individual ions in sediment interstitial waters. To avoid the temperature effect, several investigators have cooled sediments to temperatures near to those of the overlying bottom waters before extracting the interstitial water. Unless the bottom temperature is reproduced exactly, some residual error is likely even if the thermal effects are fully reversible, which is not known. The collection of interstitial water, by *in situ* methods (Sayles *et al.*, 1973b) obviates these problems and also that of any effect produced by pressure. Data from such sampling in sediments ranging from those from continental shelves to those from abyssal plains show quite conclusively that significant concentration gradients do exist in the interstitial waters of the uppermost sediments. Depletions of potassium as great as 0·3 to 0·6% relative to that found in the overlying sea water, may exist in interstitial water at depths of 30 cm below the sediment surface (Fig. 30.4). Such depletions are usually more pronounced at greater depths. In contrast, Ca^{2+} shows a 4·5–6·0% enrichment within the uppermost metre of sediment, but tends to fall again with depth. The gradients of K^+ are such that if they are ubiquitous they could produce downward diffusive fluxes which may be comparable globally to the annual imput of dissolved K^+ from rivers. The decrease in K^+ with sediment depth seems to vary geographically (Fig. 30.4), but the factor, or factors, controlling this change are unknown. Possibly sediment accumulation rates, and/or the relative amounts of weathered and unweathered material in sediments, may have a strong influence on this gradient. In the latter case one might expect unusual diffusive fluxes in sediments forming in high latitudes where the supply of sediment detritus is poorly weathered (Fig. 30.4).

In the deep-sea, both K^+ and Mg^{2+} may be taken up by both detrital clay minerals and also those formed by alteration of volcanic materials. The relative importance of these as a control on potassium and magnesium in the marine environment is, as yet, unknown. However, it is interesting to note that in deep-sea sediments on the Mid-Atlantic ridge, the most erratic variations of interstitial K^+ and Mg^{2+} occur in sediments containing volcanic material. (Bischoff and Ku, 1970).

30.4.3. DIAGENESIS OF SILICA

Considerable uncertainty remains regarding the relative importance of the various processes that remove dissolved silica from the sea (Mackenzie and Garrels, 1966; Harriss, 1966; Calvert, 1968; Burton and Liss, 1968, 1973). These processes include those regulating the silica concentration in the interstitial waters of sediments. Opaline silica tends to dominate the diagenetic geochemistry of silicon in most sediments. Quartz and aluminosilicates, including the clay minerals, seem to play a very subordinate role in this respect.

In most Recent sediments the bulk of the silicon, which is present in the form of quartz and other silicate minerals has a terrigenous origin. However, in some sediments biogenous silica (planktonic diatoms, radiolaria, silicoflagellates and some benthic sponges) may contain the major fraction of the silicon. Areas where the sediments are rich in biogenous silica include the Gulf of California (Calvert, 1966), the Southern California Borderlands (Emery, 1960), the S.W. African shelf (Calvert and Price, 1971), the Okhotsk Sea (Berzukov, 1955), the Cariaco Trench and the Antarctic regions of the

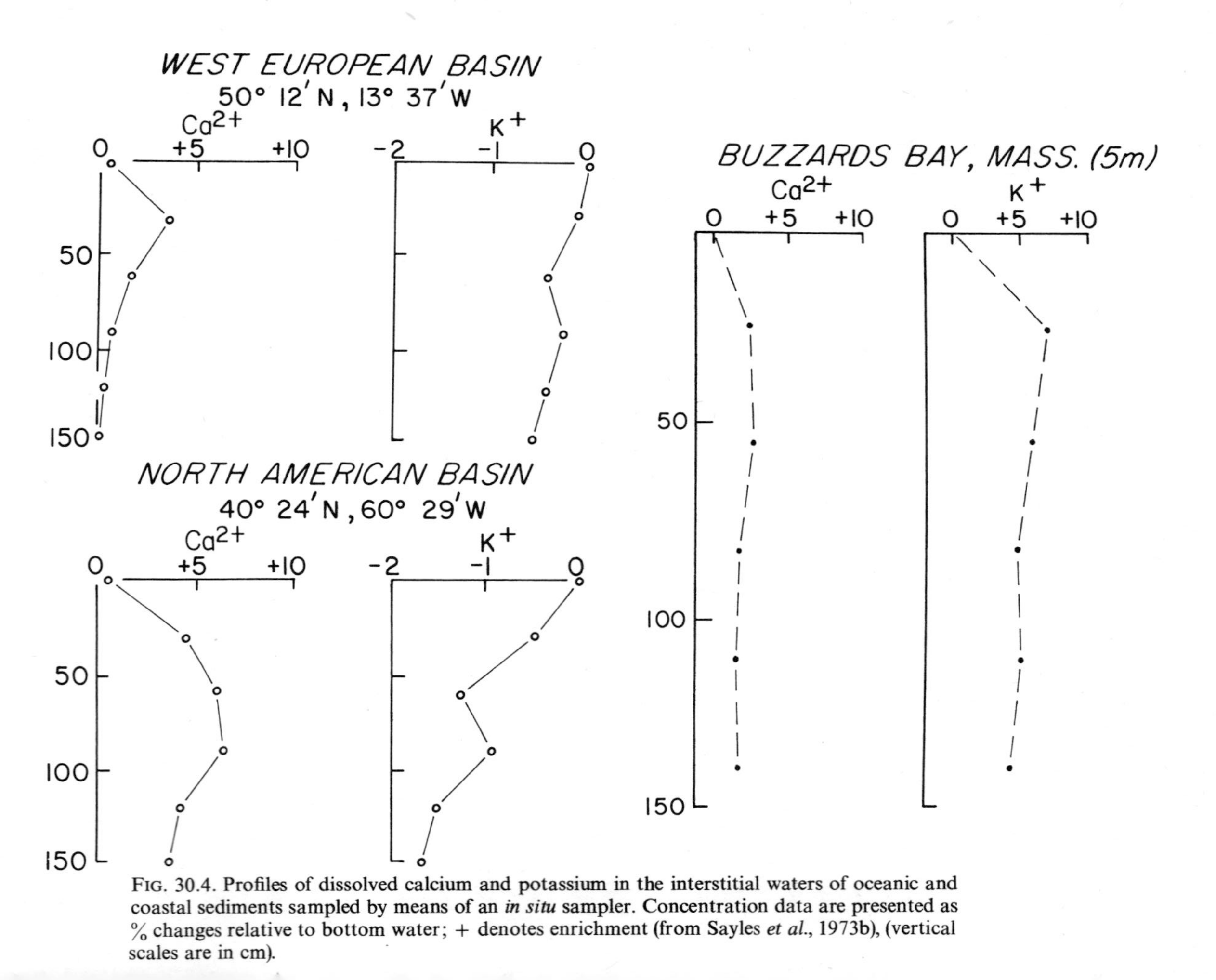

FIG. 30.4. Profiles of dissolved calcium and potassium in the interstitial waters of oceanic and coastal sediments sampled by means of an *in situ* sampler. Concentration data are presented as % changes relative to bottom water; + denotes enrichment (from Sayles *et al.*, 1973b), (vertical scales are in cm).

oceans. In the sediments from these areas, diatom frustules usually form the bulk of the biogenous silica. However, in certain areas within the Equatorial Divergence radiolarian tests may predominate in the underlying sediment (Sverdrup *et al.*, 1942; Lisitzin, 1970).

The high concentration of biogenous silica in these sediments seems to be the result of high biological productivity in the overlying waters, or of low terrigenous input (Calvert, 1966). Hence, sediments rich in biogenous silica usually contain relatively high concentrations of organic matter, especially those from areas where the bottom waters are depleted of oxygen. In these areas biogenous silica can comprise as much as 85% of the dried sediment (Calvert and Price, 1971). However, existence of biogenous silica within sediments seems to be as much a reflection of its preservation as of the biological productivity in the overlying water.

In ancient rock sequences the presence of bedded cherts is likely to be due to a late diagenetic transformation of biogenous opaline silica to microcrystalline quartz. The alteration seems to include cristobalitic porcelanite as an intermediate phase (Ernst and Calvert, 1969), and Mizutani (1970) has suggested that the change in silica minerals can be used to understand progressive changes during diagenesis. However, the processes involved in the alteration seem to be complex. Recent studies of buried deep-sea Tertiary and Mesozoic cherts (Beall and Fischer, 1969) indicate that although considerable dissolution of siliceous organisms has occurred during sediment diagenesis, the silica is redeposited largely as opaline silica and only partly as chalcedonic quartz.

The origin of chert is not necessarily confined to the alteration of opaline silica. Calvert (1971) has observed chert forming as an alteration product of volcanic glass, or as a direct precipitate from hydrothermal solutions. Moreover, Peterson and von der Borch (1965) have observed inorganic precipitation of chert in some South Australian saline lake and lagoonal sediments under very high pH conditions (10). They suppose that under these highly alkaline conditions the dissolution of detrital quartz produces silica concentrations in excess of that required for saturation at lower pH values.

Natural waters, both fresh and marine, are generally far from saturation with respect to silicic acid (Krauskopf, 1956) and are found to average 10^{-4} mol l^{-1} in waters at depths greater than a few hundred metres (Turekian, 1968). The result of this is that under these conditions silica will tend to dissolve. If it can be assumed that natural opaline silica has a solubility in distilled water or sea water similar to those of synthetic forms of silica, for example, silica gel and silica-glass, the data indicate that in waters of pH value <9 (at 25°C) its solubility is $\sim$100–140 mg SiO_2 l^{-1} for the reaction (Alexander *et al.*, 1954; Krauskopf, 1956; Siever, 1962 and Berner, 1971)

$$SiO_{2(amorph)} + 2H_2O \rightarrow H_4SiO_{4(aq)}$$

Corresponding values for the solubility of quartz, calculated from data by Siever (1962), Morey *et al.* (1962) and Stumm and Morgan (1970), are apparently much lower $\sim$4 mg $SiO_2 l^{-1}$. Morey *et al.* (1962) considered that the solubility of quartz in water seems to be influenced by the presence of a surface monolayer of silicic acid 0·03 μm thick, which is considerably more soluble than the quartz itself. Nevertheless, the dissolution rate of quartz, like its precipitation rate, is probably very slow.

Even if the temperature effects on the squeezing of interstitial water for analysis are taken into account (Fanning and Pilson, 1971; see Chapter 32) interstitial SiO_2 analysis reported in the literature almost invariably show much higher concentrations than are found in sea water. The extraordinarily high SiO_2 content ($\sim$700 mg $SiO_2 l^{-1}$) in interstitial waters from some Pacific Ocean sediments (Arrhenius, 1963) may be due to the devitrification of volcanic glass to phillipsite and other minerals. In more normal sediments, such as those of the Gulf of California and other areas (Siever *et al.* 1965), interstitial water SiO_2 contents usually lie in the range 20–50 mgl^{-1} SiO_2. The waters are thus super-saturated with respect to quartz, but at the temperature of the sediments, are undersaturated or possibly saturated with respect to amorphous silica (Siever, 1962). In such sediments there does not appear to be a typical depth profile applicable to all cores. More recent data of Fanning and Pilson (1971) and Heath (1974) seem to bear this out.

Although quartz is often recognized as overgrowths, as a replacement mineral, and as micro-crystalline cement in ancient sediments, its precipitation in Recent sediments has not been satisfactorily demonstrated. Its absence is often explained by a very slow reaction rate (Morey *et al.,* 1962). However, its laboratory synthesis, by direct precipitation from sea water onto crushed quartz, has been demonstrated under normal conditions of temperature and pressure within a period of three years (Mackenzie and Gees, 1971). Harder and Flehmig (1970) have also shown, from laboratory experiments, that quartz can precipitate in conditions comparable with that of a natural system by ageing the amorphous hydroxide precipitates of iron (Fe III and other elements produced in the presence of an aqueous solution containing silicic acid. Here, it seems, within a time as short as two weeks, euhedral quartz is exsolved or precipitated. However, quartz does not form, or forms only in very small amounts, when the silicic acid concentration of the solution containing the amorphous hydroxide-silica precipitate exceeds the solubility of amorphous silica (Fig. 30.5). Under these conditions polymerization of the silicic acid may inhibit quartz formation, and in still more highly concentrated solutions silica in the precipitate remains amorphous even after long periods of ageing. If these observations can be applied to

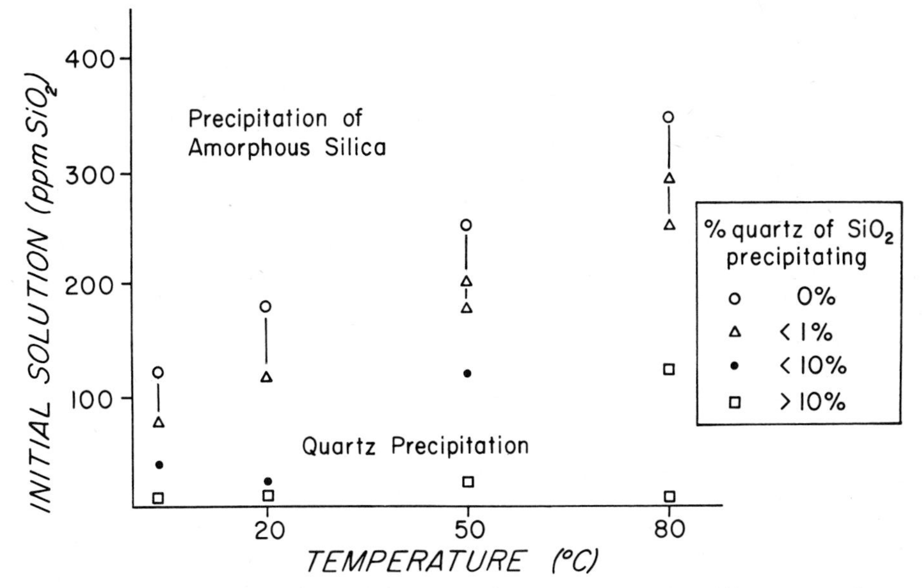

FIG. 30.5. Formation of quartz in hydroxide-silica gels as a function of the SiO_2 concentration of the initial solution and the temperature (pH = 7) (after Harder and Flehmig, 1970).

sediments it would seem that the high content of SiO_2 in most sediment interstitial waters will inhibit any precipitation of quartz, at least during the early stages of diagenesis.

Most workers agree that much of the silicon occurring in sediment interstitial waters is the result of the dissolution of biogenous silica (see Calvert, 1968; Fanning and Schink, 1969). However, Burton and Liss (1973) have opposed this argument and claim, from budget calculations, that most silica released from sediments into the overlying water comes from the alteration of unweathered aluminosilicate debris, especially in Antarctic regions.

The rate of dissolution of the silica walls of diatoms seems to vary with their water contents (Huang and Vogler, 1972), pH, and with the various ions fixed in some way to their surface (Lewin, 1961). Thus, pretreatment of diatoms with Al, Be, Fe, Ga etc. tends to inhibit dissolution. Hurd (1972) has produced further evidence to suggest that iron, magnesium and calcium aluminosilicates may form on their surfaces and reduce the rate of dissolution.

Whatever mechanism is involved in the dissolution of opaline silica in sea water (see Iler, 1955) less than 10% of the biogenous silica produced at the surface, and perhaps as little as 1%, will be found at the sediment surface. Upon burial, further dissolution takes place, with the result that only 0·05–0·15% of the original amount produced by organisms is retained within sediments (Hurd, 1972). Organic-rich sediments seem to retain more opal than do oxidising sediments. For instance, diatoms in organic rich anoxic sediments in saline lakes (pH = 9) on Mount Meru, Lahar, East Africa (Heckey, 1971) show little evidence of dissolution, even though the sediment interstitial water (pH = 8·5–9·0) contains very low concentrations of SiO_2 (<10 mg $SiO_2 l^{-1}$). The reason for this is probably the existence of an organic coating on the diatoms which inhibits dissolution. Support for this contention is given by Siever (1962) who showed that the solubility of silica gel is drastically reduced in the presence of peat water which, he presumes, produces a protective film of adsorbed organic material on the free silica surfaces. In such circumstances it is likely that amorphous silica will not attain equilibrium with the surrounding water until this film is removed.

30.4.4. DIAGENESIS OF MINOR ELEMENTS

The partitioning of elements between the terrigenous and the various authigenic and diagenetic components of marine sediments hinders our understanding of the behaviour of many elements, especially the minor ones, in sediments. Chemical treatments, designed to preferentially remove certain sediment phases, have occasionally been employed (Goldberg and Arrhenius, 1958; Chester and Hughes, 1967; Presley *et al.*, 1972) to chemically identify a particular phase, but are not always satisfactory as sometimes more than one sedimentary component is attacked. Correlation of a particular minor element, or group of minor elements, with a certain identifiable and plentiful amount of an authigenic or diagenetic component, such as iron or manganese hydroxides, certain phosphates, carbonates etc. can often produce a reasonably clear picture of the transformation and movement of minor elements during sediment burial. However, in many instances such correlations may be unreliable as a particular host component may occur only in small amounts and may be admixed with others containing the same elements.

Concentrations of many minor elements in sediment interstitial waters are often one or two orders of magnitude higher than those found in sea water (Tageyeva and Tikhomirova, 1962; Shishkina, 1964; Brooks *et al.*, 1968; Presley *et al.*, 1972). This is particularly true in shallow water oxidizing sediments. Even if the probable errors due to temperature effects upon squeezing, etc. are taken into account, such concentrations, together with the high degree of vertical variability within particular sediment profiles, imply that a considerable degree of reaction of minor metals takes place during burial. However, attempts to interpret such changes are few, and little has been done to relate them to particular and well defined reactions occurring within sediments.

Terrigenous silicate minerals are responsible for much of the minor element content of sediments. This is especially so in rapidly accumulating sediments. Thus, most of the variations in minor elements are governed by changes in terrigenous supply, and by dilution with silica and carbonates. In view of the reluctance of terrigenous silicates to alter appreciably during early diagenesis, as has been seen by the lack of change in major elements in sediment interstitial waters, it is probable that the terrigenous component plays only a small part in controlling changes in minor elements in interstitial waters.

Few of the minor elements analysed in interstitial waters are exclusively related to silicates in the sediments. Possible exceptions may include lithium and rubidium. Relative to chlorine, lithium shows little change in concentration between sediment interstitial water and the overlying sea water (Friedman *et al.*, 1969), a conclusion confirmed by Kaplan and Presley (1969) and Presley and Kaplan, (1970) for the upper parts of deep-sea cores. However, in the coastal sediments of Saanich Inlet, Presley *et al.* (1972) have noted some small and unexplained deviations between the content of lithium in some interstitial waters ($\sim$90 ppb) and the overlying sea water (115 ppb). Likewise Morozov (1969) found the rubidium content of the interstitial waters of various sediments to be similar to that of the overlying sea water. Certainly, compared with elements such as Zn, Cu, Mo, Ni, Co etc., these two elements show little variation in concentration implying that much of the change in transition and similar elements noted by many workers is not controlled by the alteration of silicates, but is probably caused by changes in the organic substances in sediments and more importantly by chemical transformations involving manganese and iron phases.

30.4.5. MINOR ELEMENTS AND ORGANIC MATTER

Marine sediments contain varying quantities of organic compounds of which humic substances, amino and fatty acids and carbohydrates constitute an important part. Various oxygen-containing and amino functional groups in such compounds especially those of low molecular weight are likely to be of considerable geochemical significance as they are known in other systems to react with various cations (see also Chapter 31). However, the problems in the identification of such products, and the assessment of their importance in the natural environment, are formidable. Much of the difficulty lies in establishing a true association between the minor elements and a specific group, or groups, of organic compounds in heterogenous substrates where elements are likely to be associated with both inorganic and organic phases. Within soils and fresh waters, and under laboratory conditions simulating these environments, the importance of exchange between metals and certain functional groupings of organic matter is reasonably well established (Kononova, 1966). In addition, the ability of organic matter to react with various metals leading to the formation of organo-metal complexes of a reasonably stable nature is also well known (Schnitzer and Skinner, 1965; Drozdova, 1968; Zunino *et al.*, 1972). The stability of these complexes under varying physico-chemical stresses, e.g. Eh and pH, and in the presence of other metals is known only from empirical assessments (Kee and Bloomfield, 1961). However, the evidence for their existence in marine sediments is largely circumstantial. Associations between metals and organic matter have often been deduced from an apparent association and correlation between the presence and the amount, and/or behaviour of, certain elements in a sediment and of specific organic groups.

The close association between many minor metals, e.g. Cu, Ni, Mo, Zn, Pb and U and the amount of organic carbon in many sediments, especially those from reducing environments (Baturin *et al.*, 1967; Price and Calvert, 1973a) is often interpreted as being due to the existence of metal-organic substances, even though the presence of co-existing sulphides and other inorganic phases may be the determining factor in these environments. Thus, the extent to which metal-organic compounds occur, and their variation, between sediments of different origin and during sediment burial is not easily appreciated.

Unlike most minor elements iodine (and bromine) is exclusively contained in sediment organic matter (Vinogradov, 1939; Sugawara and Terada, 1957; Shishkina and Pavlova, 1965). Its concentration in surface sediments is such that I/C (org) ratios appear fairly constant ($\sim 3 \times 10^{-2}$), at least on continental shelves and platforms where bottom conditions are highly oxidizing. Much of the iodine (and bromine) appears to have been incorporated into the sediment following uptake by planktonic seston (Price and Calvert, 1973), and may accumulate there through the mediation of bacteria. During burial of oxidizing sediments, iodine (and bromine) is lost relative to organic matter and often reaches I/C(org) ratios of $\sim 1 \times 10^{-3}$ at depths of 1m or less (Price *et al.*, 1970; Bojanowski and Paslawska, 1970). Further, in sediments of uniform lithology, forming under conditions where sediment accumulation rates are constant, the loss of iodine relative to carbon at depth follows a first order reaction (Fig. 30.6), and probably relates to the change of organic matter in sediments during early diagenesis. Such trends may also be affected by a certain degree of recycling of iodine between surface and buried sediments. The rates of change of I/C(org) ratios with depth appears to be dependent on sediment accumulation rates (Price *et al.*, 1970). The relatively high iodine concentrations in sediment interstitial waters ($\sim 1{\cdot}9$ ppm) from southern parts of the Baltic Sea (Bojanowski and Paslawska, 1970) also imply a release of iodine from organic matter during sedi-

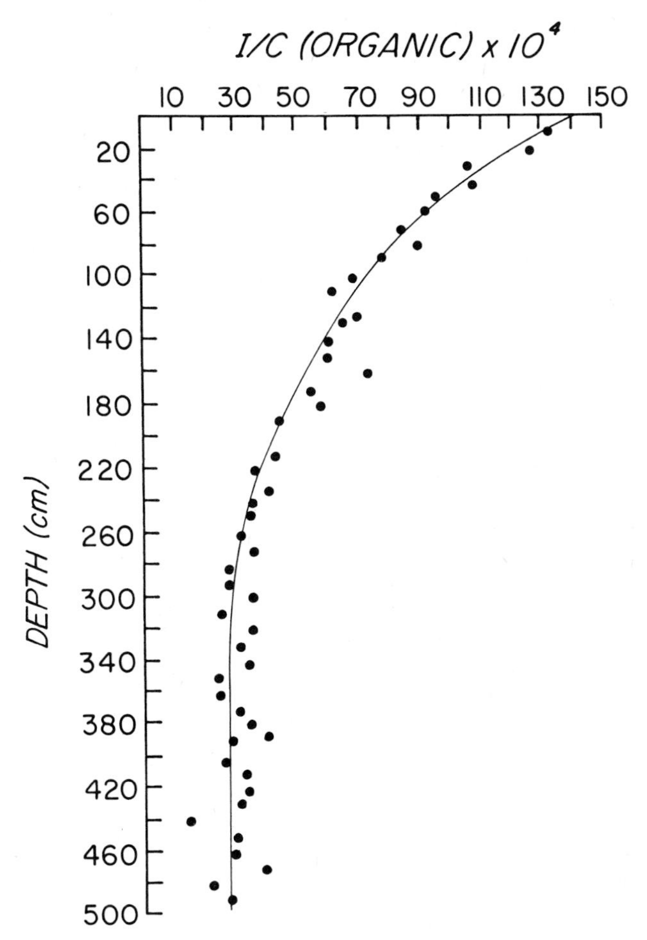

FIG. 30.6. Change in iodine/carbon (organic) ratios with depth in a sediment core (L-66) of uniform lithology from the Gulf of California (after Price *et al.*, 1970).

ment burial. In contrast, the I/C(org) ratios of surface sediments forming under reducing conditions show values one order of magnitude lower than those of oxidizing sediments, and during their burial the change in I/C(org) ratio is minimal (Price and Calvert, 1973). The differences in uptake of iodine by organic matter at the surface of oxidizing and reducing sediments, and its release or lack of release during burial, in all probability relates to differences in composition and behaviour of unspecified organic substances, formed by the degradation of organic matter, under different redox, and hence bacterial conditions.

In addition to their occurrence in silicates and inorganic precipitates many metals are known to concentrate in sediments as coordination complexes with appropriate organic molecules or in a few instances (e.g. Hg or As) as methyl compounds (see for example, Jensen and Jernelöv, 1964; McBride and Wolfe, 1971). Much of the uranium in reducing muds is also known to be organically complexed (Kolodny and Kaplan, 1973). Further, the ability of organic substances such as humates to take up metals, e.g. Ag, Co, Cu, Fe, Mn, Mo, Ni, Pb, V and Zn, in appreciable quantities is now well known (Swanson *et al.*, 1966). However, many of these complexes have not been identified and their behaviour in the marine environment is unknown. Parker (1961) and Price (1967) have observed that zinc is highly concentrated in the uppermost few centimetres of rapidly accumulating oxidizing marine sediments. Surfaces of oxidizing sediments can also show unusually high concentrations of Pb and Cu in excess of that which may be expected from their association with manganese and iron oxide or hydroxide phases. It appears that the sediment surface often acts as a reservoir for metals, many of which may occur as organically bound substances, perhaps originating through the mediation of bacteria. The depletion of Zn, Cu, Pb and I with depth in a profile of rapidly forming and lithologically identical oxidising marine sediments from Loch Fyne, Scotland (Price and Calvert, 1973a) is such that below 25 cm Zn, Cu and Pb attain constant values. These sediments are essentially free of manganese and iron hydroxides. Variations in their concentration ratios to organic carbon (Fig. 30.7) seem to be depth dependent. The changes in these ratios and also that of I(tot)/C(org) appear to be different; Zn/C(org) ratios show the most rapid and I(tot)/C(org) the least change in the uppermost 25 cm of the sediment. Relative to organic carbon the extent of the variation seems to be $Zn > Cu > Pb \gg I$. Such behaviour may imply a breakdown and release of metal-organic substances from the sediment during burial.

Investigations by Jensen and Jernelöv (1964), and Jernelöv (1969) have shown that $CH_3.Hg.CH_3$ and CH_3Hg^+ can be synthesized from inorganic mercury by bacteria in oxidizing sediments and held near the sediment surface by bonding onto various products of protein degradation. However, upon burial much of the complexed Hg is released into the overlying waters. The affinity of mercury for sulphide ion is so strong that methylation usually occurs in aerobic systems. Exceptionally, it may occur under anaerobic conditions where either the HS^- activity is low or the Hg ion concentration is abnormally high (Fägestrom and Jernelöv, 1971). Thus, in most marine anoxic environments Hg is likely to be retained in the sediment as sulphide. Much of the Zn, Cu, Pb and Mo in anoxic sediments may also occur as the sulphides, and consequently the elements will show little tendency to be lost from the sediment during and after burial.

Mobilization of both organic and inorganic forms of metals during sedi-

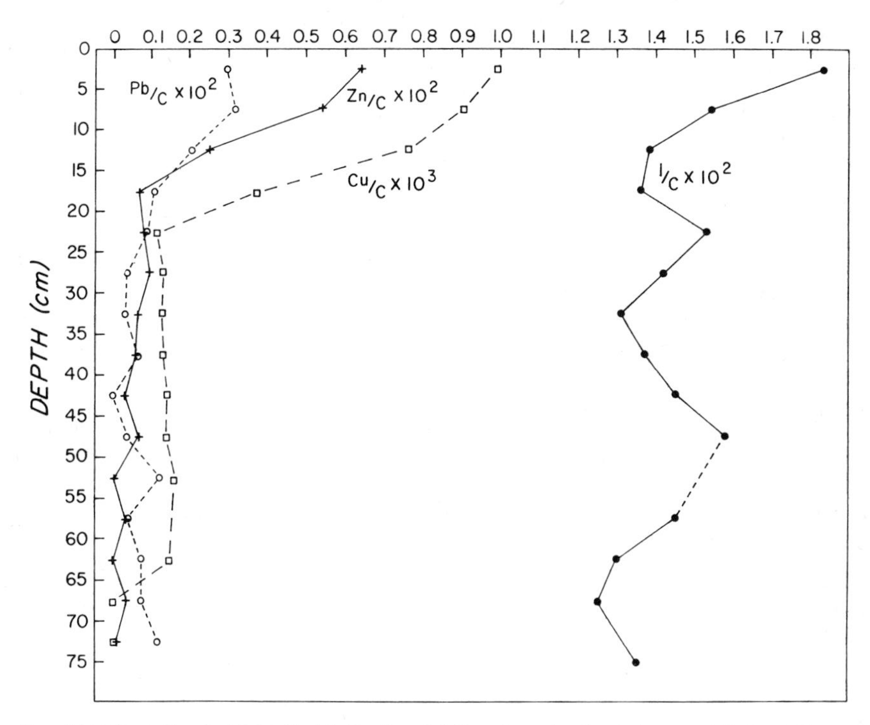

FIG. 30.7. Variation in Pb/C, Zn/C, Cu/C and I/C ratios with depth in an oxidizing sediment core from Loch Fyne, Scotland (Price and Calvert, unpublished data).

ment burial is likely to result in some elements being organically complexed in sediment interstitial waters. Often the measured concentrations of minor metals in interstitial waters exceed the solubility of phases most likely to precipitate, and the existence of metal-organic complexation in such waters has been inferred (see Presley *et al.*, 1972). However, real evidence for complexation is difficult to obtain, and the relative proportions of inorganic and organically bound forms of metals are as yet, unknown.

30.4.6. DIAGENESIS OF IRON, MANGANESE AND PHOSPHORUS

The behaviour of manganese, iron and their associated elements in both sea water and sediments has attracted considerable attention in recent years. Much of this interest concerns the dissolution, migration and precipitation of iron and manganese phases in media of different pH and, in particular, under changing redox conditions.

Many oxidizing surface sediments become reducing when they are buried. The attendant reduction of iron and manganese will dictate their chemistry in both the sediments and their interstitial waters. Reduction of these elements at some depth in a sediment can often be discerned by a marked colour change; brown coloured sediments which prevail in oxidizing conditions rapidly give way to grey coloured sediments where the conditions are reducing. Within continental shelf and platform sediments this colour change is usually well defined, and takes place within the uppermost few centimetres ($\sim$5 cm). Exceptionally, it may take place at greater depths. For instance, oxidizing sediments varying in thickness from a few millimetres to 18 cm occur beneath the Kara Sea (Zenkevitch, 1963), and even greater thicknesses are found in the highly oxidixed sediments in some part of the Gulf of Bothnia (Gripenberg, 1934). In oceanic sediments the thickness of the upper oxidized sediment layer usually increases seawards from a few centimetres in the marginal areas to a metre or more in truly oceanic sediments (Fomina, 1962; Skornyakova, 1965; Lynn and Bonatti, 1965); in central oceanic regions such colour changes are usually not seen. Interestingly, the depth to which molecular oxygen extends in coastal sediments is shallower than that to which oxidized Mn and Fe are found (Kanwisher, 1962), and this perhaps indicates the importance of sediment overturn by the benthic community to the establishment of a near surface oxidizing substrate. The penetration of ^{238}Pu to depths of 15 cm in open ocean sediments tends to support this conclusion (Noshkin and Bowen, 1973).

In general, the thickness of the oxidizing layer in hemipelagic and pelagic sediments seems to be inversely dependent on the amounts of metabolisable organic matter contained in the sediment, which in turn is controlled to a considerable extent by the total sediment accumulation rate (Price and Calvert, 1970). In addition to variations in the supply of manganese and iron from overlying waters, the latter factor also dictates the amounts of dispersed manganese and iron in surface sediments. Sediments of low accumulation rate, such as those occurring in areas of scour and in the central oceanic regions may build up relatively high levels of dispersed manganese and iron because of a lack of dilution by other sedimentary materials.

Oxidation of organic matter by biological agents will cause a downward depletion of oxygen in sediments, and as a consequence of this, and other reactions, lower Eh (pE) values may be expected at depth. In such a situation dispersed iron and manganese, which, under highly oxygenated conditions (pE > 11) occur only as Fe^{3+} and Mn^{4+} hydroxides or oxides, will at depth be reduced to Fe^{2+} and Mn^{2+} ions. Thus, concentrations of soluble iron (II) and manganese (II) will increase at the expense of Fe (III) and Mn (IV). Their highest concentrations in sediment interstitial waters will be controlled by two factors. Firstly, by the solubility of their respective carbonates, or for Fe, and less so for Mn, by sulphide precipitation. Secondly, by an upward diffusion of ions out of the reduced zone, and their reprecipitation as

Fe (III) and Mn (IV) either at, or near, the sediment water interface or in the overlying sea water.

The dissolution and reprecipitation of Fe^{3+} as hydroxide tends to occur at considerably lower pE values ($\sim$4) than the equivalent reaction involving manganese (pE $\sim$ 7·5) (See Fig. 30.8; and Stumm and Morgan, 1970). Further,

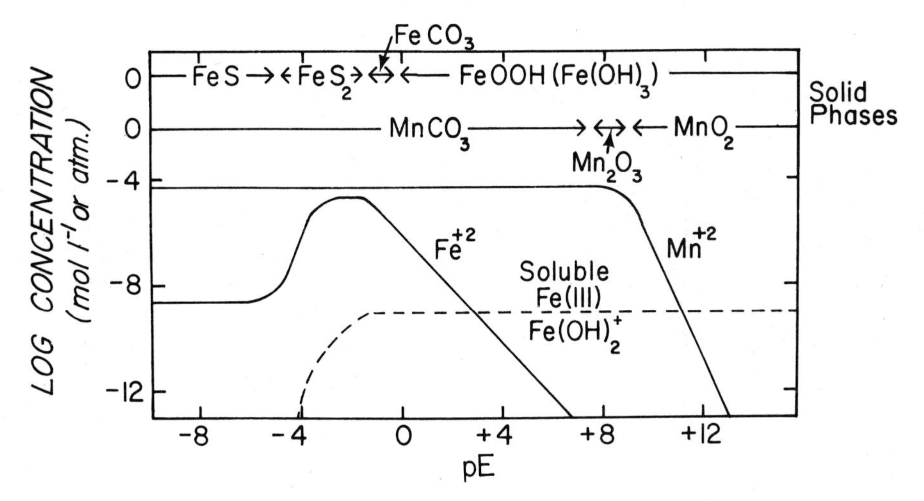

FIG. 30.8. Equilibrium concentrations of pertinent redox components of iron and manganese as a function of pE at a pH of 7·0, calculated using standard equilibrium constants (from Stumm and Morgan, 1970).

Fe^{2+} acts as a reducing agent with respect to higher oxides or hydroxides of manganese which can attract appreciable amounts of Mn(II) into their structures (Stumm and Morgan, 1970). The gross effects of such reactions during sediment burial are that Mn^{2+} will appear prior to Fe^{2+} upon a progressive lowering of pE in interstitial waters and will also reoxidize to Mn (IV) phases (e.g. δ-MnO_2) closer to the sediment water surface than will Fe (III). Hence, the apparent vertical movement of Mn in sediments is more pronounced than that of iron.

The ability of Fe (III) to complex with ortho-phosphate is an important feature in the cycle of phosphorus in sediments. Fe^{3+} can interact chemically with ortho-phosphate to form insoluble phosphate-iron (III) complexes. Furthermore, under slightly acid conditions $FePO_4$ may precipitate within certain sediments (Nelson, 1967), especially those of estuarine and non-marine types. In marine environments of more alkaline pH a metastable ferric compound containing both PO_4^{3-} and OH^- in variable proportions will precipitate; the ratio PO_4^{3-}/OH^- being dependent on the pH of the environment (Morgan and Stumm, 1965). During sediment burial the phosphate and its associated Fe (III) are likely to be redistributed. High concentrations of phosphate (up to 3·5% P_2O_5 by weight) are associated with Fe (III) in many shallow water ferro-manganese oxide concretions (Winterhalter and Siivola, 1967; Sevast'yanov and Volkov, 1966; Calvert and Price, 1971). This is likely to be caused by a release of PO_4^{3-} ions during sediment burial and subsequent reprecipitation with Fe (III) in concretionary form near the sediment surface. Phosphate contents of shallow water ferro-manganese oxide concretions seem to be considerably higher than those found in pelagic Fe–Mn concretions or nodules. This may be due to the difference in availability of phosphorus (from organic matter) in the two environments. Nevertheless, many deep-sea sediments (El Wakeel and Riley, 1961), and their contained nodules (Mero, 1965), can contain unusual levels of P_2O_5 suggesting that much of the phosphorus in them is associated with metal oxides. For instance, the phosphorus in eastern Pacific Ocean sediments (Bonatti *et al.*, 1971) can show a threefold decrease in P_2O_5 content between the surface oxidized sediment (0·4%) and the underlying reducing sediment (0·11%) (Table 30.5). Although this has been interpreted as being due to the solubilisation of calcium phosphate across a "redox interface" in the sediment (Bonatti *et al.*, 1971), it is more likely to have been caused by dissolution of ferric oxide containing phosphate.

Both arsenic and yttrium are often associated with iron and phosphorus enriched oxidized sediments and are known to be highly concentrated in Fe-rich ferro-manganese nodules (~300 ppm As and ~50 ppm Y) (Sevast'yanov, 1967; Fomina and Volkov, 1969; Calvert and Price, 1971). Arsenic probably occurs as an arsenate replacing phosphate in iron phosphate, and yttrium may exist as yttrium phosphate, its most likely form in the marine sedimentary environment (Sillén, 1961). In oxidized sediments the behaviour of arsenic closely parallels phosphorus (Sevast'yanov, 1967). Under reducing conditions reduction to $H\,AsO_2$ and $As\,S_2^-$ will take place, depending on the a_{HS^-} and pE, and will result in either its solubilization or stabilization within the reducing zone.

The fixation of phosphorus as Fe (II)-orthophosphate (vivianite) is known in many soils and non-marine sediments (Mackereth, 1965; Kjensmo, 1968; Rosenquist, 1970) and may also be important in controlling the geochemical migration and immobilization of phosphorus within marine sediments. However, its existence within oxidizing marine sediments although expected (Garcia, 1969) has not been confirmed by mineral identification. Its occurrence in anaerobic sediments is also possible where sulphide ion activities are low ($a_{HS^-} > 10^{-1\cdot5}$), and phosphate concentrations are high ($a_{HPO_4^{2-}} > 1\cdot0$; Nriagu, 1972). Where sulphide concentrations are higher, and/or phosphate is lower, the precipitation of Fe (II) as sulphide precludes fixation of PO_4^{3-} as iron phosphate. However, in situations where $CaCO_3$ is present

TABLE 30.5

Variations in major and minor elements in the uppermost 20 cm of hemipelagic sediments in core P6702–59. 2°45′N, 85°20′W, water depth 3274 m (East Pacific) (after Bonatti *et al.*, 1971)

Depth cm	$CaCO_3$ %	Al_2O_3 %	MnO %	FeO %	P_2O_5 %	Co	Ni	La	Cr	U	V	Fe^{+3}/Fe^{+2}	S %
0–1	38·5	5·4	3·6	2·5	0·35	40	240	33	21	0·45	88	19·0	0·38
1–3	40·0	5·2	4·2	2·7	0·43	33	325	35	27	0·45	77	7·8	0·36
3–4	36·5	5·9	3·6	2·5	0·42	24	315	32	26	0·52	77	7·2	
4–5	38·5	6·6	4·1	2·7	0·42	27	360	35	28	0·59	86	11·0	0·36
5–6	36·5	5·3	2·7	2·5	0·39	25	255	28	29	0·53	90	6·6	0·37
6–7·5	38·3	6·1	0·84	2·0	0·20	10	190	10	34	0·55	115	5·2	0·37
7·5–9	42·3	5·2	0·32	2·6	0·15	12	145	11	23	0·63	9	5·4	0·38
9–11	46·3	4·2	0·24	2·5	0·14	13	130	11	28	0·61	100	5·4	0·39
11–12	50·2	4·7	0·25	2·4	0·11	12	145	11	34	0·69	110	2·8	0·39
12–14	54·1	5·0	0·27	2·5	0·11	10	170	<10	35	0·82	117	4·1	0·38
14–17	54·1	4·5	0·24	2·2	0·12	7	145	<10	34	1·13	110	6·1	0·26
17–20	50·1	4·0	0·27	2·3	0·10	9	155	<10	42	1·37	110	5·9	0·34

Major elements expressed as wt %. Minor elements expressed in ppm by wt. Analyses for S and Fe^{+3}/Fe^{+2} were made on subsamples different from those used for the other elements.

in a sediment the interstitial PO_4^{3-} concentration may be controlled by the precipitation of calcium phosphate. This has recently been observed in sulphide-rich anoxic sediments and will be discussed below.

The roles of diffusion and advection of chemical species in sediment interstitial waters have been examined by Anikouchine (1967), Manheim (1970) and Tzur (1971). Such processes are important in controlling the dissolution of many elements in sediments, particularly manganese (Lynn and Bonatti, 1965; Bender, 1971; Michard, 1971). However, the theoretical approaches by these workers have only had a limited success because of the small amount of data available on the concentration of dissolved minor elements in sediment interstitial waters. In particular, most published information does not include data on vertical profiles which are most useful when examining the effects of diagenesis and the consequent diffusion of metals within sediments. From diffusion controlled models the behaviour of manganese and related elements, upon burial and reduction would be expected to follow an exponential downward increase in concentration in the sediment interstitial waters which, together with sediment compaction, will cause an upward diffusion of ions through the sediment. When conditions are favourable, their reprecipitation may occur near the sediment-water surface. However, the presence of a redox interface at a particular depth within a sediment, and the reaction of metals with anions, can cause considerable departures from the concentration profile outlined above. Li

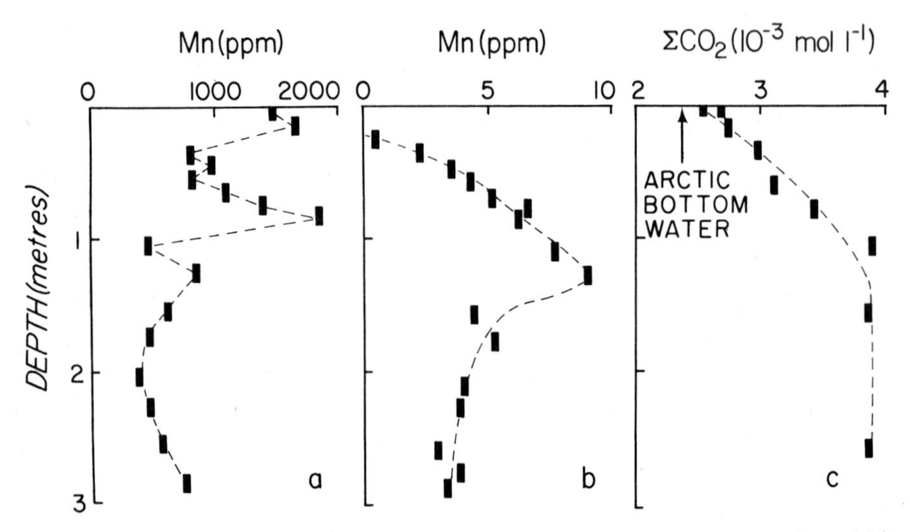

FIG. 30.9. The content of manganese in the sediment (a); its concentration in the interstitial waters (b), and the total inorganic carbon concentration in the interstitial water (c), Data for a core from the Arctic Basin (82°N, 156°W) (after Li *et al.*, 1969).

et al. (1969) have shown that although the concentration of manganese in other interstitial waters of Arctic Basin sediments increases downwards for the first metre or so, and thus follows the expected trend, at greater depth its concentration decreases significantly (Fig. 30.9). The position of maximum manganese concentration coincides with that of ΣCO_2. The decrease in manganese concentration in the lower parts of the profile is probably a reflection of equilibrium with respect to $Mn CO_3$, although mineralogical confirmation of this phase could not be shown. Calvert and Price (1972) have observed similar decreasing concentrations of interstitial manganese at depth in certain inshore marine sediments. This trend has been attributed to the precipitation of manganese carbonate, which has been identified in the sediment, and to a downward exponentially decreasing concentration of dissolved Mn, caused by a diminishing quantity of solid Mn (e.g. MnO_2) at depth. Similarly, pelagic sediments show evidence that manganese is to some extent reactive. For instance, manganese concentrations in the interstitial waters of various sediment types from marginal and pelagic regions of the Pacific Ocean range from ~1 to 6600 ppb (Presley *et al.*, 1967). The nickel (~400 ppb) and cobalt (~20 ppb) concentrations in some of these interstitial waters are considerably higher than those in sea water. However, it is interesting to note that in those sediments in which interstitial manganese concentrations are low (e.g. anoxic sediments, Presley *et al.*, 1967; Presley *et al.*, 1972), concentrations of minor elements such as Zn, Cu, Ni and Co are also low. It would seem that a major proportion of these elements is released into sediment interstitial waters as a consequence of the dissolution of hydrous manganese oxides and perhaps to a lesser extent of hydrous iron oxides.

Reprecipitation of interstitial manganese and iron at, or near, the sediment surface leads to the formation of disseminated oxides/hydroxides, or ferro-manganese oxide concretions or nodules. Manheim (1965) has suggested that ferro-manganese nodules in the sediments off continental shelves and platforms originate from a precipitation of metals from the overlying sea water. Price and Calvert (1970) have modified this view, believing that only those oceanic nodules forming on topographical highs, such as central Pacific seamounts, develop solely as a consequence of metal precipitation from sea water.

The differences in composition and mineralogy between ocean floor nodules and those occurring on seamounts have been noted by many workers (Mero, 1965; Barnes, 1967; Cronan and Tooms, 1969; Price and Calvert, 1970). The former tend to be enriched in Cu, Ni, Zn, Mo and other elements while the highest concentrations of Co, Pb, Ti and Ce occur in seamount nodules. These differences may be associated with differences in the oxidation states of Mn (i.e. Mn(II)/Mn(IV) ratios) in different nodules (Price and

Calvert, 1970). However, accurate and reliable measurements of Mn(II)/Mn(IV) ratios are difficult to obtain since post-formation oxidation of manganese often occurs (Skornyakova *et al.*, 1962).

Nodules forming in shallow water marine sediments usually show minor element concentrations one or more orders of magnitude lower than their oceanic equivalents (Goldberg *et al.*, 1963; Manheim, 1965; Winterhalter, 1966; Calvert and Price, 1971). This may be due either to the unavailability of minor elements in shallow water marine environments, or to a difference in the adsorptive behaviour of minor metals which may occur if they are organically chelated within the interstitial environment (Price, 1967). Differences in the specific surface area of Mn oxides may also be important in this respect.

The growth rates of shallow water nodules ($\sim$0·3 mm per 10^3 yrs., Ku and Glasby, 1972) are two or more orders of magnitude higher than those of open ocean nodules ($\sim$1–6 mm per 10^6 yrs; Bender *et al.*, 1966; Barnes and Dymond, 1967; Ku and Broecker, 1969). The greater degree of diagenetic remobilization of manganese in shallow water sediments is almost certainly responsible for the much higher growth rate of nodules in, and on, these sediments.

Regional variations in the composition of nodules from the Pacific Ocean floor have also been explained in terms of a complex interplay between the rates of sedimentation and assimilation of organic matter into sediments and the level of diagenetically mobilized manganese within sediment interstitial waters (Price and Calvert, 1970). That regional changes in the level of diagenetic mobility of manganese and iron occur in oceanic sediments is also demonstrated by the work of Skornyakova and Andrushenko (1964) who showed, on the basis of a carbonate and biogenous silica free analyses of oceanic sediments and ferro-manganese nodules, that the ratios Mn (nodule)/Mn (sediment) and Fe (nodule)/Fe (sediment) vary systematically throughout the Pacific Ocean. The Mn ratio is highest (80–100) in the eastern marginal Pacific, intermediate (40–60) in the equatorial oozes, low (30–35) in the north-west pelagic clays and lowest (10–40) in the South Central Pacific. The Fe ratio has a similar but less pronounced regional trend.

Manganese in the reduced state (as manganese carbonate) is known to often occur as concretions in marine sediments of both shallow water (Zen, 1959; Manheim, 1961; Calvert and Price, 1971) and oceanic origin (Lynn and Bonatti, 1965). The composition of the concretions is rarely that of rhodochrosite, but usually comprises an admixture of the 3-phase system $MnCO_3$–$CaCO_3$–$MgCO_3$, and often has almost equi-molecular proportions of Mn and Ca carbonates with subordinate amounts of Mg and Fe. Unlike ferro-manganese oxide nodules, the minor element content (e.g. Co, Ni, Cu, and Zn) of $(MnCa)CO_3$ is low and the Sr content is close to that found in inorganically precipitated calcite (Calvert and Price, 1971). The presence of this carbonate phase in sediments must imply a high activity of both Mn^{2+} and bicarbonate ions within the interstitial environment.

The occurrence, and abundance, of manganese carbonate in sediments (especially silty sediments) in ancient rock sequences has been considered by Strahkov (1969) to be due to the interaction of Mn^{2+} and Ca^{2+} ions with CO_2 between sediment layers which originally had a different porosity. The tendency towards a build-up of a high partial pressure of CO_2 and a high metal bicarbonate ion-pair concentration resulting from bacterial action in organic-rich mud horizons will be relieved by migration of CO_2 and the bicarbonate to adjacent silt horizons in which the partial pressure of CO_2 is lower. Precipitation of a mixed carbonate will consequently occur within the silts. The mechanism establishes a pattern of continued migration of bicarbonates and CO_2 from the fine-grained to the coarse-grained sediments, and hence a segregation of metal carbonates at specific horizons within a sediment profile may take place.

Siderite ($Fe\,CO_3$) is thermodynamically stable under very much more restricted conditions than is manganese carbonate, and the limitations on its precipitation are such that it is unlikely to precipitate in the marine environment. It has not yet been found to be forming in Recent marine sediments, although it is known to occur at depth in certain deep-sea drillings. For siderite to exist, both the pE and the HS^- activity must be low. As low pE values usually result from the anaerobic decompositon of organic matter, which as a matter of course nearly always involves reduction of sulphate to HS^-, the conditions necessary for siderite formation are rarely encountered in the marine system. Siderite can only form in sea water in which sulphate reduction is inhibited so that metastable concentrations of SO_4^{2-} persists (Berner, 1971). Thus, in some respects the existence of siderite in marine sediments suffers many of the constraints of vivianite. However, within nonmarine environments, in which SO_4^{2-} concentrations and hence those of HS^-, are usually low, the only factors restricting the formation of these two minerals are changes in pH and the presence of competitive concentrations of elements such as calcium. Hence, both vivianite and siderite have been found in many nonmarine environments (see Rosenquist, 1970; James, 1966). Siderite cannot form in water containing H_2S, nor can it form in waters rich in dissolved Ca^{2+}, and Berner (1971) has calculated for the reaction,

$$Fe\,CO_{3\,siderite} + Ca^{2+} \leftrightharpoons Ca\,CO_3 + Fe^{2+}\,aq$$

$$K = 0{\cdot}5 = \frac{a_{Fe^{2+}}}{a_{Ca^{2+}}}.$$

Thus, for siderite to be stable relative to calcite the concentration of iron must be greater than 5 per cent of that of Ca. In sea water it is less than 0·1 per cent (Berner, 1970a), and it is doubtful if the appropriate $a_{Fe^{2+}}/a_{Ca^{2+}}$ ratio exists in the interstitial environment of Recent marine sediments. The development of siderite which occurs in many sediments, may relate to events in the later stages of diagenesis and probably takes place only after HS^- and Ca^{2+} are diagenetically fixed in a sediment.

The occurrence of iron sulphides, e.g. pyrite (FeS_2), its dimorph marcasite, and minerals approximating to FeS (commonly known as hydrotroilite) in marine sediments, seems to be dependent both on the supply of reactive iron and on the abundance of metabolizable organic matter. The latter factor is often of greater importance as a requirement for the growth of sulphate reducing bacteria which produce H_2S by reducing the sulphate of sea water and interstitial water. The sulphur content of the organic matter, which averages about 1% of its dry weight (Kaplan *et al.*, 1963), is insufficient to contribute significantly to the sulphides seen in sediments, and this has been confirmed by studies of sulphur isotopes in various sulphur containing minerals (Kaplan *et al.*, 1963).

As a consequence of the importance of organic matter in sulphide genesis, iron sulphides are usually much more prevalent in shallow water marine sediments than in oceanic sediments. In addition, the presence of much iron sulphide in deep water organic-rich anoxic sediments, such as those which occur in the Black Sea (Ostroumov, 1953; Volkov, 9161), is perhaps more a reflection of the concentration of H_2S in the overlying water than of the organic matter content of the sediments. However, in instances where a plentiful supply of sulphide ion occurs, the abundance of iron sulphide species may be limited by the amounts of reactive iron reaching the anoxic environment (Ostroumov *et al.*, 1961). In oceanic sediments iron sulphides seem only to occur in the upper levels of sediment profiles if the content of organic matter is high. For instance, Shishkina (1964) believes that sulphides exist in sediments of certain Pacific Ocean trenches in which both the sediment accumulation rate and the content of organic matter are high relative to those in adjacent regions.

As the rate of sulphate reduction in a sediment is assumed to be directly proportional to the concentration of metabolisable organic matter (Berner, 1970b), there should be a direct relationship between the concentration of available organic matter and the fraction of total iron converted to pyrite. Berner has shown that this relationship exists in near surface coastal sediments off Connecticut (Fig. 30.10).

The iron sulphide minerals which are first formed by the reaction of reactive Fe and HS^- are mackinawite (tetragonal $Fe_{(1+x)}S$) and/or greigite (cubic Fe_3S_4) (Berner, 1962; Polushkina and Sidorenko, 1963; Jedwab,

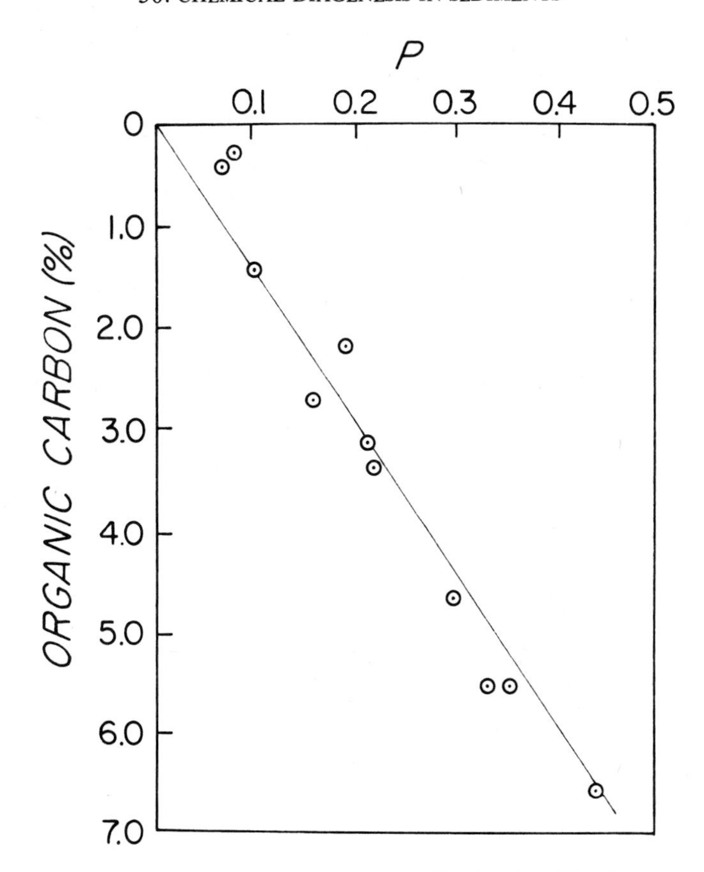

FIG. 30.10. A plot of P (=Fe pyrite/(Fe pyrite + Fe HCl), where Fe HCl refers to iron soluble in hot concentrated HCl solution) vs. percent organic carbon for the top 2 cm of some Connecticut coastal sediments (from Berner, 1970b).

1967). Because of their fine grain size they usually impart a black colour to sediments in which they occur. This is especially true in those sediments which accumulate in basins, such as some fjords and restricted (heavily polluted) bays, in which H_2S exists in the sea water. In other areas, where H_2S is found only within the interstitial water, the sediments are often olive green rather than black in colour. When it is present the black colour usually disappears with depth (van Straaten, 1954), and the boundary between the black and the underlying grey sediment is often well marked. Such a change implies that both mackinawite and greigite are metastable with respect to pyrite. However, Volkov (1961) and Berner (1970c) have noted that black sediments can exist at depth within certain areas of the Black Sea, and have attributed this to an arrested stage of diagenesis caused by a paucity of H_2S

and elemental sulphur in the interstitial environment. Berner (loc. cit) has attributed this to fluctuations in the salinity and hence in the sulphate concentration of Black Sea water during Pleistocene times (see Chapter 15).

The mechanism by which mackinawite and greigite are transformed to pyrite is not well understood, but reactions involving elemental sulphur may be involved (Ostroumov *et al.*, 1961; Volkov, 1961; Berner, 1970b), e.g.

$$Fe_3S_4 + 2\ S^0 \rightarrow 3\ FeS_2.$$

However, even though elemental sulphur can be produced by bacteria it is uncertain whether that in anaerobic sediments it has been produced in this way. Experimental studies by Rickard (1969) and Berner (1970b) have shown that pyrite, sometimes in framboidal form, will develop at low temperatures and neutral pH values by the reaction between iron monosulphides and elemental sulphur in the presence of H_2S. In the absence of elemental sulphur there is no tendency for pyrite to form.

From the results of laboratory experiments, Berner (1969) has deduced that iron and sulphur migrate within anaerobic sediments during the early stages of diagenesis. He has shown that variations in the content of organic matter in an otherwise homogenous anaerobic sediment can bring about the migration of both elements, and in some cases can lead to the localization of iron sulphide minerals

30.4.7. FORMATION OF IRON SILICATES

The geneses of chamosite in ancient ironstones and of glauconite, which is abundant in many ancient rock sequences and recent sediments, have been persistent problems. However, as a result of the work of Strahkov (1959) there is some general agreement that post-depositional processes are most important in their formation. James (1966) has concluded that the large quantity of iron required to form rock sequences containing much iron silicate was derived either from rock weathering under a variety of conditions, or from volcanic activity. These and other arguments about the origin of these two minerals have not been resolved, partly because of the puzzling absence of these types from iron minerals in modern marine environments.

Von Gaertner and Schellmann (1965), Porrenga (1966, 1967), Rohrlich *et al.* (1969) and Emelyanov (1971) have described chamosite from Recent sediments on the continental shelves of Liberia, Guinea and Sarawak, the Niger and Orinoco deltas and from sea lochs in Scotland. In most instances, especially the latter, the chamosite occurs within, and around, faecal pellets and as coatings around detrital silicates. It appears to have a poorly ordered structure, and to be iron-rich, which suggests that iron almost completely fills the octahedral site of the mineral (Rohrlich *et al.*, 1969). Sometimes faecal remains also contain hydrobiotite (Emelyanov, 1971), which may be mistaken for glauconite.

The mode of occurrence of the chamosite suggests a recent diagenetic origin; chamosite crystals are arranged tangentially around the faecal pellets, and also appear to replace detrital silicates. This concentric structure is thought to represent an early stage in the formation of oolitic chamosite which is characteristic of many ancient ironstones (Rohrlich *et al.*, 1969). It seems clear that the physiographic environment is not critical since this mineral occurs in very different situations, and factors such as sedimentation rate, bacterial activity and the content of organic matter of the sediments may play a far more important role (Rohrlich *et al.*, 1969).

It has been shown that glauconite, some of which may be of Recent origin, exists in aerobic sediments of shallow to intermediate water depths (see for example Ehlmann *et al.*, 1963; Pratt, 1961; Porrenga, 1967). It also seems that much of the glauconite initially formed is poor in potassium but that it becomes richer with later diagenetic modification (Porrenga, 1967). However, much of the glauconite observed may be an admixture of hydrogoethite, chamosite and hydrobiotite which occur in faecal remains and shell infillings off the west coast of Africa (Emelyanov, 1971). The occurrence of glauconites with high Fe^{3+}/Fe^{2+} ratios in sediments containing organic matter and whose interstitial waters contain dissolved oxygen suggests that the mineral is stable in an environment of intermediate and probably fluctuating redox potential (Berner, 1971).

30.4.8. PHOSPHORITE FORMATION

Phosphorite is a general term describing sedimentary deposits of marine origin composed mainly of carbonate apatite. Phosphorites are typically found as structureless, or layered, concretionary masses, as pellets, and crusts, and as altered skeletal remains on many continental shelves and slopes. Since their discovery by Murray and Renard (1891) they have been reported from many areas, especially the western shelves of continents (McKelvey, 1967; Tooms *et al.*, 1969), usually between the 40° parallels of latitude. They also occur on the eastern coasts of continents, for example on the Blake plateau off Florida (Pratt and McFarlin, 1966), the Atlantic coastal plain (Pevear and Pilkey, 1966), south east of New Zealand (Summerhayes, 1967), and in the Gulf of Aden (Gevork'yan and Chugunnyy, 1970). Careful investigations of these occurrences are rare, and in almost all instances palaeontological or isotope dating shows them to be old (Miocene–Pleistocene). The rarity of phosphorite as true Recent deposits has suggested to many workers that special conditions are required for their formation, and the occurrence of phosphorite in many areas where upwelling of nutrient rich waters are brought onto shallow banks or platforms leads to the conclusion that these conditions are an important factor governing the formation of this mineral complex.

The principal phase in phosphorites is usually carbonate-fluorapatite, which has the approximate formula $Ca_{10}(PO_4, CO_3)_6F_{2-3}$ (Altschuler *et al.*, 1958). When it is crypto-crystalline and virtually isotropic it is termed collophane (Dietz *et al.*, 1942). Anisotropic francolite (a carbonate–fluorapatite) and common dahlite (a carbonate-hydroxyapatite) develop chiefly by the substitition of CO_3^{2-} groups for PO_4^{3-} groups (McConnell, 1952; Altschuler *et al.*, 1958; Ames, 1959). The charge imbalance consequent on this substitution is compensated for by the introduction of F^- (Gulbrandsen *et al.*, 1966) and/or OH^- (Altschuler *et al.*, 1958). Other important substitutions are Na^+ for Ca^{2+}, Si^{4+} for P^{5+} and SO_4^{2-} for PO_4^{3-}. In addition, certain minor elements, notably U and the rare earth elements, are found to be highly concentrated and may give clues to the origin of phosphorites. However, a major problem in evaluating the composition of phosphorites is the admixture of carbonate-apatite with detrital and diagenetic minerals and with organic matter.

The considerable controversy concerning the mechanism of phosphorite formation is partly due to the difficulty of obtaining Recent material for study. In the main, arguments for their origin rest on two basic hypotheses: (i) a direct inorganic precipitation from sea water; or (ii) phosphatization of existing sediments by some diagenetic process.

The basic thesis for direct inorganic precipitation has been modelled on Kazakov's (1937) work and invokes the concept that the ascent of phosphate-rich water in regions of upwelling involves a decrease in the partial pressure of CO_2. The consequent increase in pH results in phosphate supersaturation and precipitation of carbonate fluorapatite (see McKelvey, 1967). Although certain criticisms have been levelled at Kazakov's hypothesis, colloidal precipitation of apatite has also been favoured by Dietz *et al.* (1942). Various authors (e.g. Emery, 1960; McConnell, 1965; Roberson 1966) have inferred that the bulk of ocean water is near saturation with respect to carbonate apatite. In the absence of data on the degree of saturation of calcium phosphates in bottom waters, it is impossible to evaluate the importance of direct precipitation in the formation of phosphorite.

Most of the available data indicate that phosphorites form as a consequence of diagenetic reactions. Phosphatization of skeletal carbonate was first recognized by Murray and Renard (1891) and subsequently by many other workers (see d'Anglejan, 1967), but the geological period at which phosphatisation took place is uncertain. In most instances in which the formation of carbonate-apatite has been claimed to be contemporary (Goodell,

1967) real evidence is lacking (d'Anglejan, 1968), and the dating of phosphorites using the $^{234}U/^{238}U$ disequilibrium method (Ku, 1965; Kolodny, 1969; Kolodny and Kaplan, 1970; Veeh, 1967; Veeh *et al.*, 1973) has almost invariably shown it not to be Recent. Until a few years ago the only well documented case of Recent phosphatization in sediments was that of phosphatized wood from the Gulf of Tahuantepec (Goldberg and Parker, 1960). However, Manheim *et al.* (1973) have shown that Recent Foraminifera in sediments off Chile and Peru are undergoing phosphatization. Evidence from a similar sediment type suggests that phosphorite is forming on the inner shelf on the continental margin of South West Africa (18–24° S). From studies of these phosphorites Baturin (1969, 1970), Baturin *et al.* (1972) and Price and Calvert (1973) have suggested a contemporary formation for some of the carbonate-apatite occurring in diatomaceous, organic-rich and H_2S-bearing sediments, and have evoked the well known chert-carbonaceous shale-phosphorite association of ancient rocks in their interpretation. Phosphatized clots of mud and phosphatized lumps of excrement are recognizable in the sediment at a level of a few percent by weight (8% P_2O_5, Table 30.6) (Baturin, 1969). Further examination shows phosphate to be enriched in yellow/green coloured unconsolidated clots and as thin bands, within presumably porous horizons, in laminated sediments. There appears to be a gradation from this unconsolidated material, through friable phosphorite, to compact or lithified lumps with high phosphorus contents,

TABLE 30.6

Analysis of phosphorites from Recent sediments of the shelf off South West Africa. (Price and Calvert unpublished data)

	Unconsolidated			Friable concretions		Pelletal and massive concretions	
			*		*		*
SiO_2	43·0	30·60	44·0	1·80	4·65	4·60	0·15
Al_2O_2	0·10	0·12	—	0·07	—	0·10	—
Fe_2O_3	0·67	0·81	1.12	0·95	1·00	2·53	0·20
CaO	22·00	26·50	6·96	50·90	42·42	45·50	46·24
P_2O_5	11·70	16·00	8·32	31·60	28·66	29·40	32·09
F	1·30	1·60	0·34	2·23	3·02	2·03	3·02
CO_2	—	—	1·17	—	5·34	—	6·33
Sr	1140	1200	—	2070	—	2080	—
U	70	45	—	130	—	110	—
Mo	45	30	—	30	—	15	—
Y	7	5	—	56	—	230	—
Ce	32	41	—	72	—	156	—
La	<10	10	—	10	—	90	—

* Data from Baturin (1969).

i.e. 32% P_2O_5 (Table 30.6). These various phosphorites are thought by Baturin (1969) to represent different stages in a single diagenetic process. However, measured $^{234}U/^{238}U$ ratios of the friable and indurated phosphorite are both close to those existing at radioactive equilibrium (Veeh *et al.*, 1973) suggesting that these phosphorites are not being formed in the Recent sediments in which they now occur, or that the time for consolidation is considerable (perhaps 10^5 years). The age of the unconsolidated phosphorite has been shown to be Recent (Baturin *et al.*, 1972).

Sediments seem to act as reservoirs of phosphorus in natural systems, and the phosphate concentration in waters overlying and within sediments is buffered by solubility and adsorption or ion exchange equilibria (Stumm and Morgan, 1970). Although the nature of this control is not really understood, there seems to be general agreement that chemical interactions of phosphate (orthophosphate) with Ca^{2+} are important. Such interactions appear to be dependent on the availability of Ca^{2+} ions relative to others (e.g. Fe^{3+}, Al^{3+}), on the pH of the interstitial environment, and on the susceptibility of clay minerals to take up phosphate (Weaver and Wampler, 1972).

Evidently, phosphate-rich solutions are capable of converting calcite to carbonate-apatite at the surface, and within, a sediment. Experimental work by Ames (1959), Stumm and Leckie (1970), and others, has shown that $CaCO_3$ is converted to hydroxy or carbonate-apatite by partial replacement of CO_3^{2-} groups by PO_4^{3-}. However, such experiments alone do not necessarily reflect interactions between circulating phosphate-rich sea water and calcareous organic remains. Micro-organisms may influence the extent of these interactions by maintaining a substantial fraction of the phosphorus in the form of particulate or dissolved organic phosphorus and by controlling such environmental factors as pH, which determine solubility equilibria (Stumm and Morgan, 1970). Stumm and Leckie (1970) have demonstrated that the precipitation of calcium phosphate at marine pH values is greatly accelerated by calcium carbonate, the surface of which acts as a nucleating agent for crystallization. It would appear that after an initial chemisorption of phosphate and the heterogenous formation of nuclei of amorphous calcium phosphate, these nucleii are slowly transformed into crystalline apatite (hydroxy-apatite) with the subsequent growth of the crystals. In the absence of calcium carbonate, high degrees of supersaturation can be maintained indefinitely (see Stumm and Morgan, 1970).

The phosphate for carbonate-apatite formation is likely to originate from decomposition of organic phosphorus compounds and skeletal phosphate, and for diatomaceous deposits from the diatoms themselves, (for instance, Lisitsin (in Baturin, 1969) has found 0·8% P_2O_5 by weight in certain diatom frustules). Increases in organic productivity in the euphotic zone in areas of upwelling, e.g. off S.W. Africa, will lead to a high and constant supply of phosphorus for phosphorite formation in the sediments. In such sediments interstitial water phosphate concentrations may be one or two orders of magnitude higher than those found in the overlying sea water (see Baturin, 1969). Phosphate contents in the interstitial waters of reducing sediments off California (Rittenberg *et al.,* 1955) range up to fifty times the values reported for sea water. However, within oxidizing sediments there is little evidence that the phosphate concentration differs from that in the overlying waters (Fanning and Pilson, 1971). The relative proportion of inorganic to organic phosphorus in this instance is not known.

Phosphorites are notable for concentrating many elements, e.g. F, Na, Mg, Al, K, Zn, Sr, Ba, U and lanthanides, many of which seem to substitute for Ca^{2+} in the lattice. However, hypotheses that attempt to explain the composition of phosphorites often present a confusing picture. Much of this confusion results from the complexity of the apatite structure and the presence of foreign matter. If we regard the phosphate-rich diatomaceous sediment, and the more consolidated phosphorites on the South West African shelf, as representing different stages of one diagenetic process (see above), the change in the chemistry of the minor elements relative to the phosphate content in the different phosphorites may indicate element uptake at different stages of development. The analyses of unconsolidated phosphorites given in Table 30.6 suggests that many elements, for example fluorine and uranium, are fixed in the mineral at the earliest stage of formation (see Altschuler *et al.,* 1958, Gulbrandsen *et al.,* 1966). In this context Altschuler *et al.* (1958), Gulbrandsen *et al.* (1966) and Cruft (1966) have also observed that lanthanides are highly concentrated in fluor-apatites of ancient phosphorites, although Ce can show relative depletion in some phosphorites (Altschuler *et al.*, 1958; Tooms *et al.*, 1969). With ageing, further uptake of rare earth elements may be expected (Semenov *et al.,* 1962). For example, the analyses of Ce (and La) in the different phosphorites given in Table 30.6 suggest that although much fixation occurs at the time of formation, later (diagenetic) uptake is also important especially with Y, as it is only concentrated in the consolidated phosphorites.

REFERENCES

Alexander, G. B., Heston, W. M. and Iler, R. K. (1954). *J. Phys. Chem.* **58**, 453.

Altschuler, Z. S., Clarke, R. S. and Young, E. J. (1958). *U.S. Geol. Surv. Prof. Pap.* **314–D**, 45.

Ames, L. L. (1959). *Econ. Geol.* **54**, 829.

Anikouchine, W. A. (1967). *J. Geophys. Res.* **72**, 505.

Arrhenius, G. (1950). *Geochim. Cosmochim. Acta,* **1**, 15.
Arrhenius, G. (1963). *In* "The Sea" (Hill, M. N. ed.) Vol. 3. Interscience Publishers, New York.
Bass Becking, L. G. M., Kaplan, I. R. and Moore, D. (1960). *J. Geol.* **68**, 234.
Bada, J. L., Luyendyk, B. P. and Maynard, J. B. (1970). *Science, N.Y.* **170**, 730.
Bada, J. L. and Schroeder, R. A. (1972). *Earth Planet Sci. Lett.* **15**, 1.
Bader, R. G. (1955). *Geochim. Cosmochim. Acta,* **7**, 205.
Barber, R. T. (1968). *Nature, Lond.* **220**, 274.
Barnes, S. S. (1967). *Science, N.Y.* **157**, 63.
Barnes, S. S. and Dymond, J. R. (1967). *Nature, Lond.* **213**, 1218.
Baturin, G. N. (1969). *Dokl. Akad. Nauk SSSR,* **189**, 227.
Baturin, G. N. (1970). *Oceanology,* 327.
Baturin, G. N., Kochenov, A. V. and Shimkus, K. M. (1967). *Geokhimiya,* 41.
Baturin, G. N., Merkulova, K. I. and Chalov, P. I. (1972). *Mar. Geol.* **13**, M37.
Beall, A. O. and Fischer, A. G. (1969). *Init. Rep. Deep-Sea Drilling Proj.* **1**, 521.
Bender, M. L. (1971). *J. Geophys. Res.* **76**, 4212.
Bender, M. L., Ku, T. L. and Broecker, W. S. (1966). *Science, N.Y.* **151**, 325.
Berner, R. A. (1962). *Science, N.Y.* **137**, 669.
Berner, R. A. (1963). *Geochim. Cosmochim. Acta,* **27**, 563.
Berner, R. A. (1964). *Mar. Geol.* **1**, 117.
Berner, R. A. (1969). *Amer. J. Sci.* **267**, 19.
Berner, R. A. (1970a). "Handbook of Geochemistry", Vol. II-1. Springer-Verlag, Berlin.
Berner, R. A. (1970b). *Amer. J. Sci.* **268**, 1–23.
Berner, R. A. (1970c). *Nature, Lond.* **227**, 700.
Berner, R. A. (1971). "Principles of Chemical Sedimentology." McGraw-Hill, New York.
Bezrukov, P. L. (1955). *Dokl. Akad. Nauk. SSSR* **103**, 473.
Biscaye, P. E. (1965). *Geol. Soc. Amer. Bull.* **14**, 35.
Bischoff, J. L. and Ku, T. L. (1970). *J. Sediment Petrology,* **40**, 960.
Bischoff, J. L. and Ku, T. L. (1971). *J. Sediment. Petrology,* **41**, 1008.
Bischoff, J. L., Green, R. E. and Luistro, A. O. (1970). *Science, N.Y.* **167**, 1245.
Blumer, M. (1967). *Adv. Chem. Ser.* **67**, 312.
Bojanowski, R. and Paslawska, S. (1970). *Acta Geophys. Pol.* **18**, 277.
Bonatti, E. (1963). *N.Y. Acad. Sci. Trans. Ser II,* **25**, 938.
Bonatti, E., Fisher, D. E., Joensuu, O. and Rydell, H. S. (1971). *Geochim. Cosmochim. Acta,* **35**, 189.
Bordovskiy, O. K. (1965). *Mar. Geol.* **3**, 83.
Boström, K. (1967). *Adv. Chem. Ser.* **67**, 286–311.
Brooks, R. R., Presley, B. J. and Kaplan, I. R. (1968). *Geochim. Cosmochim. Acta,* **32**, 397.
Brown, F. S., Baedecker, M. J., Nissenbaum, A. and Kaplan, I. R. (1972). *Geochim. Cosmochim. Acta,* **36**, 1185.
Bruevich, S. V. (1961). "Khim. Tikhogo Okeana Izdat," Nauka, Moscow.
Burton, J. D. and Liss, P. S. (1968). *Nature, Lond.* **220**, 905.
Burton, J. D. and Liss, P. S. (1973). *Geochim. Cosmochim. Acta,* **37**, 1761.
Calvert, S. E. (1966). *Geol. Soc. Amer. Bull.* **77**, 569.
Calvert, S. E. (1968). *Nature, Lond.* **219**, 919.
Calvert, S. E. (1971) *Contrib. Mineral Petrol.* **33**, 273.
Calvert, S. E. and Price, N. B. (1970). *Contrib. Mineral Petrol.* **29**, 215.

Calvert, S. E. and Price, N. B. (1972). *Earth Planet. Sci. Lett.* **16**, 245.
Carroll, D. and Starkey, H. C. (1960). *Clays and Clay Minerals, 7th Nat. Conf.* 80–101.
Chester, R. and Hughes, M. J. (1967). *Chem. Geol.* **2**, 249.
Cooper, J. E. (1962). *Nature, Lond.* **193**, 744.
Cooper, W. J. and Blumer, M. (1968). *Deep-Sea Res.* **15**, 535.
Cronan, D. S. and Tooms, J. S. (1969). *Deep-Sea Res.* **16**, 335.
Cruft, E. F. (1966). *Geochim. Cosmochim. Acta,* **30**, 375.
d'Anglejan, B. F. (1967). *Mar. Geol.* **5**, 15.
d'Anglejan, B. F. (1968). *Can. J. Earth Sci.* **5**, 81.
Dasch, E. J. (1969). *Geochim. Cosmochim. Acta,* **33**, 1521.
Dayhoff, M. O., Lippincott, E. R. and Eck, R. V. (1964). *Science, N.Y.* **146**, 1461.
Degens, E. T. (1965). "Geochemistry of Sediments," Prentice Hall, New Jersey.
Degens, E. T. (1967). *In* "Diagenesis in Sediments" (Larsen, G. and Chilingar, G. V. eds.) pp 343–390. Elsevier, Amsterdam.
Degens, E. T., Prashnowsky, A., Emery, K. O. and Pimenta, J. (1961). *Neues Jahrb. Geol. Palaeontal. Monatsh.* **3**, 413.
Degens, E. T., Emery, K. O. and Reuter, J. H. (1963). *Neues. Jahrb. Geol. Palaeontol. Monatsh.* **5**, 231.
Degens, E. T., Reuter, J. H. and Shaw, K. N. F. (1964). *Geochim. Cosmochim. Acta,* **28**, 45.
Dietz, R. S., Emery, K. O. and Shepard, F. R. (1942). *Bull. Geol. Soc. Amer.* **53**, 815.
Drever, J. I. (1971a). *J. Sediment Petrology,* **41**, 982.
Drever, J. I. (1971b). *Science, N.Y.* **172**, 1334.
Drozdova, T. V. (1968). *Soviet Soil Sci.* **10**, 1393.
Ehlmann, A. J., Hulings, N. C. and Glover, E. D. (1963). *J. Sediment. Petrology,* **33**, 87.
El Wakeel, S. K. and Riley, J. P. (1961). *Geochim. Cosmochim. Acta,* **25**, 110.
Emelyanov, E. M. (1971). *ICSU/SCOR Symp., Cambridge* 1970. *Inst. Geol. Sci. Rep.* **70/16**, 99–103.
Emery, K. O. (1960). "The Sea off California", Wiley, New York.
Emery, K. O. and Hoggan, D. (1958). *Bull. Amer. Ass. Petrol. Geol.* **42**, 2174.
Emery, K. O., Stitt, C. and Saltman, P. (1964). *J. Sediment Petrology,* **34**, 433.
Ernst, W. G. and Calvert, S. E. (1969). *Amer. J. Sci.* **267–A**, 114.
Fägestrom, T. and Jernelöv, A. (1971). *Water Res.* **5**, 121.
Fanning, K. A. and D. R. Schink (1969). *Limnol. Oceanogr.* **14**, 59.
Fanning, K. A. and Pilson, M. E. Q. (1971). *Science, N.Y.* **173**, 1228.
Farrington, J. W. and Quinn, J. G. (1971a). *Geochim. Cosmochim. Acta,* **35**, 735.
Farrington, J. W. and Quinn, J. G. (1971b). *Nature, Phys. Sci.* **230**, 67.
Fomina, L. S. (1962). *Tr. Inst. Okeanol., Akad. Nauk SSSR,* **54**, 158.
Fomina, L. S. and Volkov, I. I. (1969). *Dokl. Akad. Nauk SSSR,* **185**, 188.
Foree, E. G. and McCarty, P. L. (1970). *Environ. Sci. Technol.* **4**, 842.
Friedman, G. M., Fabricand, B. P., Imbimbo, E. S., Brey, M. E. and Sanders, J. E. (1969). *J. Sediment Petrology,* **38**, 1313.
Garcia, A. W. (1969). U.S. Coast Guard Rep. 25.
Gevork'yan, V. Kh. and Chugunnyy, Yu. G. (1970). *Oceanology,* **10**, 233.
Goldberg, E. D. and Arrhenius, G. (1958). *Geochim. Cosmochim. Acta,* **13**, 153.
Goldberg, E. D. and Parker, R. H. (1960). *Geol. Soc. Amer. Bull.* **71**, 631.

Goldberg, E. D., Koide, M., Schmitt, R. A. and Smith, R. H. (1963). *J. Geophys. Res.* **68**, 4209.
Goodell, H. G. (1967). *J. Geol.* **75**, 665.
Griffin, G. M. (1962). *Geol. Soc. Amer. Bull.* **73**, 737.
Griffin, J., Windom, H. and Goldberg, E. P. (1968). *Deep-Sea Res.* **15**, 433.
Gripenberg, S. (1934). *Havsforsk. Skr.* **96**, 1.
Gulbrandsen, R. A., Kramer, J. R., Beatty, L. B. and Mays, R. E. (1966). *Amer. Mineral,* **51**, 819.
Harder, H. (1971). *Min. Soc. Japan,* **Spec. Pap. 1**, 106.
Harder, H. and Flehmig, W. (1970). *Geochim. Cosmochim. Acta,* **34**, 295.
Hare, P. E. (1969). *In* "Organic Geochemistry" (Eglinton, G. and Murphy, M. T. J. eds.), pp. 438–463. Longman Springer-Verlag, New York.
Hare, P. E. (1972). *Carnegie Inst. Washington Year Book,* **71**, 1615, 592–596.
Hare, P. E. and Mitterer, R. M. (1969). *Carnegie Inst. Washington, Year Book.* **67**, 205.
Harriss, R. C. (1966). *Nature, Lond.* **212**, 275.
Hart, R. (1970). *Earth Planet Sci. Lett.* **9**, 269.
Heath, G. R. (1974). *In* "Palaeo-Oceanography" (Hay, W. W. Ed.), Soc. Econ. Palaeontol. Mineral., Spec. *Pub.*
Heckey, R. E. (1971). Ph.D. Thesis, Duke University, Durham, U.S.A.
Holland, H. D. (1965). *Nat. Acad. Sci.* **53**, 1173.
Huang, W. H. and Vogler, D. L. (1972). *Nature, Lond.* **235**, 157.
Hurd, D. C. (1972). "Interactions of Biogenic Opal, Sediment and Sea Water in the Central Equitorial Pacific." Ph.D. Thesis, University of Hawaii.
Hurley, P. M. (1966). *In* "Potassium Argon Dating" (Schaeffer, O. A. and Zahringer, J., eds.), Springer-Verlag, New York, pp. 134.
Iler, R. K. (1955). "The Colloidal Chemistry of Silica and Silicates." Cornell, Ithaca, New York.
James, H. L. (1966). *U.S. Geol. Surv. Prof. Pap.* **440**W.
Jannasch, H. W. and Wirsen, C. O. (1973). *Science, N.Y.* **181**, 641.
Jannasch, H. W., Eimhjellen, K., Wirsen, C. O. and Farmanfarmaian, A. (1971). *Science, N.Y.* **171**, 672.
Jedwab, J. (1967). *Soc. Belge, Geol. Bull.* **76**, 1.
Jensen, S. and Jernelöv, A. (1964). *Nature, Lond.* **223**, 753.
Jernelöv, A. (1969). *In* "Chemical Fallout" (Miller, M. W. and Berg, G. G., eds.) C. C. Thomas, Springfield, U.S.A.
Jørgensen, H. (1945). *Gjellerup, Copenhagen,* 10.
Kanwisher, J. W. (1962). *Proc. Symp. Univ. R.I., Narragansett Mar. Lab., Occas. Publ.* **1**, 13–19.
Kaplan, I. R. and Presley, B. J. (1969). *Init. Rep. Deep-Sea Drilling Proj.* **1**, 411.
Kaplan, I. R. and Rittenberg, S. C. (1963). *In* "The Sea" (Hill, M. N., ed.), Vol. 3, pp. 583–619. Interscience, New York.
Kaplan, I. R. and Emery, K. O. and Rittenberg, S. C. (1963). *Geochim. Cosmochim. Acta,* **27**, 297.
Kazakov, A. V. (1937). *Trans USSR Sci. Inst. Fert. Insect Fungicides,* **142**, 95.
Kee, N. S. and Bloomfield, C. (1961). *Geochim. Cosmochim. Acta,* **24**, 206.
Keller, W. D. (1963). *Proc. 11th Nat. Conf. Clays Clay Min.,* Pergamon Press, New York.
Kjensmo, J. (1968). *Arch. Hydrobiol.* **65**, 125.
Kolodny, Y. (1969). *Nature, Lond.* **224**, 1017.

Kolodny, Y. and Kaplan, I. R. (1970). *Geochim. Cosmochim. Acta,* **34**, 3.
Kolodny, Y. and Kaplan, I. R. (1973). *Proc. Int. Symp. Hydrogeochem. Biogeochem. Tokyo 1970.*
Kononova, M. M. (1966). "Soil Organic Matter." Pergamon Press, London.
Krause, H. R. (1959). *Arch. Hydrobiol.* **24**, 297.
Krauskopf, K. B. (1956). *Geochim. Cosmochim. Acta,* **10**, 1.
Ku, T. L. (1965). *J. Geophys. Res.* **70**, 3457.
Ku, T. L. and Broecker, W. S. (1969). *Deep-Sea Res.* **16**, 625.
Ku, T. L. and Glasby, G. P. (1972). *Geochim. Cosmochim. Acta,* **36**, 699.
Kvenvolden, K. A. (1967). *J. Amer. Oil Chem. Soc.* **44**, 628.
Kvenvolden, K. A., Peterson, E. and Brown, F. S. (1970). *Science, N.Y.* **169**, 1079.
Lewin, J. C. (1961). *Geochim. Cosmochim. Acta* **21**, 182.
Li, Y-H., Bischoff, J. and Mathieu, G. (1969). *Earth Planet Sci. Lett.* **7**, 265.
Lisitzin, A. P. (1970). *In* "Sedimentation in the Pacific Ocean." (Bezrukov, P. L., ed.). Nauka, Moscow.
Lynn, D. C. and Bonatti, E. (1965) *Mar. Geol.* **3**, 457.
Mackenzie, F. T. and Garrels, R. M. (1965). *Science, N.Y.* **150**, 57.
Mackenzie, F. T. and Garrels, R. M. (1966). *Amer. J. Sci.* **264**, 507.
Mackenzie, F. T. and Gees, R. (1971). *Science, N.Y.* **173**, 533.
Mackereth, F. J. H. (1965). *Trans. Roy. Soc. Lond.* **250**, 165.
Mangelsdorf, P. C., Wilson, T. R. S. and Daniell, E. (1969). *Science, N.Y.* **165**, 171.
Manheim, F. T. (1961). *Geochim. Cosmochim. Acta* **25**, 52.
Manheim, F. T. (1965). *Univ. R.I., Narragansett Mar. Lab., Occas. Publ,* **3**, 217.
Manheim, F. T. (1970). *Earth Planet. Sci. Lett.* **9**, 307.
Manheim, F. T. and Rowe, G. (1973). In preparation.
Manheim, F. T. and Sayles, F. L. (1971). *Initial Rep. Deep-Sea Drilling Proj.* **8**, 857.
McBride, B. C. and Wolfe, R. C. (1971). *Biochim. J.* **10**, 4312.
McConnell, D. (1950). *J. Geol.* **58**, 16.
McConnell, D. (1952). *Bull. Soc. Fr. Mineral Cristallogr.* **75**, 428.
McConnell, D. (1965). *Econ. Geol.* **60**, 1059.
McKelvey, V. E. (1967). *Bull. U.S. Geol. Surv.* **1252–D**, 1.
Melson, W. G. and Thompson, G. (1973). *Geol. Soc. Amer. Bull.* **84**, 703.
Mero, J. L. (1965). "The Mineral Resources of the Sea." Elsevier, Amsterdam.
Michard, G. (1971). *J. Geophys. Res.* **76**, 2179.
Mizutani, S. (1970). *Sedimentology,* **15**, 419.
Mohamed, A. F. (1949). *Amer. J. Sci.* **247**, 116.
Mopper (1973). Ph.D. Thesis, Woods Hole–M.I.T.
Morey, G. W., Fournier, R. O. and Rowe, J. J. (1962). *Geochim. Cosmochim. Acta,* **26**, 1029.
Morgan, J. J. and Stumm, W. (1965). *Proc. 2nd Conf. Water Poll. Res., Tokyo* 1964. Pergamon Press, London.
Morozov, N. P. (1969). *Oceanology,* **9**, 291.
Murray, J. and Renard, A. F. (1891). Report on Deep-Sea Deposits. *In* "Scientific Report of H.M.S. *Challenger* Expedition", 525 pp. H.M.S.O., London.
Nayudu, Y. R. (1964). *Volcanol. Bull.* **27**, 1.
Nelson, B. W. (1963). "Clays and Clay Minerals." Pergamon Press, New York.
Nelson, B. W. (1967). *Science, N.Y.* **158**, 917.
Nicholas, D. J. D. (1967). *Mineralium Deposita.* **2**, 169.

Nissenbaum, A., Presley, B. J. and Kaplan, I. R. (1972). *Geochim. Cosmochim. Acta,* **36**, 1007.
Noshkin, V. E. and Bowen, V. T. (1973). *Proc. Int. Atomic Energy Agency Symp., Seattle, Washington*, July 1971, (In press).
Nriagu, J. O. (1972). *Geochim. Cosmochim. Acta,* **36**, 459.
Oppenheimer, C. H. (1960). *Geochim. Cosmochim. Acta,* **19**, 244.
Ostroumov, E. A. (1953). *Tr. Inst., Okeanol. Akad. Nauk. SSSR,* **7**, 70.
Ostroumov, E. A., Volkov, I. I. and Formina, L. C. (1961). *Tr. Inst. Okeanol. Akad. Nauk. SSSR.* **50**, 93.
Parker, P. L. (1961). *Inst. Mar. Sci., Univ. Texas, Progr. Rep.*
Parker, P. L. (1969). *In* "Organic Geochemistry" (Eglington, G. and Murphy, M. I. T. eds.). Longman Springer-Verlag, Berlin.
Parker, P. L. and Leo, R. F. (1965). *Science, N.Y.* **148**, 373.
Perry, E. A. and Hower, J. (1970). *Clays Clay Miner.* **18**, 165.
Peterson, M. N. A. and Griffin, J. J. (1964). *J. Mar. Res.* **22**, 13.
Peterson, M. N. A. and von der Borch, C. C. (1965). *Science, N.Y.* **149**, 1501.
Pevear, D. R. and Pilkey, O. H. (1966). *Bull. Geol. Soc. Amer.* **77**, 849.
Poluschkina, A. P. and Sidovenki, G. A. (1963). *Zafiski Vses. Mineral. Obshch.* **42**, 547.
Porrenga, D. H. (1966). 14th Nat. Conf. Clays and Clay Minerals (Swineford, Ada, ed.), pp. 221–233. Pergamon Press, New York.
Porrenga, D. H. (1967). *Mar. Geol.* **5**, 495.
Powers, M. C. (1959). *In* "Clays and Clay Minerals" (Swineford, Ada., ed.). Pergamon Press, New York.
Prashnowsky, A., Degens, E. T., Emery, K. O. and Pimenta, I. (1961). *Neues Jahrb. Geol. Palaeontol Montash.* **8**. 400.
Pratt, W., (1961). 1st Nat. Coastal and Shallow Water Res. Conf., pp. 656–658.
Pratt, R. M. and McFarlin, P. F. (1966). *Science, N.Y.* **151**, 1080.
Presley, B. J., Brooks, R. R. and Kaplan, I. R. (1967). *Science, N.Y.,* **158**, 906.
Presley, B. J. and Kaplan, I. R. (1970). *Init. Rep. Deep-Sea Drilling Proj*, **4**, 415.
Presley, B. J., Kolodny, Y., Nissenbaum, A. and Kaplan, I. R. (1972). *Geochim. Cosmochim. Acta,* **36**, 1073.
Price, N. B. (1967). *Mar. Geol.* **5**, 511.
Price, N. B. and Calvert, S. E. (1970). *Mar. Geol.* **9**, 145.
Price, N. B. and Calvert, S. E. (1973a) ICSU/SCOR Symp., Cambridge 1970. Inst. Geol. Sci. Rep. 70/16, 185
Price, N. B., Calvert, S. E. and Jones, P. G. W. (1970). *J. Mar. Res.* **28**, 22.
Price, N. B. and Calvert, S. E. (1973b). *Geochim. Cosmochim. Acta.*
Price, N. B. and Calvert, S. E. (1973c). (In preparation)
Rashid, M. A. and King, L. H. (1970). *Geochim. Cosmochim. Acta,* **34**, 193.
Rateev, M. A., Gorbinova, Z. N., Lisitzin, A. P. and Nosov, G. L. (1969). *Sedimentology,* **13**, 21.
Redfield, A. C., Ketchum, B. H. and Richards, F. A. (1963). "The Sea: Composition of Sea Water, Comparative and Descriptive Oceanography," Vol. 2, pp. 26–27. Interscience Publishers, New York.
Rhead, M. M., Eglington, G., Draffan, G. H. and England, P. J. (1971). *Nature, Lond.* **232**, 327.
Richards, F. A. (1965). *In* "Chemical Oceanography." (Riley, J. P. and Skirrow, G., eds.), Vol. 1. Academic Press, London and New York.
Rickard, D. T. (1969). *Stockholm Contr. Geol.* **20**, 67.

Rittenberg, S. C. (1940). *J. Mar. Res.* **3**, 191.
Rittenberg, S. C., Emery, K. O. and Orr, W. L., (1955). *Deep-Sea Res.* **3**, 23.
Rittenberg, S. C., Emery, K. O., Hulsemann, J., Degens, E. T., Fay, R. C., Reuter, J. H., Grady, J. R., Richardson, S. H. and Bray, E. E., (1963). *J. Sediment Petrology,* **33**, 140.
Roberson, C. E. (1966). *U.S. Geol. Surv. Prof. Pap.* **550–D**, 178.
Rohrlich, V., Price, N. B. and Calvert, S. E. (1969). *J. Sediment. Petrology,* **39**, 624.
Rosenfeld, W. D. (1948). *Arch. Biochem. Biophys.* **16**, 263.
Rosenquist, I. Th. (1970). *Lithos,* **3**, 327.
Russell, K. L. (1970). *Geochim. Cosmochim. Acta,* **34**, 893.
Savin, S. M. and Epstein, S., (1970). *Geochim. Cosmochim. Acta,* **34**, 43.
Sayles, F. L., Manheim, F. T. and Waterman, L. S. (1973*a*). *Init. Rep. Rep. Deep-Sea Drilling Proj.* **20**, 783.
Sayles, F. L., Wilson, T. R. S., Hume, D. N. and Mangelsdorf, P. C., Jr. (1973*b*). *Science,* **181**, 154.
Schnitzer, M. and Skinner, S. M. (1965). *Soil. Sci.* **99**, 278.
Semenov, E. I., Kholodov, V. N. and Barinskii, R. L. (1962). *Geochemistry,* **5**, 501.
Sevast'yanov, V. F. (1967). *Dokl. Akad. Nauk. SSSR,* **176**, 191.
Sevast'yannov, V. F. and Volkov, I. I. (1966). *Dokl. Akad. Nauk. SSSR,* **166**, 172.
Shishkina, O. V. (1964). *Geochem. Int.* **3**, 522.
Shishkina, O. V. (1966). (Bruevich, S. V., ed.). *Nauk USSR Gzd. Akademiia,* 289.
Shishkina, O. V. and Bykova, V. S. (1962). *Tr. Inst. Morsk. Gidrofiz., Akad Nauk SSSR,* **25**, 187.
Shishkina, O. V. and Pavlova, G. A. (1965). *Geochemistry,* **2**, 559.
Siever, R. (1962). *J. Geol.* **70**, 127.
Siever, R., Beck, K. C. and Berner, R. A. (1965). *J. Geol.* **73**, 39.
Sigvaldason, G. E. (1959). *Beitr. Mineral. Petrol.* **6**, 405.
Sillén, L. G. (1961). *In* "Oceanography" (Mary Sears, ed.). Amer. Assoc. Adv. Sci. Publ. 67. Washington, D.C.
Sillén, L. G. (1966). *Ark. Kemi,* **25**, 159.
Skopintsev, B. A. (1964). *Sci. Pap., Inst. Chem. Technol.* Prague, Technol. Water, **8**, 1.
Skornyakova, N. S. (1965). *Int. Geol. Rev.* **7**, 2161.
Skornyakova, N. S., Andrushenko, P. F. and Fomina, L. S. (1962). *Okeanologia,* **2**, 264.
Skornyakova, N. S. and Andrushenko, P. F. (1964). *Litol. Polez. Iskop.* **5**, 21.
Smith, K. L. and Teal, J. M. (1973). *Science, N.Y.* **179**, 282.
Sorokin, Y. I. (1962). *Mikrobiologiya,* **3**, 402.
Sorokin, J. I. (1964). *J. Cons., Cons. Perm. Int. Explor. Mer.* **29**, 41.
Stevenson, F. J. and C.-N. Cheng. (1972). *Geochim. Cosmochim. Acta,* **36**, 653.
Stevenson, F. J. and Tilo, S. N. (1970). *Proc. 3rd Int. Meet. Org. Geochem., London* 1966. p. 227.
Strahkov, N. M. (1959). *Eclogae Geol. Helv.* **51**, 753.
Strahkov, N. M. (1969). "Principles of Lithogenesis," Vol. 2, Oliver and Boyd, Edinburgh.
Stumm, W. (1966). *Proc. 3rd Int. Conf. Water Pollut. Res.* **13**
Stumm, W. (1973). *Water Res.* **7**, 131.
Stumm, W. and Leckie, J. O. (1971). *Proc. 5th Int. Conf. Water Pollut. Res.* Pergamon Press, Oxford.
Stumm, W. and Morgan, J. J. (1970). "Aquatic Chemistry." Interscience Publishers, New York.

Sugawara, K. and Terada, K. (1957). *J. Earth Sci., Nagoya Univ.* **5**, 81.
Summerhayes, C. P. (1967). *N.Z. J. Mar. Freshwater Res.* **1**, 267.
Sverdrup, H. U., Johnson, M. W. and R. H. Fleming (1942). "The Oceans," Prentice-Hall, New York.
Swanson, V. E., Frost, I. C., Rader, C. F. and Huffman, C. (1966). *U.S. Geol. Surv. Prof. Pap.* **550-C**, 174.
Tageeva, N. V. and Tikhomirova, M. M. (1962). *Akad. Nauk SSSR*, 246.
Thimann, K. V. (1963). "The Life of Bacteria," 2nd Ed. Macmillan, New York.
Thorstenson, D. C. (1970). *Geochim. Cosmochim. Acta,* **34**, 745.
Tooms, J. S., Summerhayes, C. P. and Cronan, D. S. (1969). *Oceanogr. Mar. Biol. Ann. Rev.* **7**, 49.
Turekian, K. K. (1968). In "Handbook of Geochemistry" (Wedepohl, K. H. ed.) Vol. 1. Springler-Verlag. Berlin.
Tzur, Y. (1971). *J. Geophys. Res.* **76**, 4208.
van Straaten, L. M. J. U. (1954). *Leidse Geol. Meded.* **19**, 1.
Veeh, H. H. (1967). *Earth. Planet. Sci. Lett.* **3**, 145.
Veeh, H. H., Calvert, S. E. and Price, N. B. (1974). *Mar. Chem.* **2**, 189.
Velde, B. and Hower, J. (1963). *Amer. Mineral.* **48**, 1239.
Vinogradov, A. P. (1939). *Tr. Biogeokhim Lab., Akad. Nauk SSSR,* **5**, 33.
Volkov, I. I. (1961). *Tr. Inst. Okeanol., Akad. Nauk SSSR,* **50**, 68.
Von Gaertner, H. R. and Schellmann, W. (1965). *Mineral. Petrogr. Mitt.* **10**, 349.
Weaver, C. E. and Wampler, J. M. (1972). *Geochim. Cosmochim. Acta,* **36**, 1.
Weaver, C. E., Beck, K. C. and Pollard, C. O. (1971). *Geol. Soc. Amer. Spec. Pap.* **134**, 296.
Wehmiller, J. and Hare, P. E. (1971). *Science, N.Y.* **173**, 907.
Whitehouse, U. G. and McCarter, R. S. (1958). *Nat. Acad. Sci.-Nat. Res. Counc. Publ.* **566**, 81.
Winterhalter, B., (1966). *Geotek. Julk.* **69**, 1.
Winterhalter, B. and Siivola, J. (1967). *Comp. Rend. Soc. Geol. Finlande,* **39**, 161.
Zen, E. (1959). *J. Sediment. Petrology,* **29**, 513.
Zenkevitch, L. (1963). "Biology of the Seas of the USSR". George Allen and Unwin, London.
Zobell, C. E. (1942). *J. Sediment Petrology,* **12**, 127.
Zobell, C. E. (1946). *Bull. Amer. Ass. Petrol. Geol.* **30**, 477.
Zobell, C. E. (1964). *In* "Advances in Organic Geochemistry" (Ingesson, E. ed) Macmillan, New York.
Zobell, C. E. (1968). *Bull. Misaki Mar. Biol. Inst., Kyoto Univ.* **12**, 77.
Zunino, H., Galindo, G. Peirano and Aguilera, M. (1972). *Soil. Sci.* **114**, 229.

Chapter 31

Factors Controlling the Distribution and Early Diagenesis of Organic Material in Marine Sediments*

EGON T. DEGENS† and KENNETH MOPPER

Woods Hole Oceanographic Institution,
Woods Hole, Massachusetts 02543, U.S.A.

* Woods Hole Oceanographic Institution Contribution No. 3075.

† Present address: Geologisch-Paläontologisches Institut der Universität Hamburg, Von-Melle-Park 13, 2 Hamburg 13, Federal German Republic.

31.1. Introduction

The amount of organic matter in both modern and ancient sediments varies from almost 100% in some peat and marsh deposits, to as little as a few parts per million in some sand and turbidite deposits. Commonly, the organic content is between 0·1 and 5%; in general, deep-sea sediments contain ca. 0·2% organic matter, geosynclinal sediments ca. 2%, and shelf sediments 1–5%. The ultimate source of this material is photosynthetically fixed carbon (primary production) in the biosphere. Many varied and interrelated biochemical and geochemical processes transform the organic matter of living cells into *kerogen*, the inert organic residue which is widely dispersed in ancient sedimentary rocks.

The various organic compounds present in sediments during early stages of diagenesis interact not only among themselves, but also with metal ions and mineral surfaces. As a consequence, the complexity of the reaction sequences and the resultant organic products becomes immense. Thus, Blumer (1973) has estimated that for fossil porphyrins alone the number of structural permutations exceeds 2×10^6. If so many compounds can arise from chlorophyll and haemoglobin, their possible precursors, then the number of molecular species that can be generated from the main building blocks of life in the course of diagenesis must be almost infinite. So far, over 1000 different organic molecules have been identified in sediments. The number of compounds isolated from crude oil is of the order of 10 000. It is evident from these figures that theoretically there is a wide field for future research. However, from a practical standpoint, the question arises: will a knowledge of the compounds present enhance our understanding of the nature and diagenetic history of sedimentary organic matter? In view of the limited conclusions that have been drawn from the distribution patterns of the more familiar molecules, and considering the vast number of types of molecules which are present in sediments, it seems likely that this type of research represents an unfruitful avenue of approach.

In the present chapter, the fate of a few major compounds is traced from the biological cell, through the water column, and into sediments. It is believed that this approach can bring to light, and possibly solve, major problems in the field of organic geochemistry.

The problems considered here include:

(1) What major factors control the cycling of organic matter in the World Ocean?

(2) What factors determine the quantity and quality of organic matter that is incorporated into sediments?

(3) How does the environment at the sediment-water interface affect the composition and structure of the organic input?

(4) What happens to organic matter during early diagenesis?
(5) Can fossil organic matter be used as a means to elucidate palaeo-environments?

These problems will be investigated in terms of amino acids and sugars. Collectively, these compounds comprise 40–80% of the organic matter of most organisms and, hence, represent a large proportion of the organic input to aquatic sedimentary environments.

31.2. Selection of Data

This chapter presents data which has been collected over the past few years in the authors' laboratories (Degens, 1970; Mopper and Degens, 1972, 1973), together with that of other investigators. Variations in the analytical techniques used by different investigators in the field of organic geochemistry make it difficult to compare results from different laboratories, except in a qualitative or at best semi-quantitative manner. For this reason, it has been necessary to be selective in choice of literature. There is no implication that the authors' results are better or more accurate than those of other investigators, but they are consistent for all material investigated.

The emphasis of this chapter will be placed on the elucidation of factors which affect the distribution and early diagenesis of organic matter in various marine habitats. Reviews of the general nature of organic matter and organic interactions in the marine environment may be found elsewhere (e.g. Breger, 1964; Degens, 1965, 1967; Durrsma, 1965; Wangersky, 1965; Eglinton and Murphy, 1969; Wagner, 1969; Williams, 1971).

31.3. Environmental Factors Regulating the Input of Organic Matter to Sediments

31.3.1. Introduction

Environmental factors controlling the input of organic matter to sediments fall into four general categories; (1) biological, (2) physical, (3) chemical, and (4) geological. However, attempts to use categories such as these to classify organic matter lead to difficulties because these divisions are so broad and interrelated that they become almost arbitrary. For this reason the chapter is organized in terms of well-defined segments of the oceanic biogeochemical cycle.

31.3.2. PRIMARY PRODUCTION

The total annual primary productivity (expressed on a dry weight basis) in the ocean is ca. $5{\cdot}5 \times 10^{13}$ kg or 155 g m^{-2}. The figure for total land productivity is twice this amount, namely $10{\cdot}9 \times 10^{13}$ kg (dry weight). The standing crop of the ocean and the land are about 0·009 and 12·5 dry kg m^{-2} respectively on a dry weight basis; therefore, marine phytoplankton photosynthesis is more efficient by a factor of almost 1500 relative to average land plants (Whittaker, 1970). As a consequence it might be expected that there would be a high concentration of metabolic waste products in the marine environment.

The world-wide distribution of oceanic productivity is shown in Fig. 31.1. From this it can be seen that vast regions of the sea are almost barren, whereas others are densely populated. The most productive areas are found where upwelling brings nutrients to the surface. These areas include the equatorial region of the Pacific, parts of the Arctic and Antarctic Oceans, and some continental margins.

31.3.3. TERRESTRIAL CONTRIBUTIONS

The present transgressive stage (eustatic rising of sea level) in oceanic evolution favours deposition of terrigenous detritus in near-shore environments. Manheim *et al.* (1970) and Meade (1973) concluded that most of the suspended material in surface waters of the outer shelves off the east coast of North America is composed of organic matter and skeletal debris, and that the amounts of terrigenous minerals rapidly decrease away from the continents (see Fig. 31.2). Microscopic investigation of marine suspended matter reveals that a substantial portion of it consists of marine organisms. However, it is not known whether all the organic debris is of marine origin. Furthermore, in addition to particulate organic matter, there is in the sea a vast reservoir of dissolved organic matter, the source of which is even more uncertain. The question then arises: is there an input of terrigenous organic material to the open ocean and, if so, how large is it relative to marine contributions? Stable carbon isotope analyses may be used to elucidate this problem.

The marine plankton of temperate regions have a $\delta\,^{13}C$ of about $-20‰$ (relative to PDB standard) with no apparent differences between zoo- and phytoplankton populations. Common land plants have $\delta^{13}C$ values 5 to 10‰ lower (Craig, 1953; Degens, 1969). A number of studies have shown that the isotopic composition of organic matter in recent marine sediments is identical to that of marine plankton (see e.g. Sackett and Thompson, 1963; Hunt, 1968), and local variations in the $\delta^{13}C$ values of plankton arising from temperature and respiration effects are reflected in the $\delta^{13}C$ values of the underlying sediment. Only sediments deposited in river estuaries and close to shorelines reveal a terrigenous influence (Fig. 31.3). It may be concluded,

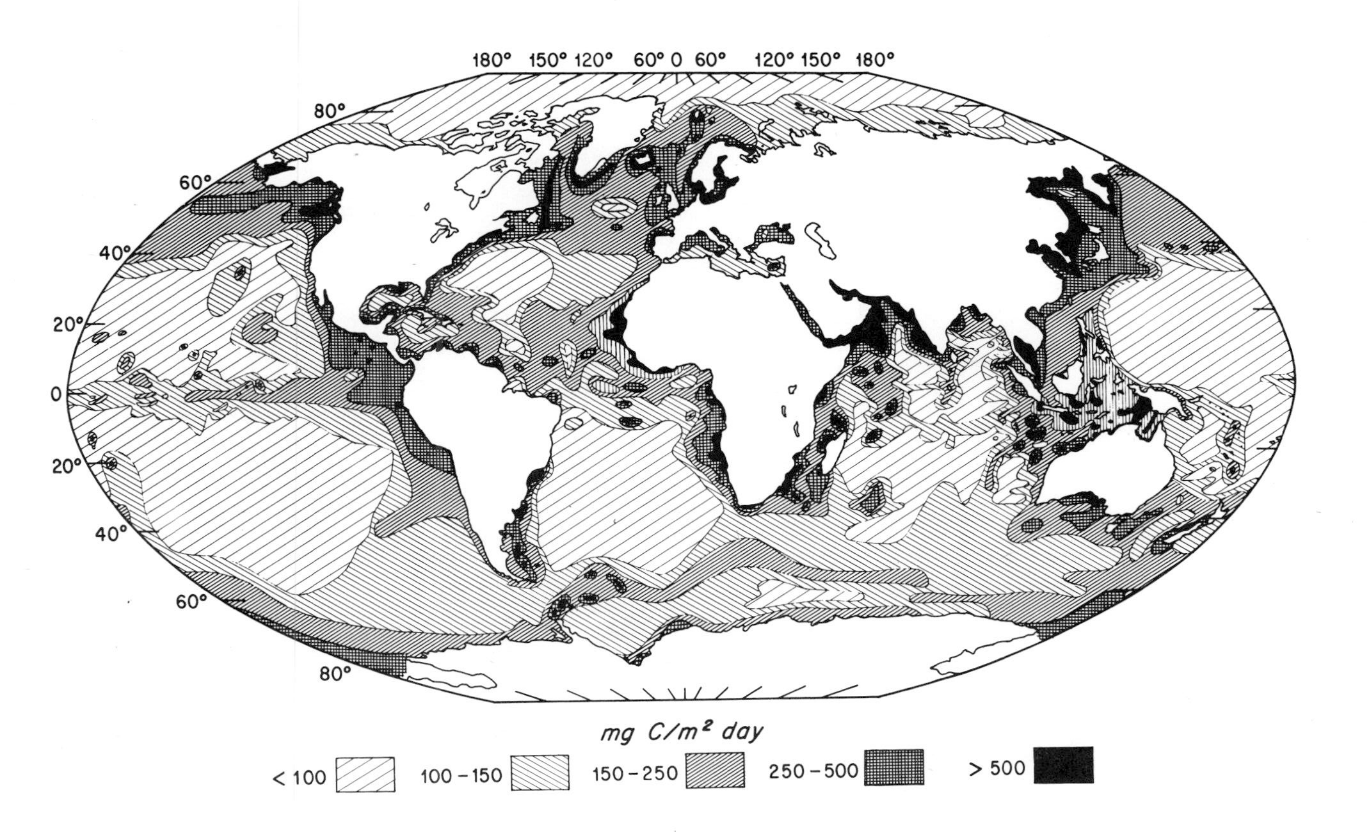

FIG. 31.1. Distribution of primary production in the World Ocean (after Koblentz-Mishke, *et al.*, 1968).

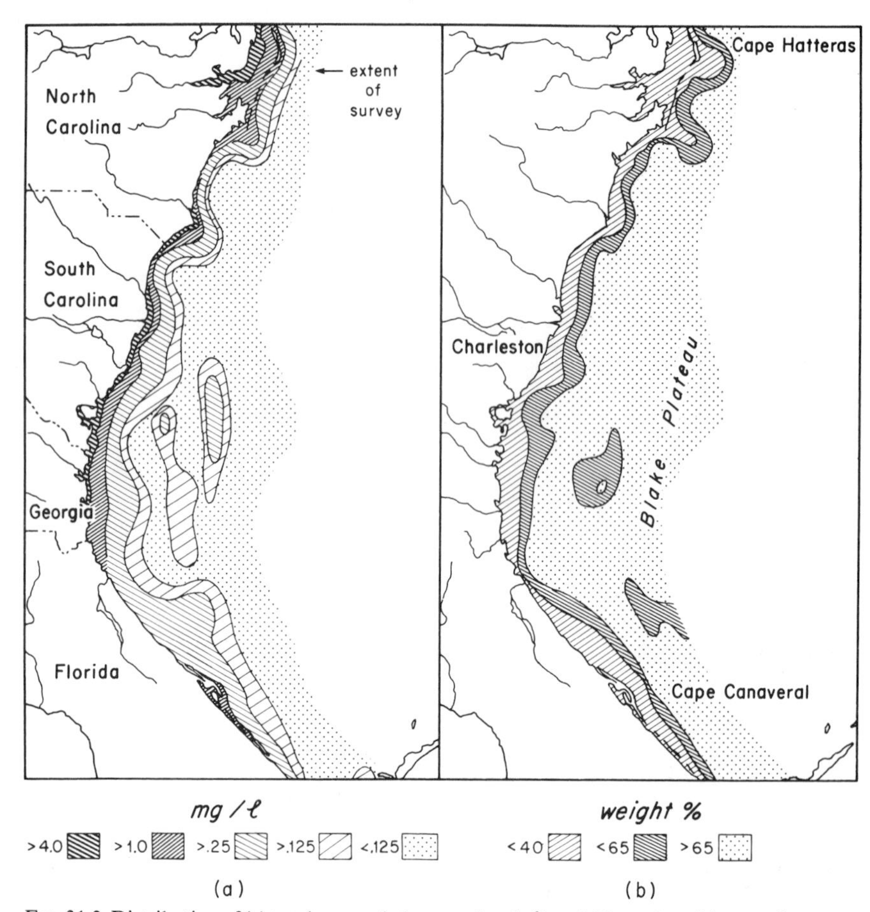

FIG. 31.2. Distribution of (a) total suspended matter (mg l^{-1}) and (b) combustible organic matter (weight %) in the surface waters of the Atlantic continental margin from Cape Hatteras to the Florida Keys (after Manheim *et al.*, 1970).

therefore, that the bulk of the organic debris contained in recent marine deposits is derived from the local biomass.

In contrast, during a regressive stage (eustatic lowering of sea level) it is possible that a larger percentage of terrigenous organic matter may have been swept into the sea. However, carbon isotope studies of deep-sea cores covering the last 100 000 years give no indication of the presence of land-derived organic matter, with the exception of that originating from occasional turbidity flows. Geochemical evidence based on carbohydrate analyses of Pleistocene sediments (see page 107) also support this conclusion. It appears that, regardless of whether oceanic evolution is in a transgressive or regressive

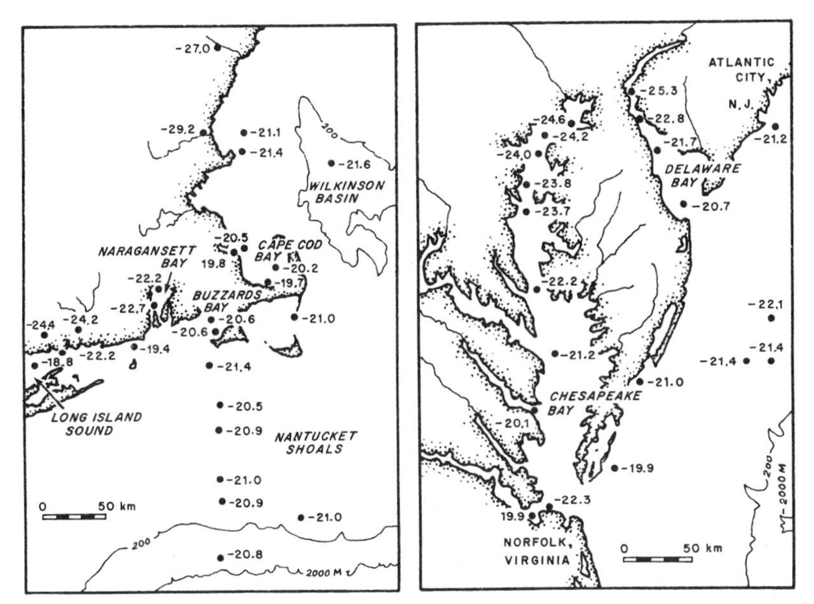

FIG. 31.3. $\delta^{13}C$ values of organic matter of surface sediments from the continental shelf of the Northeastern United States (after Hunt, 1968).

stage, organic contributions from land are small relative to those from the marine biomass. The pieces of wood which occur sporadically in marine sediments should therefore be considered more in the nature of artifacts rather than as indications of a large scale organic input from land.

31.3.4. EFFICIENCY OF THE RECYCLING OF ORGANIC MATTER IN THE EUPHOTIC ZONE

Most organic matter produced in the euphotic zone of the ocean is recycled through processes such as grazing and respiration, which makes it difficult to calculate the flux of organic matter leaving this zone. An indirect calculation can be made since this total flux is approximately equal to the sum of the following: (1) the flux of organic matter to the sediment which is subsequently permanently trapped; (2) the flux of organic matter liberated from the sediment; (3) the flux of organic matter biologically oxidized at the sediment-water interface; (4) the flux of organic matter oxidized by deep water metabolism; and (5) the flux of organic matter solubilized in the water column.

The flux of organic matter trapped in the sediment can be calculated if the average sedimentation rate and the average organic carbon contents of the sediments are known. For example, the average sedimentation rate for one Argentine Basin core is 6 cm per 1000 yrs. Assuming a density of 2·7 g ml^{-1}

for the inorganic phase and an average water content of 40 %, the wet sediment has a density of 1·67 g ml^{-1}. Thus, the sediment flux is about 10 g cm^{-2} per 1000 yrs. The average carbon content is 0·67 % on a dry weight basis or 0·36 % on a wet basis. Therefore, the organic carbon flux is 0·036 g-C cm^{-2} per 1000 yrs, or about 0·4 g-C m^{-2} yr^{-1}. The average primary productivity in the water column in the region of the core is 100 g-C m^{-2} yr^{-1}; therefore, the flux of organic matter permanently entering the sediment is about 0·4 % of the primary productivity in the euphotic zone.

When the sedimentation rate and depth of the water column (see Section 31.3.5) in an area are uniform, the amount of organic matter in the sediment would be expected to reflect primary productivity in the euphotic zone. For example, the organic content of the sediments directly below the Peru Coastal Current (productivity ~300–500 g-C m^{-2} yr^{-1}) ranges from 2 to 7 % (Ryther *et al.*, 1970). In contrast the organic content of the sediment to the west of the current (productivity 100–200 g-C m^{-2} yr^{-1}) falls to about 1 %.

Assuming an average age of deep water D.O.M. (dissolved organic matter) of 3400 yrs, (Williams *et al.*, 1969) and that steady-state conditions prevail, Williams (1971) calculated that 0·5 % of the mean annual primary productivity (corresponding to a flux of 0·5 g-C m^{-2} yr^{-1}) must enter the deep ocean from the euphotic zone in order to maintain the present reservoir of D.O.M. If the steady-state assumption is correct, then the percent of the total organic flux leaving the euphotic zone to form D.O.M. (~0·5 %) must be compensated by both the flux of D.O.M. oxidized by deep water metabolism and that of D.O.M. adsorbed or condensed onto P.O.M. (particulate organic matter). The latter flux is unknown; however, estimates of the former have been published recently. Using the D.O.C. (dissolved organic carbon) data of Menzel and Ryther (1970) for the North Atlantic, Craig (1971) calculated that the rate of consumption of organic matter in deep water is of the order of $3·5 \times 10^{-4}$ mg-C l^{-1} yr^{-1}. This rate corresponds to a consumption of approximately 1–2 g-C m^{-2} yr^{-1} for a water column of 3000 m.

It is important to note that Craig's calculations are based on the assumption that simple vertical diffusion-advection models can explain the vertical profile of the ΣCO_2-^{14}C-^{13}C-O_2-alkalinity system in the ocean. On the basis of linear correlations between salinity and dissolved oxygen, Menzel and Ryther (1970) have suggested that lateral diffusion and advection are mainly responsible for the vertical profile observed for this system. The question of consumption vs. lateral mixing is still unresolved.

Comparison of the results of Craig's calculation with those of Williams reveals that the consumption rate of D.O.M. is larger than the apparent supply rate by about a factor of two to four. In order to reconcile this discrepancy it is necessary to postulate two reservoirs of D.O.M.: (1) young, labile D.O.M. (*e.g.* metabolic exudates such as those described by Hellebust (1965) together with D.O.M. which is released by lysis of cells), and (2) aged, refractory D.O.M. If this *extreme* case is considered, in which refractory D.O.M. consists exclusively of radioactively "dead" carbon, then young, labile D.O.M. (<200 years B.P.) must constitute 60 to 70% of the total D.O.M. in order to yield the average ^{14}C "age" of 3400 years B.P. found by Williams *et al.* (1969). If as a more realistic approach it is assumed that the average age of the refractory D.O.M. is of the order of 4000–6000 years B.P., then the young, labile D.O.M. need only constitute 15–25% of the total D.O.M. in order to yield the "age" in question. It is interesting to note that only 10% of the total D.O.M. has been identified, and that the bulk of the remainder probably consists of refractory heteropolycondensates (Duursma, 1965; Degens, 1970). Although the reservoir of labile D.O.M. may be considerably smaller than that of refractory D.O.M., the flux of labile D.O.M. must be at least two to four times larger in order to account for the measured consumption rate. A small amount of the labile D.O.M. (~0·5% of primary productivity) is probably converted to refractory D.O.M. in order to maintain both the present reservoir of refractory D.O.M. and its average residence time. A summary of these different inter-relationships in the open ocean is presented schematically in Fig. 31.4.

The total flux of P.O.M. (inert + labile) to the sediment is unknown; however, its minimum value must approximate to the sum of that part of the total flux of organic matter which is subsequently permanently trapped in the sediment (i.e. ~10 cm below the sediment-water interface) and that part which is ultimately biologically oxidized at the sediment surface. The latter has been recently determined by *in situ* measurements (Smith and Teal, 1973), and is of the order of 1–4 g-C m^{-2} yr^{-1} for sediments at a depth of about 2000 m on the continental slope south of New England.

TABLE 31.1

Summary of organic fluxes in the ocean

Flux	g-Cm^{-2} yr^{-1}	Reference
Permanently trapped in sediment	0·2–0·6	Mopper and Degens (1972)
Organic matter solubilized (D.O.M.)	0·3–0·4	Williams (1971)
Biological oxidation in sediment	1·0–4·0	Smith and Teal (1973)
Deep water oxidation of organics	1·0–2·0	Craig (1971)
Total range*	2·5–7·0	

* Total range = flux of organic matter leaving the euphotic zone; thus, assuming an average surface productivity of 100 g-Cm^{-2} yr^{-1}, the efficiency of recycling is about 93 to 97%.

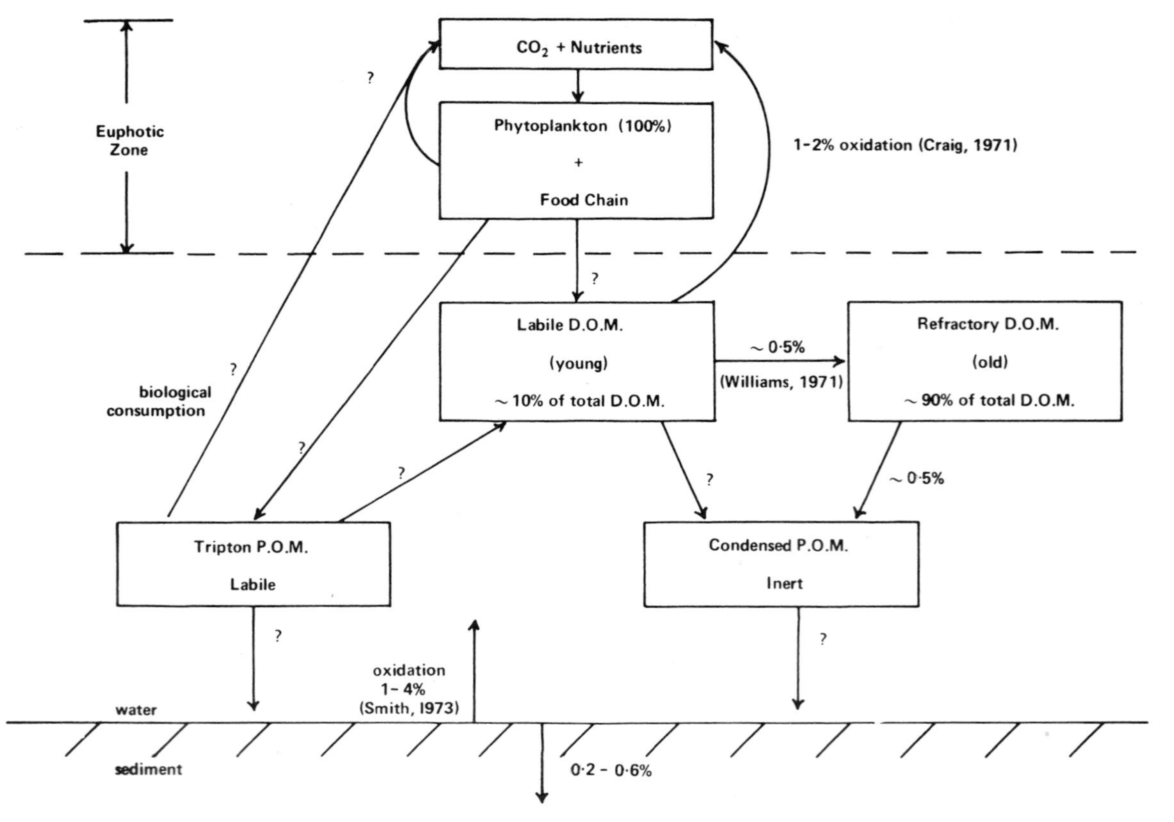

FIG. 31.4. Turnover of organic matter in the open ocean (after Mopper and Degens, 1972).

Deuser (1971) calculated that 80 to 95% of the organic carbon produced in the euphotic zone of the Black Sea is recycled in the upper 200 m (just above the O_2–H_2S interface). His calculations are based on *in situ* production rates of hydrogen sulfide and on changes in $^{13}C/^{12}C$ ratios in the dissolved inorganic carbon (ΣCO_2). Because the Black Sea is a restricted anoxic basin these calculations cannot be applied to the oceans in general in which, as oxygen is present throughout, a greater efficiency of recycling (90–98%) is expected (see Table 31.1.)

31.3.5. DEPTH OF THE WATER COLUMN

Organic detritus which survives degradation and grazing in the euphotic zone and settles into the abyssal regions of the oceans may be subjected to direct biological consumption as suggested by Ogura (1970); however, no direct measurements exist to support this hypothesis.

In long water columns the combined factors of condensation reactions, metal complexation, microbial degradation and deep-sea oxidation may have the effect of eliminating the more labile constituents and thereby increasing the degree of meta-stability and the overall molecular weight of P.O.M. and D.O.M. An increase in molecular weight of D.O.M. with depth is, in fact, observed. Thus, Degens (unpublished data) has observed that the bulk of D.O.M. in the euphotic zone has a molecular weight <5000, with a maximum at 1500. In contrast, most of the D.O.M. at a depth of 5000 m has a molecular weight >10000. Although an increase in the degree of metal complexation with depth may also occur, this parameter has not been measured.

From the schematic presentation given in Fig. 31.5 it seems likely that the P.O.M. originating from seston which reaches the sediment will be of zero age. The organic input to shallow water sediments is probably dominated by this fraction because of the high primary productivity and the low degree of degradation which occurs in the short water column. In deep-sea sediments however, the organic input derived from labile seston is probably less significant as it has undergone extensive degradation in the long water column and also because it underlies water of lower productivity.

Under the conditions represented by Fig. 31.5 the dissolved organic matter (D.O.M.) may condense to form P.O.M.; although the exact mechanism of this conversion is somewhat obscure (e.g. Menzel, 1966), Barber (1966) has observed that if filtered sea water is bubbled with air, surface active organic compounds (molecular weight <100000) concentrate at the surface of bubbles and may then precipitate as P.O.M. under the action of bacteria. Ogura (1970) and Barber (1968) have demonstrated that much of the surface and deep-sea D.O.M. is inert to microbial attack and chemical oxidation. These observations do not contradict the data of Craig (1971), who calculated

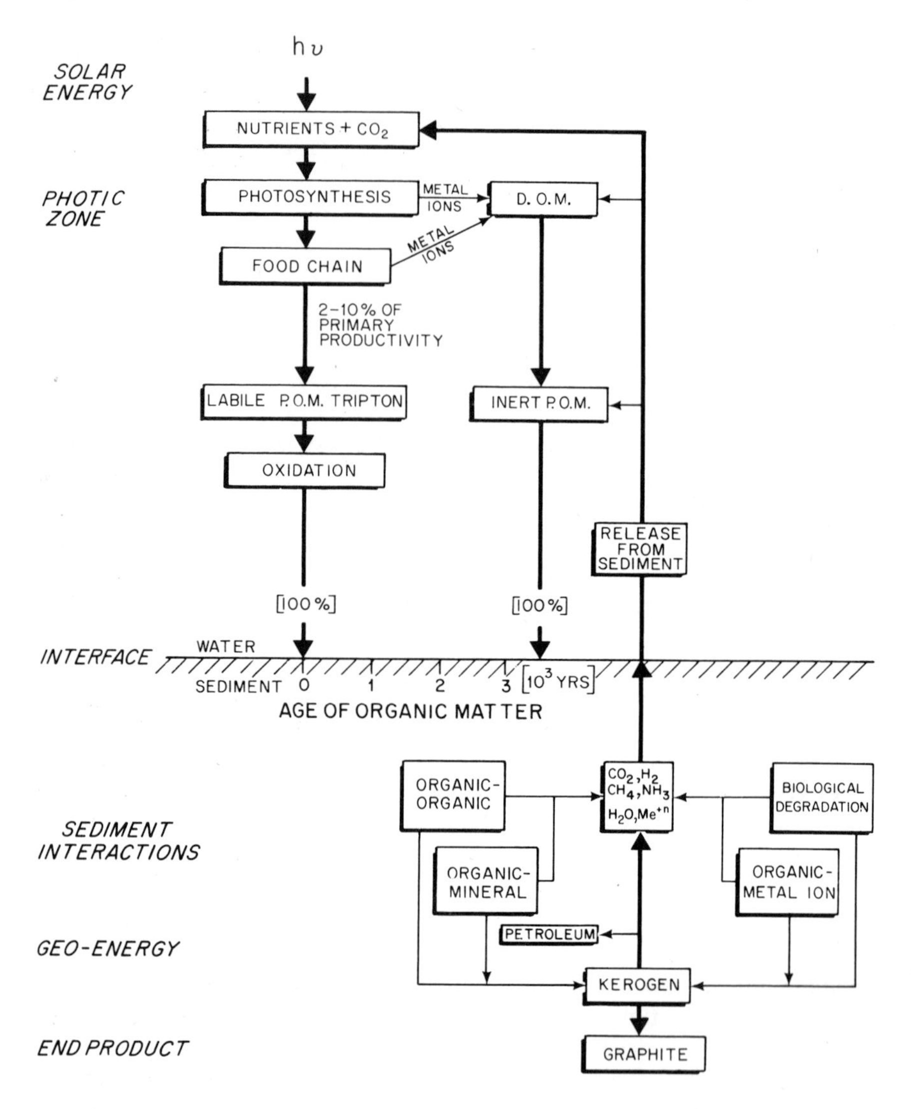

FIG. 31.5. Some aspects of the biogeochemical cycle in the ocean. Two major organic inputs to sediment are emphasized: (1) labile particulate organic matter (P.O.M.) or tripton, and (2) inert P.O.M. which forms from aged dissolved organic matter (D.O.M.). The organic input to shallow water sediments is probably dominated by labile P.O.M. and is therefore of zero age. The organic input to deep-sea sediments is probably dominated by inert P.O.M. and is therefore considerably older. Diagenesis of organic matter results in the release of nutrients into the environment, thereby closing the cycle.

that deep water consumption of D.O.M. proceeds at a rate of about $3{\cdot}5 \times 10^{-4}$ mg-C l^{-1} yr^{-1} because this rate is exceedingly slow and may be beyond the capabilities of present laboratory measurements.

The inertness of D.O.M. may be attributed both to cross-linking resulting from complexation with metal ions (Barber and Ryther, 1969; Siegel, 1971), and to the high degree of intermolecular condensation within the D.O.M. (Degens, 1970). Furthermore, the average age of deep-sea D.O.M. is of the order of 3400 years (Williams *et al.*, 1969). Thus, P.O.M. formed from D.O.M. may not only be more inert than seston, but may also be significantly older. This refractory P.O.M. may constitute a major fraction of the organic input to deep-sea sediments since the input of seston is probably considerably reduced relative to that to shallow water sediments (see above). It is of interest to note that the ages of the organic matter in surface sediments of deep-sea and shallow water environments are about 2000–3000 years and 200–400 years respectively (Degens, 1965). However, this age difference may be partially attributed to differences in sedimentation rates in the two areas and to the homogenization of the sediments by burrowing organisms. For example, in the Argentine Basin, where the sedimentation rate is about 6 cm per 1000 yrs, homogenization of the upper 5–10 cm will yield an apparent age of about 600–800 years.

It must be emphasized that, like D.O.M., P.O.M. appears to consist of a mixture of older refractory material and younger labile seston material, and that the relative proportions of these fractions varies spatially as well as temporally, *e.g.* with depth. By determining the content of labile substances, such as organo-phosphates and chlorophyll, the relative importance of these fractions may be estimated (Menzel and Ryther, 1964).*

Handa and Yanagi (1969) observed that the order of decay of some components of P.O.M. in the upper 50 m of the Western Pacific Ocean is as follows: water-extractable carbohydrates > organic nitrogenous compounds > total organic matter > water-insoluble carbohydrates.

The chemical nature of the deep-sea D.O.M. is largely unknown and only 10% of it has been identified and found to consist of amino acids, fatty acids, carbohydrates, phenols, etc. (see e.g., Wagner, 1969 and Duursma, 1965). As much as 20 to 50% of deep-sea P.O.M. is hydrolyzable (Gordon, 1970; Degens, 1970) and is composed largely of proteins and other nitrogeneous compounds. Thus, if deep-sea P.O.M. forms in part from D.O.M., then it follows that it must also contain some nitrogenous compounds. This hypo-

* P.O.M. is conventionally collected by filtering large volumes of sea water. An alternative method for collecting this material would involve scraping the upper 1 mm of deep-sea sediment by means of a submersible. The fact that microbial activities are greatly reduced at water depths greater than 1000 m (see discussion on micro-organisms, p. 78), and that the oxidation rates are extremely low, makes the organic matter at the immediate sediment-water interface representative of the particulate fraction present in the water column above.

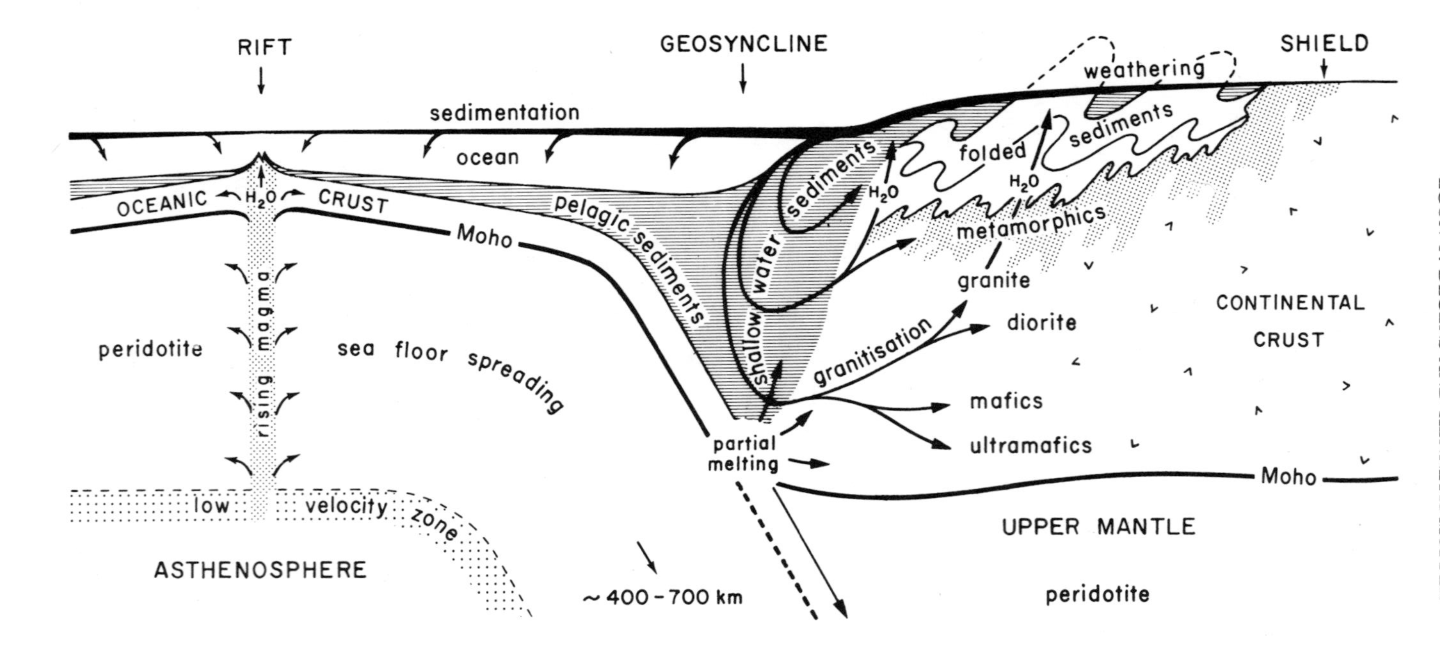

FIG. 31.6. Some major oceanic provinces, and the recycling of sediments during the process of sea floor spreading and mountain building.

thesis is supported by the C/N ratio, ca. 3, found for D.O.M. which is lower than that of 5–6 in planktonic material. Degens (1970) has postulated that D.O.M. consists predominantly of nitrogenous heteropolycondensates which are held together by organic–organic condensations, urea clathrate complexes, and metal ion coordination polyhedra. The percentage of land-derived humic material (e.g. lignin) in D.O.M. is more significant close to land than in the open ocean.

In conclusion, the depth of the water column will affect the quantity and quality of the organic input to the sediments. The reasons for this are as follows; (1) in shallow regions, where surface productivity is high, the input is probably dominated by labile sestonic material the composition of which approximately reflects that of its source (plankton); and, (2) in deep regions, where surface productivity is generally lower, the input is probably dominated by aged refractory material whose composition may be that of a composite of highly degraded seston and P.O.M. which formed from D.O.M.

31.3.6. TOPOGRAPHICAL FACTORS

According to Menard and Smith (1966), oceanic topography falls into nine different physiographic provinces, e.g. shelf and slope, basin, and rise and ridge. The distribution of these provinces in the oceans is principally determined by geological processes, such as large-scale crustal phenomena, (which produce features such as the mid-ocean ridges and fracture zones), and surface processes of erosion and redeposition (which result in the formation of continental shelves and slopes). The structural and spatial relationships between various provinces are depicted in Fig. 31.6 in which the dynamic aspects controlling the distribution of the physiographic provinces have been emphasized, although the morphology of the sea floor is only shown in a schematic fashion. The diagram reveals that areas close to continents are major sinks of terrigenous material which, during orogenic periods, is incorporated into the continental crust.

Rates of deposition of sediment may vary from as low as 1 mm per 1000 years on some abyssal plains to as high as several metres per day in tidal zones. Meade (1973) showed that along the east coast of the United States the rate of deposition of sediment is approximately equal to the present rise in sea level which is 2–3 mm per year. He also determined that episodic flooding of some rivers, due to hurricane activity, can yield in one week as much detritus as is normally supplied by the same river in several years.

The average deposition rate for the deep-sea is of the order of 1 cm per 1000 years. In the geosynclinal areas of the world, which are located in the troughs and basins bordering the continents (see Fig. 31.7), the average sedimentation rate is usually between 10 and 100 cm per 1000 years. Com-

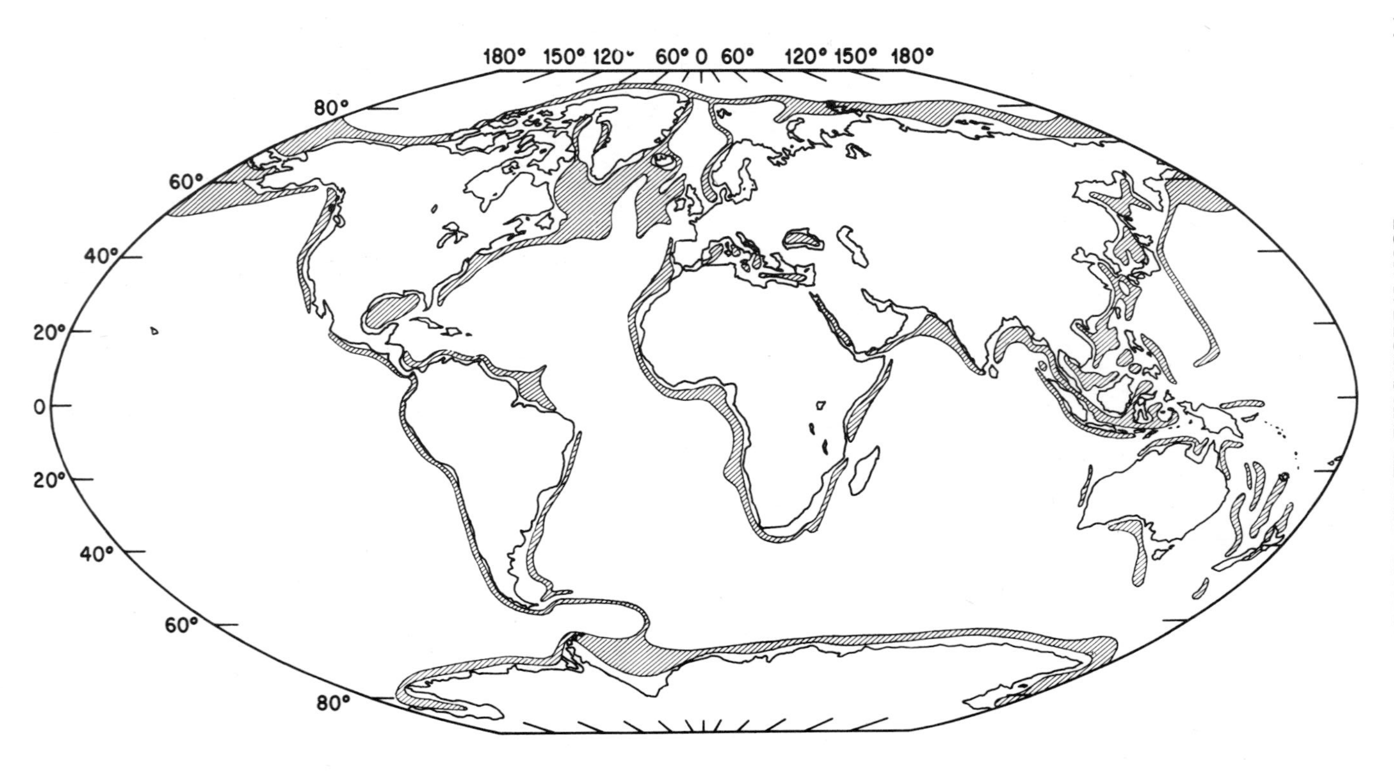

FIG. 31.7. Geosynclinal belts of the World Ocean (after Manheim and Sayles, 1973).

parison of Figs 31.1 and 31.7 reveals that regions of high organic productivity overlap geosynclinal areas. As a result, organic matter has a better chance to survive the early stages of diagenesis in these regions than it does in other deep-sea environments where surface water productivity and sedimentation rates are generally much lower.

In deltaic regions substantial quantities of terrigenous detritus are deposited in the adjacent deep-sea areas (e.g. the Indus Cone, the Mississippi Delta etc.). This type of sedimentation is enhanced during regressive stages of oceanic evolution when an increase in turbidity current activity sweeps sediment from delta deposits and outer continental shelves into deep-sea troughs and basins.

31.3.7. REDOX CONDITIONS

Most marine environments are oxic and only in few areas, such as the Black Sea, Cariaco Trench and Walvis Bay, are anoxic conditions found at the sediment-water interface. It may be significant that the interstitial waters of most marine deposits are devoid of free oxygen a few centimetres below the sediment-water interface (Kanwisher, 1962), even though they are overlain by oxygenated water. The marine benthos generally recycles the upper 2–10 cm of sediment, thereby exposing to oxygenated conditions material which has already been covered several times. Eventually, the sediment will be buried below the reworking zone and will remain in this anoxic environment. As a result, reducing micro-environments may develop even in oxic sediments, thereby giving rise to minerals such as glauconites and sulphides.

31.4. FACTORS DETERMINING THE DIAGENESIS OF THE ORGANIC MATTER IN SEDIMENTS

In this section sugar and amino acid data (Mopper and Degens, 1972) will be employed primarily as indicators of the diagenetic history of sedimentary organic matter. Since the main source of the organic matter in marine sediments is plankton, the discussion will be prefaced with a description of the amino acid and sugar compositions of representative plankton samples.

31.4.1. MARINE PLANKTON

Data on the amino acid and sugar contents of marine plankton are presented in Tables 31.2 and 31.3. The samples represent mixtures of zoo- and phytoplankton collected by an oblique tow from a depth of 200 m to the sea surface with a No. 10 net. The protein composition (Table 31.2) is fairly uniform, whereas that of the carbohydrate (Table 31.3) is highly variable.

TABLE 31.2

Distribution of amino acids and hexosamines in marine plankton (residues per 1000)*

	Iceland	Oyster pond, Woods Hole	Cariaco Trench	Walvis Bay	Peru-Callao‡	Galapagos Islands
Aspartic acid	98	90	102	104	115	109
Threonine	51	49	58	69	57	55
Serine	55	66	62	71	61	60
Glutamic acid	120	106	127	79	138	124
Proline	55	58	69	64	48	54
Glycine	133	82	155	112	159	132
Alanine	106	97	113	118	105	104
Cystine	9	1	2	1	11	13
Valine	63	52	59	76	53	58
Methionine	27	20	22	19	29	23
Isoleucine	44	37	41	54	35	38
Leucine	76	70	76	85	76	76
Tyrosine	8	38	18	40	26	33
Phenylalanine	31	31	26	42	35	32
Lysine	58	93	55	33	25	45
Histidine	6	24	4	6	13	9
Arginine	61	86	11	29	13	34
Total amino acids ($mg\ g^{-1}$ dry weight)	313	419	266	140	312	357
Hexosamines ($mg\ g^{-1}$)	18·7	36·0	14·3	9·8	7·9	55·6
Ammonia ($mg\ g^{-1}$)	2·3	3·0	3·2	1·4	0·97	10·8
AA/HA Ratio	16·7	11·6	18·6	14·3	35·5	6·4
Organic C %	54·4	54·53	43·90	29·43	45·33	48·46
N %	5·97	7·28	7·34	4·12	7·69	7·34
H %	7·74	7·45	6·24	3·18	6·19	6·98
C/N Ratio	9·1	7·5	6·0	7·1	5·9	6·6

* To facilitate comparison between samples, the concentrations are given as amino acid residues per 1000 (= mole ‰). This is the expression commonly used in the biochemical literature.
† ~50% mineral residue (mainly diatoms).
‡ 79% $CaCO_3$ (coccoliths) removed prior to analysis.

The general order of abundance of the latter is: galactose > glucose > mannose > ribose > xylose > fucose > rhamnose > arabinose. The function of galactose in plankton is unknown; however, the high degree of variability of glucose is probably related to the biochemical function of this sugar since β-glucans are the commonest food storage products of algae (Meeuse, 1962; Handa and Yanagi, 1969). The dominant amino acids in plankton are:

TABLE 31.3
Carbohydrate composition of marine plankton (mol %)

	Iceland	Oyster pond, Woods Hole	Cariaco Trench	Walvis Bay	Bermuda	Peru-Callao	Galapagos Islands
Rhamnose	2·7	5·4	7·4	3·0	4·6	4·0	5·1
Fucose	6·2	6·3	7·8	4·2	6·4	3·0	7·1
Ribose	37·4	22·1	4·9	2·8	5·6	14·8	7·1
Arabinose	2·5	0·8	2·1	0·5	1·1	3·0	6·0
Xylose	3·2	6·0	5·7	4·3	5·5	3·7	19·4
Mannose	16·0	18·0	15·0	7·5	11·9	13·1	13·2
Galactose	16·8	23·8	39·7	12·2	37·3	18·0	22·5
Glucose	15·3	17·7	17·3	65·6*	27·6	40·5	19·4
Total sugars ($\mu mol\,g^{-1}$)	13·1	113·2	144·9	217·9	86·9	40·5	44·8

* Cellulose fibre contamination present.

glycine, alanine, and glutamic and aspartic acids. The above sugars and amino acids were extracted by acid hydrolysis.*

A substantial part of the acid extractable amino acids and sugars is present as monomers within metal–organic associations. For example, 60% of the aspartic acid and 20% of tyrosine and phenylalanine are present in the unhydrolyzed EDTA extract.† The high degree of metal binding by aspartic

* *Total acid extractable sugars* were determined following hydrolysis in 1·8 N-HCl for 3 h at 100°C; *total acid extractable amino acids* were determined after hydrolysis in 6·0 N-HCl for 22 h at 105°C.

† The degree of metal binding of amino acids and sugars was determined by treating the sample material with an aqueous slurry of EDTA for 8 h at 100°C. By addition of EDTA, a strong chelating agent, to the samples metals ions (especially polyvalent ones) were removed from *in situ* metal-organic complexes having lower stability constants; metal-free organic compounds, if soluble, were then released into the aqueous solution. In unbuffered solutions EDTA exhibits minor hydrolysis effects. Because of this, sugars and amino acids detected in the EDTA extract represent an approximation to the *in sigu* metal-bound monomers. This fraction will be referred to as the *unhydrolysed EDTA extract.* Acid hydrolysis of this extract will convert EDTA-solubilized carbohydrates and proteins to monomers; this fraction will be referred to as the hydrolyzed EDTA extract. For a few sediment samples Chelex 100 (a strong chelating resin with EDTA-like functional groups) was substituted for EDTA. The yields obtained with this resin following hydrolysis agreed well with those found when the EDTA extract was hydrolysed; thus providing independent verification for the significance of the EDTA extracts. Sediments appear to be buffered by inorganic phases, whereas plankton samples are not. EDTA extraction of plankton in a buffered system would eliminate interference of hydrolysis. The hydrolyzed EDTA extract generally contained 25–70% of the sugars and amino acids released by acid hydrolysis. In contrast, the hydrolyzed water extract (100°C for 8 h followed by hydrolysis) generally released less than 20% of the acid extractable sugars and amino acids. The unhydrolyzed water extract usually amounted to less than 1% of the extract. For details of these extraction procedures see Mopper and Degens (1972, 1976) and Mopper (1973).

acid may be interpreted in the light of recent research on calcification which indicates that this amino acid provides structural sites for the coordination of calcium and magnesium ions (Ghiselin *et al.*, 1967; King and Hare, 1972; Degens, 1976).

The hydrolyzed EDTA extract contains about half of the total acid extractable amino acids. Hexosamines, which constitute 1 to 6% (dry weight) of the total plankton, are not metal-bound and are probably present as chitinous material (see page 84, Table 31.5).

Approximately half of the acid extractable sugars are released by EDTA; about half of the EDTA extract is present as monomers (see page 81, Table 31.4, line a). The metal-bound pentoses (ribose, arabinose, and xylose) occur principally as monosaccharides (line a); the metal-bound hexoses (mannose, galactose, and fucose) occur principally as polysaccharides (line b). Line c shows that hexoses are dominantly present as EDTA-*non*-extractable sugars: deoxyhexoses (rhamnose and fucose) and pentoses are dominantly present as EDTA-extractable sugars (lines a and b).

The EDTA-insoluble sugars are probably present as long-chain structural polysaccharides (hemicellulose). It is of interest to note that Handa and Yanagi (1969) observed that the sugars (xylose, mannose, galactose and glucose) usually found in such long-chain structural polysaccharides, are the dominant sugars in the water-insoluble carbohydrate fraction of particulate organic matter in the surface layer of the ocean (0–50 m). It can be calculated that mannose, galactose and glucose constitute about 75% of the EDTA-non-extractable residue of plankton (Mopper, 1973).

31.4.2. BENTHOS AND MICROORGANISMS

Benthic samples have provided data for a comparison of animal densities and biomass under various ecological conditions (Sanders *et al.*, 1965; Sanders and Hessler, 1969; Rowe, 1971). The benthic metazoan biomass density off New England decreases by a factor of about 500 from the shallow coastal zone (10–200 m) to the deep-sea (4000–5000 m) with the greatest change occurring at the edge of the continental shelf (100–200 m). Furthermore, within the depth range 200–2500 m the biomass density decreases by a factor of 25 although surface primary productivity remains fairly constant at around 100 g-C m^{-2} yr^{-1}. The differences in hydrostatic pressure over this interval may affect the diversity of species but not the actual biomass. Similar trends are developed in other regions of the ocean. However, at comparable depths the benthic animal density is significantly higher in regions of higher surface primary productivity than it is in regions of low surface productivity. The rates of decrease of metazoan biomass density as a function of water depth can be expressed by statistically significant least

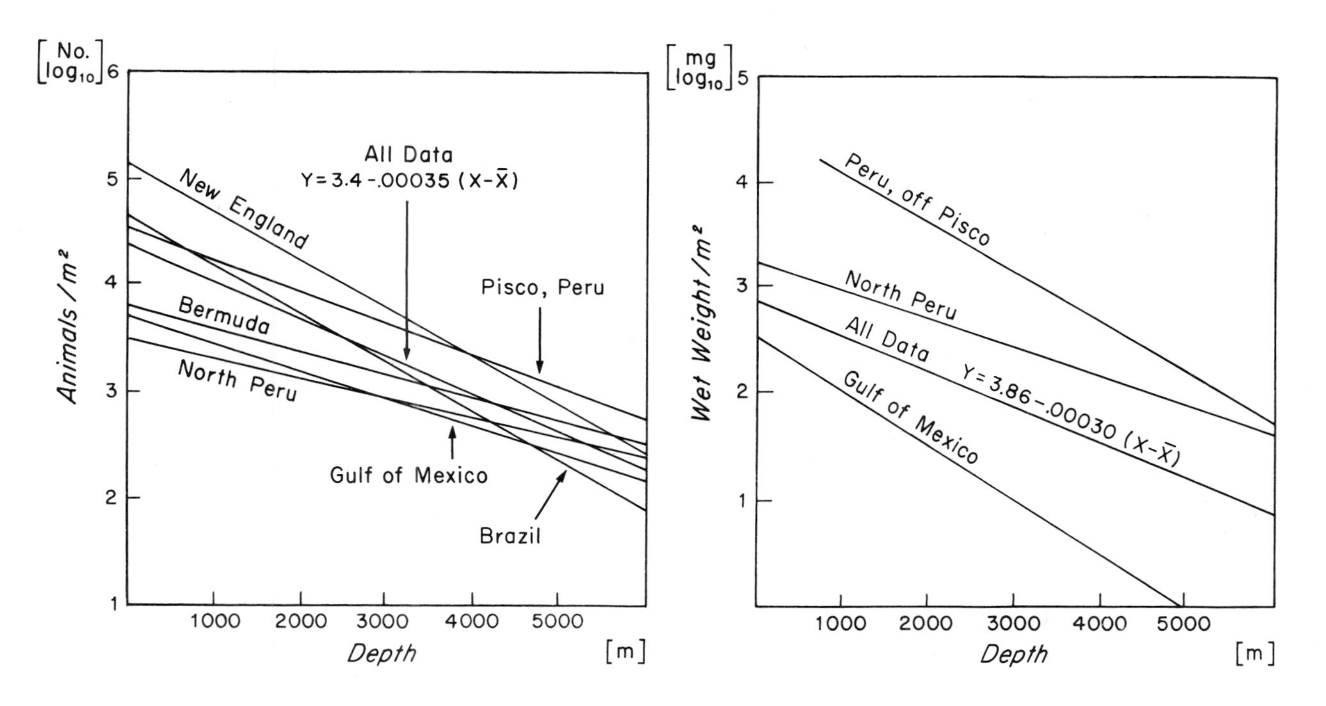

FIG. 31.8. Rates at which animal density and biomass decrease with depth down the water column. The relationships between the logarithm (base 10) of animal density or biomass, and the water depth are described by statistically significant least square linear regressions (after Rowe, 1971).

squares linear regressions as shown in Fig. 31.8. These regressions suggest that although depth exerts the most significant effect, surface productivity is the next most important control on benthic biomass. The decrease in biomass with depth is probably due to the increasingly refractory nature, and hence indigestible nature, of the organic input. A deficiency of certain nutrients or growth hormones in the organic detritus used as food may also limit biomass density in the deep-sea.

It is often said that microorganisms rapidly degrade sedimentary organic matter, even in deep-sea environments and microbial counts of up to several million bacteria per gram of deep-sea sediment can be found (see e.g. Rittenberg *et al.*, 1963). However, the work of Jannasch *et al.* (1971) and Jannasch and Wirsen (1973) who examined the *in situ* microbial degradation of organic substrates at the sediment-water interface, has shown that this is not so. Food stored at a water depth of 1500 m for several months showed negligible bacterial degradation. Organic substrates (e.g. agar, polysaccharides) inoculated at depth and embedded for several months in deep-sea sediment were utilized at rates up to several hundred times lower than control samples kept under refrigeration. In contrast, organic substrates placed in shallow-water sediments were utilized at comparable or slightly lower rates than were the controls. When clay was added to the substrates, decomposition rates did not increase even though the number of bacteria increased by a factor of 100 to 1000 relative to clay-free samples.

This decline of microbial activity with depth may be related to the increase in pressure. However, an alternative hypothesis is that with increased water depth the organic matter brought to the sediments undergoes considerable degradation and structural modification, making it less palatable for, or less degradable by, microorganisms. The fact that common microbial substrates remain virtually intact in the deep-sea suggests that physiological effects exert a greater control on the activity of bacteria than "palatability" effects. It is conceivable that microbial cell walls become "stained" by heavy metals present in the sediment pore waters, and this inhibits the activity of the microorganisms and results in a type of "suspended animation". The microbial organisms would regain their activity if chelators capable of removing the inhibiting metals from the cell walls were present in sufficient quantities, as they probably are in shallow-water sediments. Experiments are being conducted to test these hypotheses. The idea that there is a general slow-down of life processes in the deep-sea is further supported by *in situ* measurements of biological oxidation rates, which range from 1–4 g-C m^{-2} yr^{-1} for deep-sea sediments to 50–150 g-C m^{-2} yr^{-1} for the sediments of shallow coastal regions (Smith and Teal, 1973).

31.4.3. ORGANIC WASTE PRODUCTS

Since organic matter represents a small fraction of the total sediment, it is difficult to distinguish between compounds that arise from biological degradation processes and those which originate from mineral-organic interactions. Therefore, in order to distinguish clearly the characteristics of one of these mechanisms, i.e. microbial degradation, two systems were studied in which this was the principal process involved in the production of organic components, viz. sewage sludge treatment and the rumination in a cow. The data presented in Tables 31.4 and 31.5 permit the following trends to be recognized:

31.4.3.1. *Sugars* (*Table* 31.4)

(a) The degree of metal association (hydrolyzed EDTA extract) increased from 17% in the Deer Island (near Boston, Mass.) primary sludge to 77% in the Deer Island secondary sludge. This trend is not as clear in the Cranston (Cranston plant, Rhode Island) data because cellulose determinations were not performed.

(b) The principal metal-bound sugars are rhamnose, fucose, ribose, arabinose and galactose. Arabinose is the dominant metal-bound monomer.

(c) The unhydrolyzed water extracts released <0·3% of the total acid extractable sugars, whereas the unhydrolyzed EDTA extracts released about 5% in the primary sludge and about 15% in the secondary sludge. The presence of fragile monomeric sugars in these extracts confirms that metal association has a protective effect.

(d) Glucose showed the largest decrease in concentration after microbial (secondary) digestion. This was to be expected because cellulose is a major component of the primary sludges.

(e) Cow manure is also the product of microbial digestion. The rumen in a cow can be considered analogous to a secondary sludge tank of a sewage treatment plant. The carbohydrate patterns of cow manure are very similar to those of sewage sludge. For example, about 40% of the carbohydrate is metal-bound; rhamnose, arabinose and galactose are the major metal-bound species (ribose was not detected), and arabinose is again the dominant metal-bound monomer. The unhydrolyzed water extract released <0·2% of the total acid extractable sugars, while the unhydrolyzed EDTA extract released 8%. Glucose is reduced in comparison to other sugars and especially relative to the straw used as feed (not shown).

TABLE 31.4
Total percentages of sugars extractable by EDTA treatment of sediments and organic waste products

Sugar*		Rh	Fu	Ri	A	X	M	Ga	Gl	Total
Cariaco Trench										
<10 m	a†	9	25	93	30	19	2	10	3	14
(6)	b	34	34	—§	21	23	20	28	28	28
	c	57	41	7	49	58	78	62	69	58
Cariaco Trench										
>10 m	a	12	25	49	19	20	5	10	5	12
(3)	b	48	34	—	19	25	24	29	30	28
	c	40	41	51	62	55	71	61	65	60
Plankton above										
Cariaco Trench	a	15	54	93	97	31	4	7	13	20
	b	31	19	—	9	21	27	20	37	21
	c	54	27	7	—	48	69	73	50	59
Black Sea,										
(marine) 15 cm	a	9	23	43	26	15	1	11	1	12
	b	91	43	0	36	10	17	42	32	28
	c	0	34	57	38	75	82	47	67	60
Black Sea,										
(transition) 65 cm	a	20	30	79	22	31	3	18	2	20
	b	50	35	—	20	30	45	29	48	35
	c	30	35	21	58	39	52	53	50	45
Black Sea,										
(fresh) 125 cm	a	13	25	76	32	15	3	11	3	14
	b	71	103	89	41	61	56	44	45	54
	c	16	—	—	27	24	41	45	52	32
Walvis Bay,										
sediment surface	a	8	15	74	22	19	5	8	1	11
	b	33	18	—	12	28	48	27	29	28
	c	59	67	26	66	53	47	65	70	61
Lake Kivu,										
130 cm	a	19	38	53	51	33	2	23	1	14
	b	52	31	—	26	36	48	16	20	23
	c	29	31	47	23	31	50	61	79	63
Argentine Basin										
<5 m (2)	a	30	43	94	39	38	9	20	13	28
	b	52	37	—	14	30	44	50	39	36
	c	18	20	6	47	32	47	30	48	36
Oyster pond,										
sediment surface	a	10	24	63	34	15	2	8	2	11
	b	44	43	37	31	34	29	39	24	33
	c	46	33	0	35	51	69	53	74	56
Oyster pond,										
plankton	a	12	46	48	5	19	5	37	25	30
	b	58	39	28	40	18	35	20	38	30
	c	30	15	24	55	63	60	43	37	40

TABLE 31.4—*continued*

Sugar*		Rh	Fu	Ri	A	X	M	Ga	Gl	Total
N.Y. Bight, sediment surface	a	16	39	123	54	36	5	18	5	22
	b	60	48	—	27	27	49	50	44	43
	c	24	13	—	19	37	46	32	51	35
Wood in Cariaco trench, sediment, 0·64 m	a	16	63	0	88	23	5	19	1	15
	b	26	7	100	—	—	6	9	0	1
	c‡	58	30	0	12	77	89	72	99	84
Cow manure	a	5	33	—	44	2	3	9	1	8
	b	53	—	—	25	26	11	57	27	28
	c	42	67	—	31	72	86	34	72	64
Deer Island sewage sludge, primary	a	19	52	120	102	8	2	13	1	6
	b	54	27	—	—	17	17	56	4	11
	c‡	27	21	—	—	75	81	31	95	83
Deer Island sewage sludge, digested	a	23	71	194	101	12	3	14	2	17
	b	128	119	45	55	52	52	107	16	60
	c‡	—	—	—	—	36	45	—	82	23
Cranston sewage sludge, primary; collected 1972	a	6	19	45	61	2	1	4	2	6
	b	70	51	36	22	6	22	67	46	41
	c	24	30	19	17	92	77	29	52	53
Cranston sewage sludge, digested; collected 1972	a	7	37	93	60	2	2	7	1	11
	b	76	56	—	36	11	28	76	58	45
	c	17	7	7	4	87	70	17	41	44
Cranston sewage sludge, digested; collected 1971	a	11	22	150	45	4	1	5	1	8
	b	97	87	0	67	40	52	65	53	60
	c	—	—	—	—	56	47	30	46	32

* Rh = rhamnose; Fu = fucose; Ri = ribose; A = arabinose; X = xylose; M = mannose; Ga = galactose; Gl = glucose.

† a = (EDTA/total) × 100% = EDTA extracted monomers as total.
b = [(EDTA + HCl)/total] × 100% − a = EDTA extracted polymers as % of total.
c = 100% − (a + b) = non-EDTA extractable residue as % of total.

‡ Pretreated with concentrated H_2SO_4.

§ Concentration is less than in other extracts.

() No. of samples represented.

Note: because of limitations in the extraction procedures not all results total 100%.

TABLE 31.5

Total percentages of amino acids and hexosamines extractable by EDTA treatment of sediments and organic waste products

		Asp	Thr	Ser	Glu	Pro	Gly	Ala	Val	Met	Iso	Leu	Tyr	Phe	Lys	His	Arg	Total	Hex
Cariaco Trench <10 m	a*	5	0	0	0	0	9	0	0	0	0	0	0	0	3	0	0	1	0
	b	13	13	9	7	3	14	4	6	2	3	2	2	2	8	0	0	6	2
	c	82	87	91	93	97	78	96	94	98	97	98	98	98	89	100	100	93	98
>10 m	a	12	3	6	0	0	58	0	0	0	0	0	0	0	4	0	0	6	0
	b	52	8	23	12	7	76	8	20	2	2	1	0	0	82	0	0	26	0
	c	36	89	71	88	93	—	92	80	98	98	99	100	100	14	100	100	68	100
Black Sea 15 cm (marine)	a, b	31	12	14	7	2	13	7	5	0	5	5	2	4	14	0	0	10	10
	c	69	88	86	93	98	87	93	95	100	95	95	98	96	86	100	100	90	90
Black Sea (interm.) 65 cm	a	14	0	4	0	nd.	3	0	1	0	0	0	0	nd.	0	0	0	2	0
	b	1	25	24	9	nd.	10	8	4	6	4	2	14	nd.	13	0	15	10	10
	c	85	75	72	91	nd.	87	92	95	94	96	98	86	nd.	87	100	95	88	90
Black Sea (fresh) 125 cm	a, b	10	22	25	5	nd.	83	2	1	0	0	0	0	0	234	0	0	21	2
	c	90	78	75	95	nd.	17	98	99	100	100	100	100	100	—	100	100	79	98
Walvis Bay sediment surface	a	14	1	1	0	0	2	0	0	0	1	0	0	0	11	70	0	4	2
	b	16	17	14	14	16	15	13	16	0	6	4	2	3	1	0	16	11	6
	c	70	82	85	86	84	83	87	84	100	93	96	98	97	88	30	84	85	92
Lake Kivu 130 cm	a	17	0	0	0	0	2	0	0	0	0	0	0	0	0	0	0	1	0
	b	5	4	10	5	4	4	4	4	1	3	2	6	5	8	14	2	4	8
	c	78	96	90	95	96	94	96	96	99	97	98	94	95	92	86	98	95	92

240 cm	a, b	6	17	28	10	3	35	6	7	0	4	4	1	1	16	8	5	12	3
	c	94	83	72	90	97	65	94	93	100	96	96	99	99	84	92	95	88	97
Argentine Basin 0–1 m	a	65	4	20	2	5	41	2	1	0	0	0	nd.	nd.	192	nd.	nd.	18	0
	b	1	25	26	30	—	88	11	10	15	9	6	nd.	nd.	—	nd.	nd.	36	8
	c	34	71	54	68	95	—	87	89	85	91	94	nd.	nd.	—	nd.	nd.	46	92
3–5 m	a	71	5	12	1	5	105	29	11	33	12	12	nd.	nd.	12	nd.	nd.	21	0
	b	19	38	28	30	—	387	—	—	—	—	—	nd.	nd.	102	nd.	nd.	57	8
	c	10	57	60	69	95	—	71	89	67	88	88	nd.	nd.	—	nd.	nd.	22	92
Oyster pond, Plankton	a	61	13	12	3	14	13	14	9	26	16	15	20	21	17	6	17	18	0
	b	12	41	32	54	36	47	38	22	22	42	37	28	28	34	32	10	33	5
	c	27	46	56	43	50	40	48	69	52	42	48	52	51	48	62	73	49	95
Cow manure	a	13	1	1	1	0	2	2	2	0	1	1	1	1	3	0	0	2	0
	b	11	16	16	13	9	15	15	12	3	11	10	15	8	36	7	15	16	34
	c	76	83	83	86	91	83	83	86	97	88	89	85	91	71	93	85	82	66
Deer Island sludge, primary	a	9	5	3	1	3	5	5	6	2	5	3	6	2	5	0	2	4	0
	b	20	14	22	23	22	26	21	20	18	19	17	24	22	34	12	14	21	3
	c	71	81	75	76	75	69	74	74	80	76	80	70	76	61	88	84	75	97
Cranston sludge, digested	a	11	1	1	0	0	2	2	2	0	1	0	0	0	1	0	1	2	0
	b	8	18	12	9	13	19	14	21	2	18	12	14	16	31	19	21	15	34
	c	81	81	87	91	87	79	84	77	98	81	88	86	84	68	81	78	83	66

† a = (EDTA/total) × 100% = EDTA extracted monomers as total.
b = [(EDTA + HCl)/total] × 100% − a = EDTA extracted polymers as % of total.
c = 100% − (a + b) = non-EDTA extractable residue as % of total.
nd. = not detected.

31.4.3.2. *Amino Acids* (*Table* 31.5)

(a) Cow manure contains one percent of the total acid extractable amino acids in the free state (unhydrolyzed water extract). The unhydrolyzed EDTA extract, however, contains considerable quantities of aspartic acid, (Table 31.5). Acidification of these extracts releases 10 and 20 percent of the total acid extractable amino acids respectively. Lysine, aspartic acid and hexosamines are the dominant metal-bound amino acids in the cow manure.

(b) Comparison of Cranston sludge with cow manure reveals that sewage treatment induces metal binding to the same degree as the digestive tract in a cow. For example, lysine and hexosamines are the major metal-bound metal species in both systems. These trends strongly suggest that, independent of the composition of the original organic input into either a sewage plant or the digestive system of a cow, the final product will probably reflect the amino acid pattern of the metabolic imprint of microbes, which in both these systems appears to be similar.

(c) The metal association of amino acids, in contrast to that of sugars, does not increase during microbial degradation, suggesting that carbohydrates are used in preference to proteins. This can be understood if it is remembered that carbohydrates, especially cellulose, constitute the bulk of raw sewage sludge. Biological consumption within sediments takes place in tens to thousands of years instead of the few weeks which are required for sewage digestion. Furthermore, in sediments the protein content is at least as high as that of carbohydrate. Thus, in sediments, one would expect a significantly higher degree of protein consumption and a concomitant increase in metal association relative to sewage sludges (see Section 31.4.4).

Table 31.6 lists metal analyses of EDTA extracted and acid digested sewage sludges. Bound metals constitute up to 10% (dry weight) of the total; a few metals, such as lead and zinc, appear to be almost completely associated in metal-organic complexes.

From the above trends it can be concluded that microbial degradation of organic matter, such as sewage sludge and straw, coincides with an increase in metal binding in the residue. This increase strongly suggests that metal-bound organic matter is biologically inaccessible, even though it includes a large fraction of easily degradable monomers (up to 15% of the total acid extractable fraction). Such inaccessibility could be due simply to the low solution levels of free compounds as shown by the unhydrolyzed water extract, or could be due to inhibitory effects causes by the metal content of the bound organics (Mopper, 1973).

31.4.3.3. *Heavy Metals Extracted from Sewage Sludge*

Table 31.6 compares heavy metal analyses of both EDTA and nitric acid extracts of sewage sludge. Metals constitute up to 10% of the total (dry weight); most of the metals appear to be almost totally extracted with EDTA. The metal recoveries for the EDTA extraction of the Deer Island secondary sludge are generally higher than for the Deer Island primary sludge. This trend indicates that digestion of sewage sludge renders metal ions more

TABLE 31.6
Metal analyses of nitric acid digested and EDTA extracted sewage sludges ($\mu g\ g^{-1}$ *dry wt.*)

Metal		Nitric acid Digestion	EDTA* Extracted	% Extractable with EDTA
	Cranston (s)†	365	345	95
Lead	Deer Island (p)	868	975	112
	Deer Island (s)	722	750	104
	Cranston (s)	12	5	42
Cadmium	Deer Island (p)	36	28	78
	Deer Island (s)	55	55	100
	Cranston (s)	1490	480	32
Nickel	Deer Island (p)	153	115	75
	Deer Island (s)	197	170	86
	Cranston (s)	280	103	37
Chromium	Deer Island (p)	895	900	101
	Deer Island (s)	1376	1675	123
	Cranston (s)	19	—	0
Cobalt	Deer Island (p)	11	—	0
	Deer Island (s)	12	—	0
	Cranston (s)	6704	4700	70
Zinc	Deer Island (p)	1747	1950	112
	Deer Island (s)	2537	3200	126
	Cranston (s)	2860	70	2
Copper	Deer Island (p)	870	15	2
	Deer Island (s)	1200	15‡	1

* Analysis of the EDTA reagent revealed only trace quantities of heavy metals. A water extraction (8 h, 100°C) of the Deer Island (s) sample released negligible amounts of heavy metals relative to the nitric acid digestion, e.g. 0·25% Cu, 0·70% Cd, 0·34% Zn, and 0·22% Ni.
† s = secondary treated sludge; p = primary treated sludge. Analyses by A. Jacobs, W.H.O.I.
‡ Extraction of the Deer Island (s) sample with cysteic acid (a sulphur-containing amino acid) released 88 $\mu g\ g^{-1}$ Cu, or 7% of that removed by the nitric acid digestion, which is about seven times that released by EDTA. The low copper yield of the EDTA extract is not clearly understood.

readily available for extraction with EDTA. The reason for this greater availability is unknown.

The low copper recovery of the EDTA extractions (Table 31.6) is not clearly understood. It is possible that the copper is bound in *in situ* complexes which have a stronger affinity for that metal (e.g. sulphydryl complexes) than does EDTA. Further research is necessary to clarify this point.

Despite their relatively high concentrations, potentially toxic heavy metals do not appear to interfere with microbial degradation. This suggests that these metals are complexed with organic matter which, in turn, greatly reduces their activity (and hence their toxicity). Tan *et al.* (1971) found a stability constant of 6·8 (at pH 7) for the complexes formed between zinc and extracts prepared from sewage sludge.

31.4.4. METAL BINDING AND BIOLOGICAL CONSUMPTION IN SURFACE SEDIMENTS

In Section 31.3.5 it was suggested that the organic input of material to deep-

TABLE 31.7
Mol % composition of carbohydrates in sediment, HCl hydrolysis

Sugar	Rh	Fu	Ri	A	X	M	Ga	Gl
Argentine Basin <5 m (7)*	8·4	9·6	4·8	10·2	10·5	13·7	24·2	18·4
Bermuda, surface	10·7	9·8	4·1	8·9	12·6	16·8	23·8	13·1
N.Y. Bight, surface	9·1	8·1	1·6	8·6	12·9	15·1	21·5	23·1
Black Sea, marine 0·15 m	4·8	7·6	9·6	4·6	16·3	23·7	20·0	13·3
Black Sea, transition 0·65–70 m	15·1	10·9	5·5	8·9	12·6	10·0	18·3	18·8
Black Sea, fresh 1·30 m	12·5	6·3	1·8	10·4	10·0	14·6	22·9	21·9
Cariaco Trench <40 m (7)	11·0	9·2	3·5	5.5	9·7	19·3	28·0	13·6
Cariaco Trench 67–130 m (2)	8·9	4·8	0·7	9·9	12·3	23·9	23·2	17·2
Lake Kivu 1·25 m	5·3	3·0	0·5	2·9	3·7	5·3	36·9	42·5
Lake Kivu 2·0–9·3 m (3)	11·3	8·0	1·0	7·0	12·0	15·6	20·3	25·1
Oyster pond, surface	11·1	6·8	0·9	10·2	11·3	13·3	20·2	26·1
Walvis Bay, surface	15·7	12·7	2·3	6·1	9·4	8·2	26·9	18·6

* No. of samples represented.

TABLE 31.8

*Distribution of amino acids and hexosamines in surface sediments, humic acids and kerogens**

	Bermuda- Woods Hole (14)†	New York Bight (3)	Cariaco Trench (3)	Walvis Bay (1)	Argentine Basin (4)	Black Sea (4)	Arabian Sea (5)	Humic HCl (3)	Kerogen HCl + EDTA (2)	Lake Kivu (3)
OH-Proline	22	—	nd.	—	nd.	—	19	—	—	—
Aspartic Acid	83	108	101	121	125	109	78	142	51	82
Threonine	75	81	36	55	66	64	57	53	42	67
Serine	87	96	47	97	84	61	68	86	138	61
Glutamic acid	67	95	76	73	57	86	71	110	6	79
Proline	39	55	71	50	31	60	74	25	2	65
Glycine	228	173	144	171	156	138	247	179	671	130
Alanine	111	90	112	95	128	115	107	110	16	131
Cystine	7	7	6	2	3	7	18	9	—	2
Valine	62	53	54	45	124	67	58	43	12	78
Methionine	11	10	23	20	15	12	9	22	—	13
Isoleucine	27	29	38	23	33	46	33	14	5	56
Leucine	35	46	81	49	44	76	45	47	2	92
Tyrosine	9	29	23	24	8	26	9	6	—	35
Phenylalanine	15	43	59	31	19	42	15	10	—	53
β-alanine	25	—	—	—	—	9	19	3	—	1
Ornithine	24	nd.	—	—	nd.	—	24	14	—	—
Lysine	47	41	81	89	80	61	37	78	55	32
Histidine	14	3	10	3	2	1	9	22	—	6
Arginine	13	41	39	51	29	21	2	26	—	18

TABLE 31.8—*continued*

	Bermuda-Woods Hole (14)†	New York Bight (3)	Cariaco Trench (3)	Walvis Bay (1)	Argentine Basin (4)	Black Sea (4)	Arabian Sea (5)	Humic HCl (3)	Kerogen HCl + EDTA (2)	Lake Kivu (3)
Total amino acids ($mg\ g^{-1}$) dry wt.	0·92	6·13	6·31	17·03	0·25	16·70	0·45	6·14	0·23	46·1
Glucosamine ($mg\ g^{-1}$)	0·031	0·42	4·16 }	3·56	0·38 }	1·81 }	0·059 }	0·73	—	1·70 }
Galactosamine ($mg\ g^{-1}$)	0·034	0·32		1·65				0·83	—	
Ammonia ($mg\ g^{-1}$)	0·062	0·16	0·21	0·34	0·035	0·22	0·014	0·70	0·091	0·23
Organic C(%)	0·38	1·67	5·10	5·90	0·50	7·10	0·84	60·1	88·3	12·6
Organic N(%)	0·073	0·15	0·50	0·81	0·065	0·62	0·093	2·81	3·73	1·06
C/N	5·2	11·1	10·2	7·3	7·7	11·5	9·0	21·4	23.7	11·9
AA/HA ratio	14·1	8·4	1·5	3·3	0·67	9·0	7·6	3·9	—	27
N recovered %	25	67	28	41	17	38	9	6	0·8	59

* Residues per 1000.
† Number of samples.

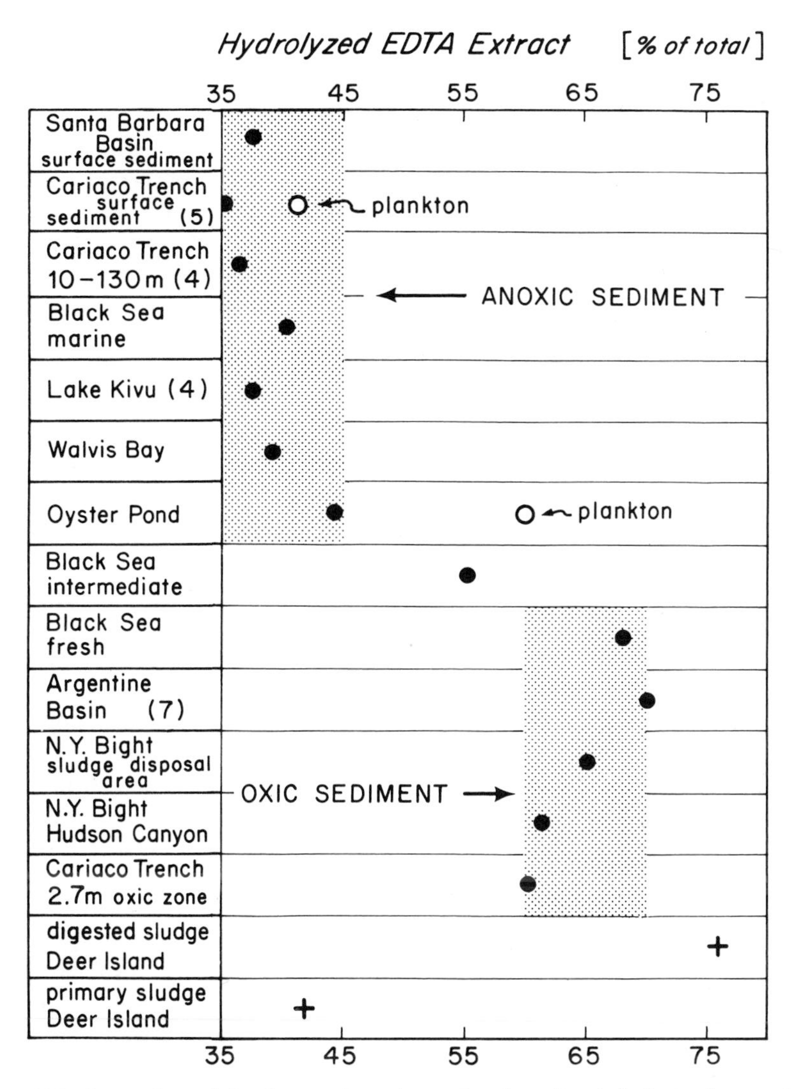

FIG. 31.9. Comparison of the degree of metal complexation of carbohydrates from various sedimentary environments. The percentage of HCl extractable sugars which is released by EDTA is 25–45% for reducing sediments, and 55–75% for oxidizing sediments. Oyster Pond is intermittently oxic (although dominantly anoxic). The "Black Sea Intermediate" sample represents a period of transition from oxidizing to reducing conditions. Digestion of primary sewage sludge has the effect of drastically increasing the degree of metal complexation. The number in parentheses is the number of samples examined.

sea sediments is less in quantity and biologically more refractory in nature than is that entering shallow-sea sediments. Despite the differences between these two inputs the sugar and amino acid compositions of surface sediments from both these environments (e.g. Argentine Basin, deep-sea; N.Y. Bight, shallow-sea) are almost identical (Tables 31.4, 31.5, 31.7 and 31.8). Thus, the concentration of EDTA-extractable sugars in the organic matter from the sediments of both environments comprises 60–70% of the total sugars. Furthermore, the total sugars represent 4–5% of the total organic matter in both sediments.

This uniformity of organic matter composition is probably attributable to the benthic bio-activity in these environments. In deep-sea areas in which the organic material probably undergoes considerable degradation in the water column, "edible" organic matter will be scarce. In contrast, in shallow regions in which the organic material probably consists predominantly of partially degraded plankton detritus (seston), "edible" organic matter will be present in considerably larger quantities. The observed benthic metazoan biomass distribution appears to reflect this pattern of edibility.

Analyses of primary and digested sewage sludges suggest that metal-bound organic matter is refractory since it is the end product of biological degradation. For example, it is shown in Fig. 31.9 that the proportion of metal-bound (EDTA-extractable) sugars in the total acid extractable sugars increases from 17% in the primary sludge to 77% in the microbially digested sludge. The degree of metal association induced by digestion of sewage sludge is similar to that of carbohydrates in oxic sediments. By analogy, the high degree of metal association of carbohydrates in shallow-sea sediments is probably attributable to a high biological consumption rate and the end-product of this degradation is a biologically resistant, metal-bound humic substance.

The high degree of metal association of organic matter and possibly microbial cell walls in deep-sea sediments is attributable to: (1) the low sedimentation rate which is conducive to a thorough reworking of the organic material by the benthic community, and (2) the refractory nature of the organic input. Therefore, although the organic matter in both shallow and deep sediments is probably qualitatively different in terms of lability, the end product of biological degradation at the sediment-water interface is remarkably similar in both composition and degree of metal association in the two environments.

Evidence to further establish these trends can be obtained from a comparison of the degree of metal binding of the organic matter in reducing sediments (Cariaco Trench, Lake Kivu and Walvis Bay) with that in oxidizing sediments (Argentine Basin). Sugars and amino acids are predominantly unassociated with metals in the surface sediments of reducing environments

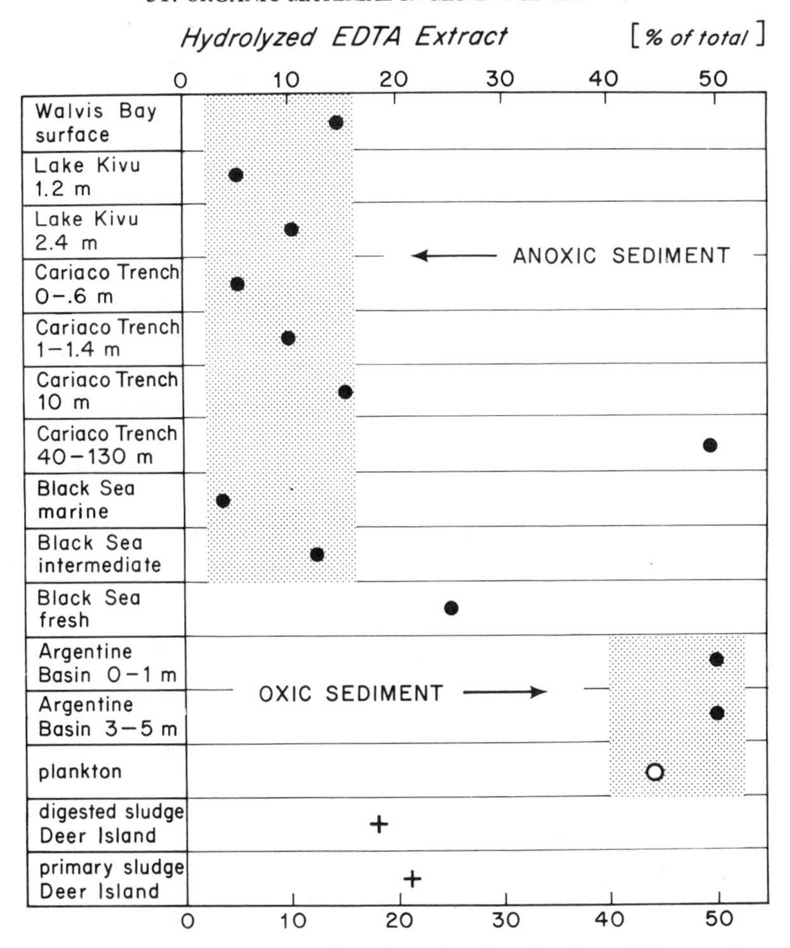

FIG. 31.10. Comparison of the degree of metal complexation of amino acids from various sedimentary environments. Reducing sediments are less complexed than oxidizing sediments. Within a reducing sediment, however, the degree of metal complexation increases rapidly with depth as shown for the Cariaco Trench sediments.

(Figs. 31.9 and 31.10), implying that a large proportion of the organic matter in reducing sediments is potentially biodegradable. The low rate of consumption of organic matter in reducing environments reflects the low biological activity there (i.e. the presence of an anaerobic population, but the absence of a metazoan biomass). Foree and McCarty (1970) have demonstrated that anaerobic and aerobic consumption rates are nearly identical under laboratory conditions. This suggests that the bulk of the consumption at oxic sediment-water interfaces is carried out by the metazoan population, not the microbial population. It is possible, however, that an active microbial community within the digestive tracts of the metazoan population is largely responsible for the degradation (Allen and Sanders, 1966), as in the rumen of a cow.

The availability of heavy metal ions in the sediment may also affect the degree of association. On the basis of K_{sp} data and the known concentrations of heavy metals in sulphide-free and sulphide-bearing seawater, it is possible to show that the thermodynamic solubility of heavy metals in reducing environments is exceedingly low in comparison to that in oxidizing environments. However, the results found by Presley *et al.* (1972), who analyzed heavy metals in the interstitial waters of reducing and oxidizing Saanich Inlet sediments, contradict this expected trend. They observed that there is very little difference in heavy metal concentrations between the oxidizing and reducing sediment, except for manganese which shows a 10-fold higher concentration in the reducing sediment. Generally, the heavy metals showed a 2 to 5-fold enrichment relative to sea water (Fe, Mn, and Zn had even higher enrichments). Thus, thermodynamic equilibria apparently do not control heavy metal concentrations in these interstitial solutions. The authors attribute the enrichments to complexation by soluble organic matter and/or equilibration with unidentified mineral phases. No evidence was given in support of either hypothesis, however. The former explanation is more likely.

The high hexosamine content of deep water sediments relative to shallow water sediments (Degens, 1970) may also be attributable to biological processes. Thus, Jannasch *et al.* (1971) suggested that in a low temperature environment an increase of pressure will tend to eliminate the growth and biochemical activity of certain types of bacteria successively as their minimum growth temperatures are shifted toward, and ultimately fall below, the environmental temperature. Consequently, it is possible that the organisms responsible for degradative hydrolysis of chitinous substances (hexosamines) may be inactive at the high pressures and low temperatures prevailing in deep-sea environments and, hence, these substances may accumulate in preference to other nitrogenous materials (e.g. proteins).

31.4.5. ORGANIC-METAL ION AND ORGANIC–ORGANIC INTERACTIONS

The end-product of the biological degradation of the organic matter of oxic sediments, appears to be biologically inert, metal-complexed humic material. Once this material is buried below the zone of biological activity, further diagenesis proceeds through slow abiotic processes, such as metal complexation and condensation reactions. The heterogenous class of humic substances contains a variety of functional groups including: aliphatic amino, aromatic amino, carboxylic, enolic, alkoxy, phenoxy, mercapto, phosphate, phospho-

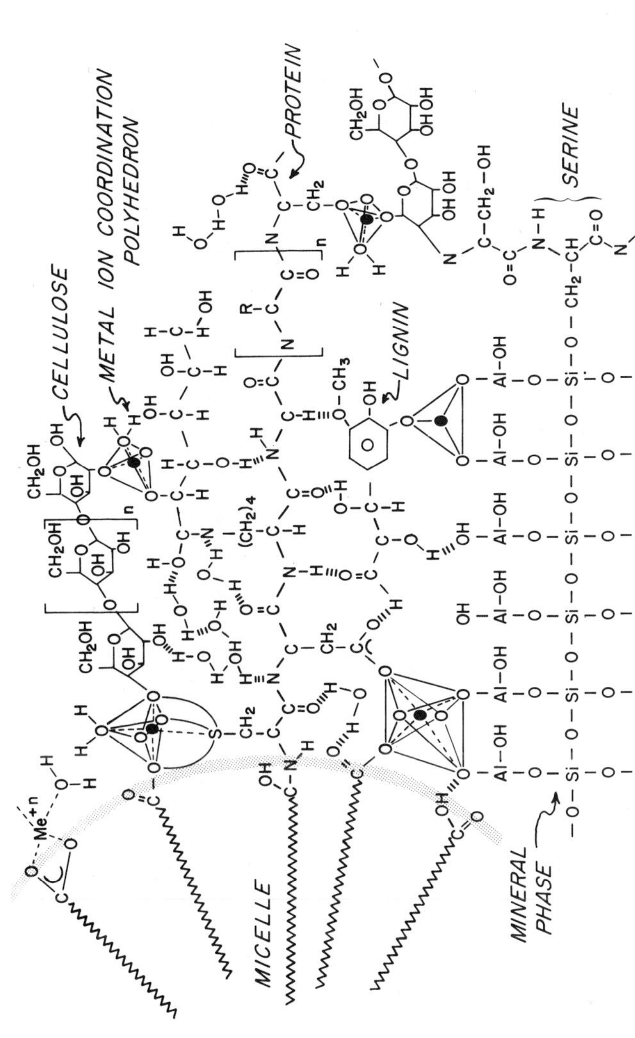

FIG. 31.11. A schematic representation of a sedimentary humic substance. The various types of interactions which occur are described below. (1) Organic-organic condensations between protein, carbohydrate, lipid, and lignin substances (the dashed lines represent hydrogen bonds). (2) Organic-mineral interactions such as condensation and adsorption of organic compounds onto the surface of a kaolinite-type mineral; the interaction of serine with silica has been demonstrated by Hecky *et al.* (1973) who showed that it is enriched in the mineralized tissues of diatom cell walls. (3) Organic-metal ion interactions result in an increase in the structural order of the humic material through the formation of metal-ion coordination polyhedra; the functional groups of different organic molecules may participate in this coordination and in one of the polyhedra shown a sulphide group (from cysteine) replaces an oxygen, and the larger ionic radius of the sulphide ion distorts the polyhedron. The incorporation of organic compounds into metal complexes appears to protect these substances from biological degradation. (4) Mineral-metal ion-organic interactions result when oxygen atoms on the mineral surface participate in metal-ion coordination polyhedra. (5) Micelle formation; molecules which contain both hydrophobic and hydrophilic portions (*e.g.* fatty acids) tend to aggregate so that the hydrophilic portion interacts with the aqueous environment, and the hydrophobic portion is protected from the aqueous environment. Indirect proof of the existence of micelles in fulvic acid is given by Ogner and Schnitzer (1970).

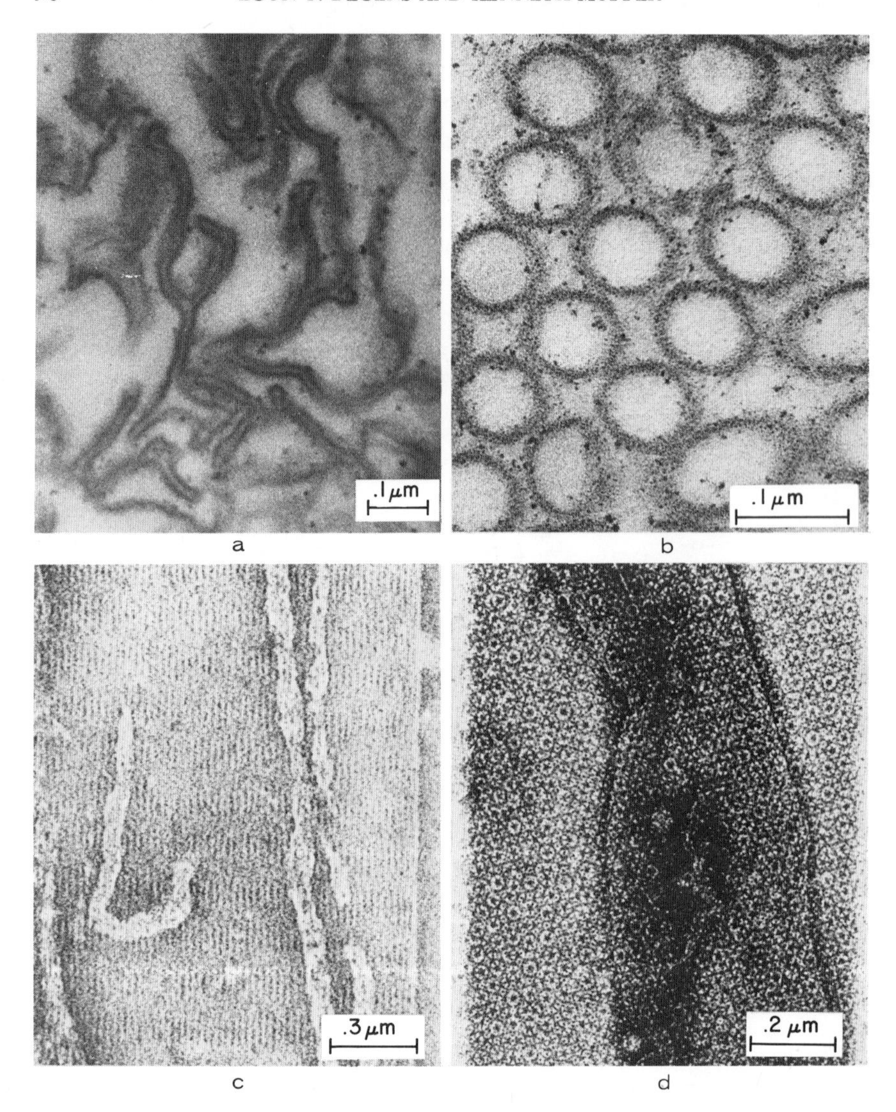

FIG. 31.12. Electron micrographs of organic materials from Black Sea sediments. (a) Branched, tubular membrane aggregated together and bound by a limiting membrane.
(b) Large tubular membranes having a diameter of 700–800 Å and consisting of unit-membranes having a width of 80 Å.
(c) Organic crystal resembling a bacterial cell wall and showing a unique pattern of sub-units which have a 40 Å periodicity.
(d) Portion of an organic crystal resembling a bacterial cell wall and consisting of a crystalline arrangement of sub-units.
All four micrographs are of the sapropel layer of Black Sea sediment. Core 1474 K, 20–70 cm. No prestaining techniques were used because *in situ* heavy metal staining made visualization possible (after Degens *et al.*, 1970).

nate, carboxyl, ether, ester, amide, thioether and hydroxyalkyl. The dehydration of the sediment during compaction is a major factor bringing about the condensation reactions between functional groups. Because of the vast number of reactions possible between these groups any discussion of them here would be unprofitable. Some degree of structural order within humic material is achieved by co-ordination of these functional groups with metal ions which results in the formation of metal ion coordination polyhedra (Matheja and Degens, 1971).

A schematic hypothetical "structure" of sedimentary humic material is given in Fig. 31.11; in order to simplify the diagram only a few of the possible interactions are shown. An electron micrograph of actual organic matter from a Black Sea sediment is shown in Fig. 31.12. The *in situ* binding of heavy metals to the organic matter made visualization possible without the use of pre-staining techniques which are usually necessary as organic tissue cannot be directly visualized. This may be regarded as direct evidence for organic-metal ion interactions in sedimentary organic matter. Similar photographs have been obtained from other sediments with a high organic content.

Changes in the degree of metal complexation of amino acids and sugars with time in sediments from oxic and anoxic environments are shown in Fig. 31.13. A high degree of metal-organic interaction is characteristic of both amino acids and sugars in the oxic environment; some significant variations with depth occur. These results suggest that for sugars biodegradation at the oxic sediment-water interface yields a material in which the metal is strongly bound and which is resistant to further alteration within the time span studied. In contrast, there is a temporal increase in metal complexation of the amino acids in the anoxic sediments (Fig. 31.13), this is to be expected since the biological activity at anoxic-sediment interfaces is low. In fact, as time passes the degree of metal binding by the amino acids in the sediments of the anoxic environment approaches that found for those of oxic environments. The degree of complexation of the sugars in Cariaco Trench sediments generally fluctuates about a mean value lower than that of the sediments from oxic environments. This implies that the functional groups of amino acids are either more reactive, or more available for reaction, than are those of the sugars. The peak at 2·7 m for the Cariaco Trench sugar data (Fig. 31.13) coincides with a band of oxic sediment and this suggests that the degree of metal association of sugars may be used as an indicator of the existence of reducing or oxidizing conditions at the sediment-water interface when the sediment was laid down.

For several amino acids, notably aspartic acid, glycine, and lysine, metal binding becomes so strong with depth (Fig. 31.13) that extraction with EDTA increases the yield by a factor of 10 relative to 6 N-HCl hydrolysis. Thus, the apparent instability of amino acids with depth which has been reported by

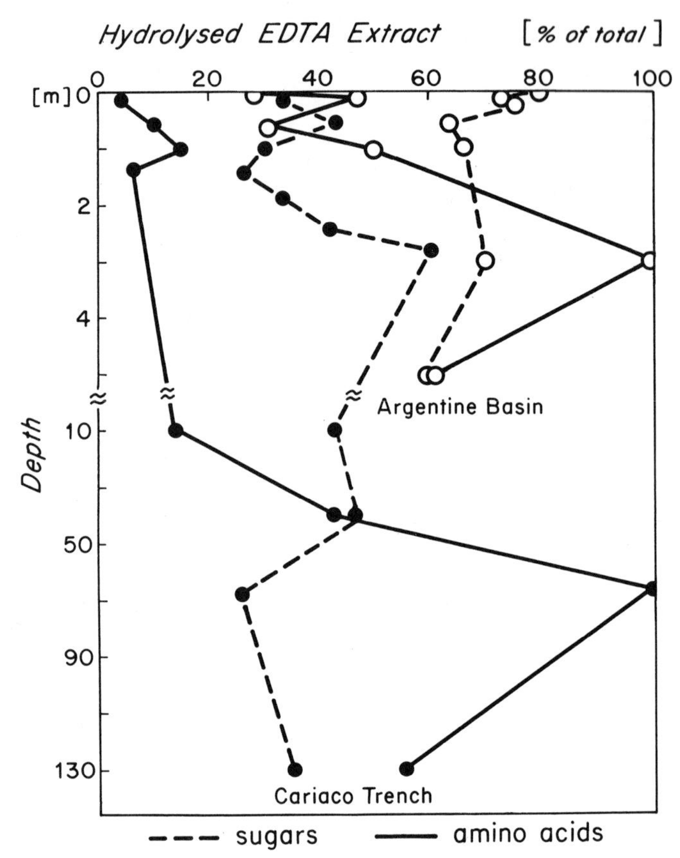

FIG. 31.13. Percentages of sugars and amino acids present as metal complexes in sediments of the Argentine Basin (oxic) and the Cariaco Trench (anoxic).

several investigators (e.g. Hare, 1969; Brown *et al.*, 1973) may be partially due to the inhibition of hydrolysis as a result of increasing metal binding.

The presence of relatively high concentrations of metal-bound monomeric sugars (10–20% of the total acid extractable organics) in all the sediments sampled (Table 31.4) is noteworthy. Incorporation of these normally easily degradable molecules into metal-organic associations apparently protects them from biotic and abiotic diagenesis in recent sediments. Thus, 15% of the total acid extractable sugars in a 5000-year-old wood sample buried in the Cariaco Trench sediment was present in a monomeric metal-bound form (Table 31.4). However, the carbohydrate composition of the overlying sediment was completely unaffected by the presence of this carbohydrate-rich wood (Mopper and Degens, 1972). This observation suggests that complexation of soluble carbohydrates by metals has fixed this fraction *in situ*, thereby inhibiting both its upward diffusion and its availability for reaction.

Metal ions tend to act as Lewis acids* (e.g. Martell, 1971) and, consequently, may catalytically promote condensation reactions between organic ligands and other organic compounds. Metal-organic complexes may therefore act as "seeds" for the precipitation of humic material from water-soluble fulvic substances. A decrease in concentration of fulvic substances together with a concurrent increase in the extractable organic matter with depth are, in fact, observed in the sediments of the Saanich Inlet (Nissenbaum *et al.*, 1972; Brown *et al.*, 1973). On a dry weight basis, metals constituted about 10% of the soluble polymerized organic matter in these sediments (Nissenbaum *et al.*, 1972); calcium, sodium and silicon were the predominant species; transition metals, such as iron, copper and zinc, were also significant.

The alkali and alkali earth metals responsible for coordination and complexation in the original cellular material (i.e. plankton) are replaced to some extent by heavy metals originating from the sedimentary environment. The degree of replacement is probably related to the availability of the heavy metal concerned. For example, in an anoxic environment the degree of replacement is low because heavy metals tend to be bound in insoluble sulphides. This observation has significant implications for the origin of life. In a prebiotic environment, which was almost certainly anoxic, the predominant metals available for coordination with organic molecules were those that remained unaffected by changes in the redox potential, e.g. sodium, potassium, calcium, magnesium, etc. and which hence did not form insoluble sulphides.

31.4.6. ORGANIC-MINERAL INTERACTIONS

Functional groups of organic molecules may coordinate to surfaces and edges of minerals as well as to metal ions (Fig. 31.11). In addition, certain clay minerals may accommodate organic molecules in interlattice positions and hence protect them from degradation (e.g. Weiss, 1969; Sørensen, 1972).

In addition to their role in producing coordination and interlayering effects, minerals may act as catalysts in the fragmentation, synthesis and polymerization of organic molecules (e.g. Burton and Neuman, 1971; Degens and Matheja, 1968; Weiss, 1969). In the authors' laboratories it has been recently demonstrated that heterocyclic nitrogen compounds are synthesized from CO_2 and NH_3 in the presence of an aqueous slurry of kaolinite (Harvey *et al.*, 1971). It was also shown that in the presence of kaolinite polysaccharides are

* A Lewis acid is capable of accepting and sharing a lone pair of electrons from a donor (Lewis base).

synthesized from aqueous solutions of paraformaldehyde, and that fatty acids are esterified to glycerides (Harvey *et al.*, 1972).

Catalytic fragmentation, synthesis and polymerization reactions undoubtedly occur in recent sediments; however, their importance appears to be negligible in comparison to biological degradation. Reactions catalyzed by minerals probably become more important as the temperature and the pressure increase during the later stages of diagenesis (e.g. Whitehead and Breger, 1950; Mulik and Erdman, 1963; Welte, 1969).

In the context of the interactions between organic compounds and minerals, it is significant to note that the organic matter within biologically precipitated minerals (referred to as mineralized tissues) is not subject to degradation either in the water column or at the sediment-water interface. Consequently, it probably represents the only unaltered organic matter entering the sediment. In unrecrystallized sediments mineralized tissues may follow different diagenetic pathways from those taken by the organic matter in the external environment (e.g. Mitterer, 1972).

31.4.7. EFFECT OF TEMPERATURE ON THE EARLY STAGES OF DIAGENESIS

The rate of abiotic diagenesis of organic matter is principally controlled by temperature. Since the sediments described up to this point have been from areas characterized by either normal or low heat flow values it is opportune now to discuss the diagenesis of organic matter in reducing sediments from an East African rift lake (Lake Kivu) which is surrounded by active volcanoes and hydrothermal springs (Degens *et al.*, 1973).

(1) The C/N ratios of the sediments from Lake Kivu increase systematically with age, whereas the ratio remains fairly constant in sediments from the other areas studied. Deaminative reactions are apparently accelerated in Lake Kivu as a result of the heat influx, the ammonia thus produced leaves the sediment and accumulates in the water column above it. At the sediment-water interface the ammonia concentration can reach values as high as 7000 μg-at N l^{-1}.

(2) The ratio of total sugars to total organic carbon in the sediments of Lake Kivu decreases rapidly down the sediment column to 15000 years B.P., below this level the fall in the ratio occurs more slowly (Fig. 31.14). In contrast, the ratios in cores from the other areas examined were fairly constant with time. The higher rate of loss in the upper part of the Kivu sediment column is probably the result of the thermal oxidation of carbohydrates to CO_2 and H_2O, and/or their reduction to furfural-type derivatives. The slower rate of loss in the lower part of the core is probably attributable to incorporation of the residual sugars into highly condensed, inextractable humic substances.

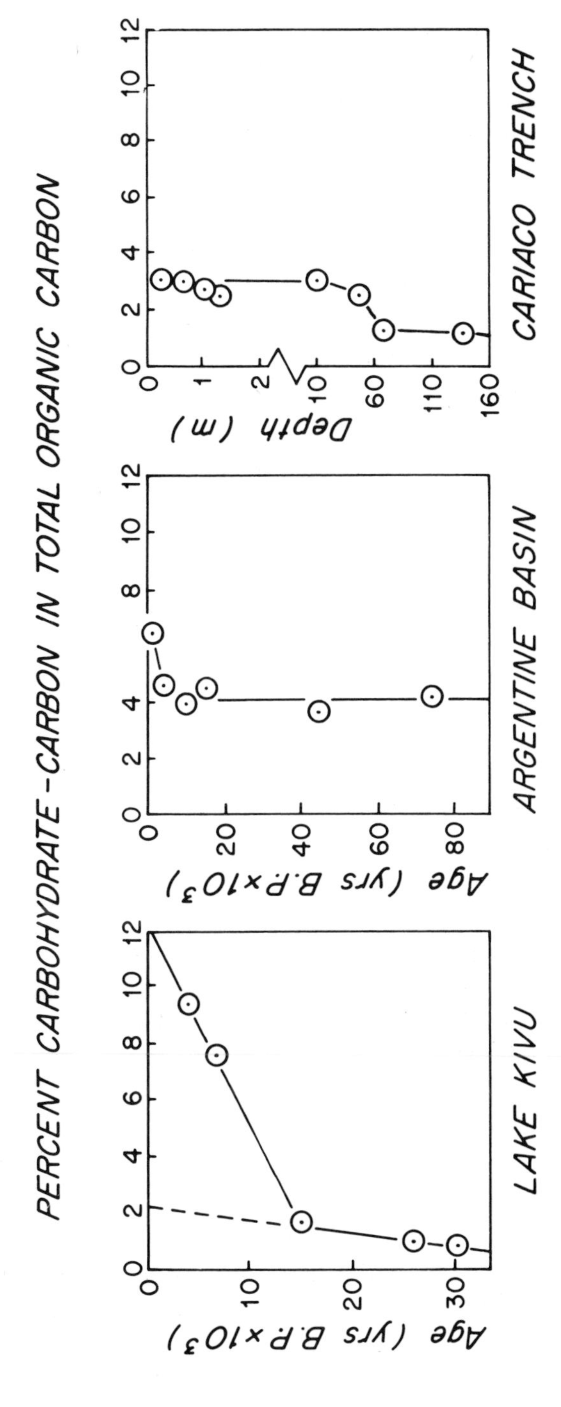

FIG. 31.14. Concentration of total sugars relative to total organic carbon in the sediments of Lake Kivu, the Cariaco Trench and the Argentine Basin.

(3) The degree of metal ion-amino acid interaction in the sediments of Lake Kivu appears to be enhanced by an order of magnitude relative to that in areas with a "normal" geothermal gradient, such as the Cariaco Trench in which the environment is also reducing. Sugars do not show this trend as clearly as the amino acids.

31.5. Palaeoenvironmental Criteria

31.5.1. Organic Carbon and Nitrogen

In the previous section it was noted that the C/N ratio is fairly constant during the early stages of diagenesis except in areas of high heat flow. The total organic content, however, can fluctuate widely in response to variations in environmental parameters, such as primary productivity, redox potential, rates of deposition, mineralogy etc. Attempts to interpret the variation of organic carbon with depth are thus frequently beset by many uncertainties. However, two instances are described below in which changes in the organic carbon content can be correlated with specific environmental factors:

(1) In the Black Sea, the environment at the sediment-water interface has changed several times from oxic fresh water to anoxic saline water during the last 25 000 years (Degens and Ross, 1972). The frequency of marine spills and fresh water dominance can be deduced from the fossil assemblages, and changes in the redox environment at the sediment-water interface and are reflected in the mineralogy and the organic carbon content of the sediments (Fig. 31.15). The distribution of the organic carbon in the sediments suggests that between 17 000 and 9300 B.P. the Black Sea was an oxygenated fresh water body down to the bottom. Around 9300 years B.P. the first invasion of Mediterranean water occurred, and this produced reducing conditions at the sediment-water interface for about 200 years. The frequency of saline spills with concurrent reducing conditions at the sea bottom over the next 2000 years is reflected in the fluctuation of organic carbon in the sediments. Finally, about 7300 years B.P. the influx of Mediterranean water became so pronounced that saline (brackish) and reducing conditions became permanently established at the sediment-water interface. The shape of the organic carbon curve from 7000 years B.P. to the present is principally the result of changes in the rate of deposition which, in turn, was controlled by variations in the productivity of the calcareous organisms (coccoliths) living in the surface waters.

(2) Argentine Basin sediments are impoverished in organic carbon relative to those of the Black Sea. However, in comparison to "normal" open ocean sediments, which generally contain ca. 0·2% organic carbon (Bordovskiy, 1965), those of the Argentine Basin are rich in organic carbon (usually between 0·5 and 1·0%). Their relatively high organic content is probably related to the

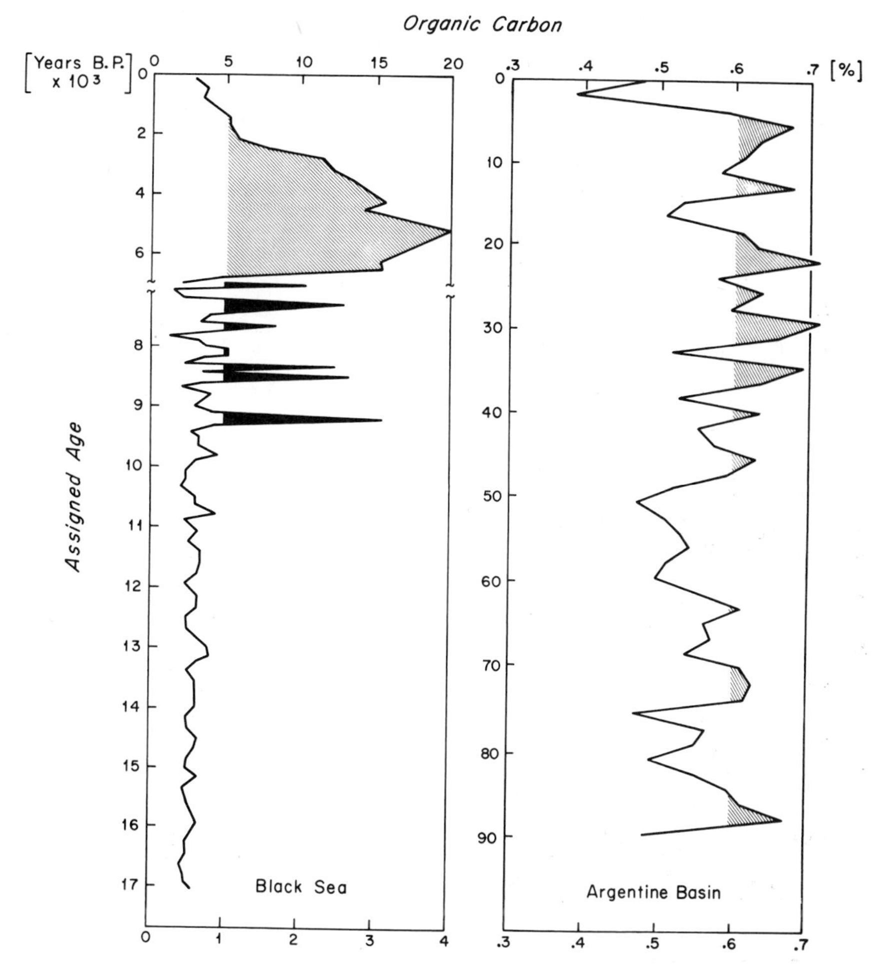

FIG. 31.15. Organic carbon in Black Sea and Argentine Basin sediments. Note that the lower portion of the Black Sea curve is plotted on an expanded scale to enhance the fluctuations. A zippeton pattern is used to emphasize concentration differences (after Mopper and Degens, 1972).

high primary productivity of the surface water resulting from upwelling in the area of the Subtropical Convergence near the edge of the continent (Ryther, 1963; Deacon, 1963). The Subtropical Convergence separates northward moving, nutrient-rich, Sub-Antarctic water from southward moving, nutrient-poor, subtropical water. The productivity of the sub-Antarctic water is 150–250 g-C m^{-2} yr^{-1} whereas that of the subtropical water is ~50 g-C m^{-2} yr^{-1} (Ebeling, 1962).

In contradiction to the suggestion by Stevenson and Cheng (1972), the

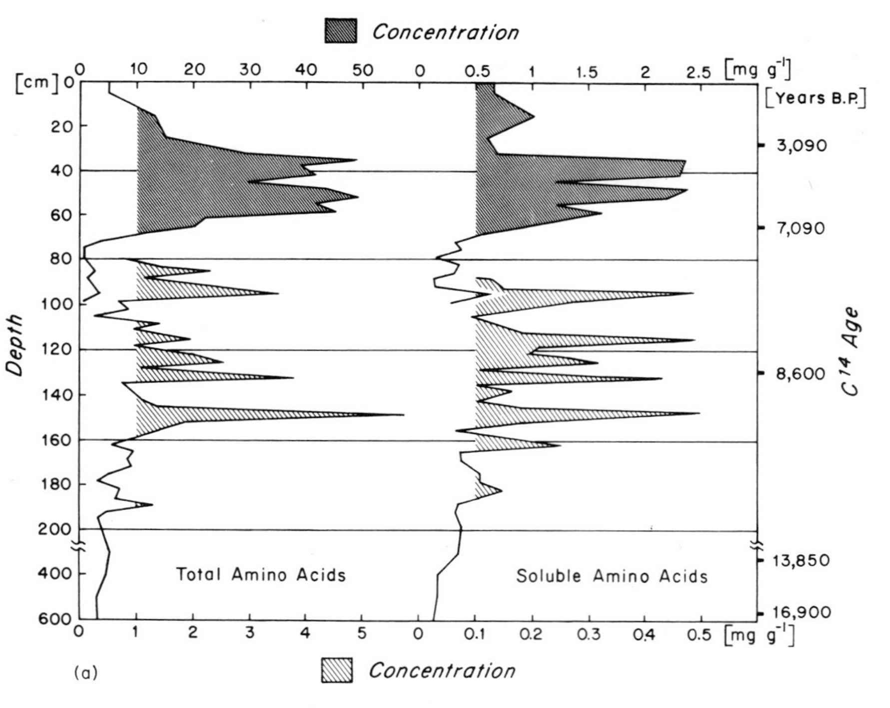

FIG. 31.16. (a) Distribution of total and soluble amino acids in a Black Sea core (Degens and Ross, 1972). The spacing between the individual samples is about 3 cm. The lower values are plotted on an expanded scale. The observed fluctuations in the concentration levels of amino acids are principally determined by changes in deposition rates and the redox potential at the sediment-water interface at the time of deposition.

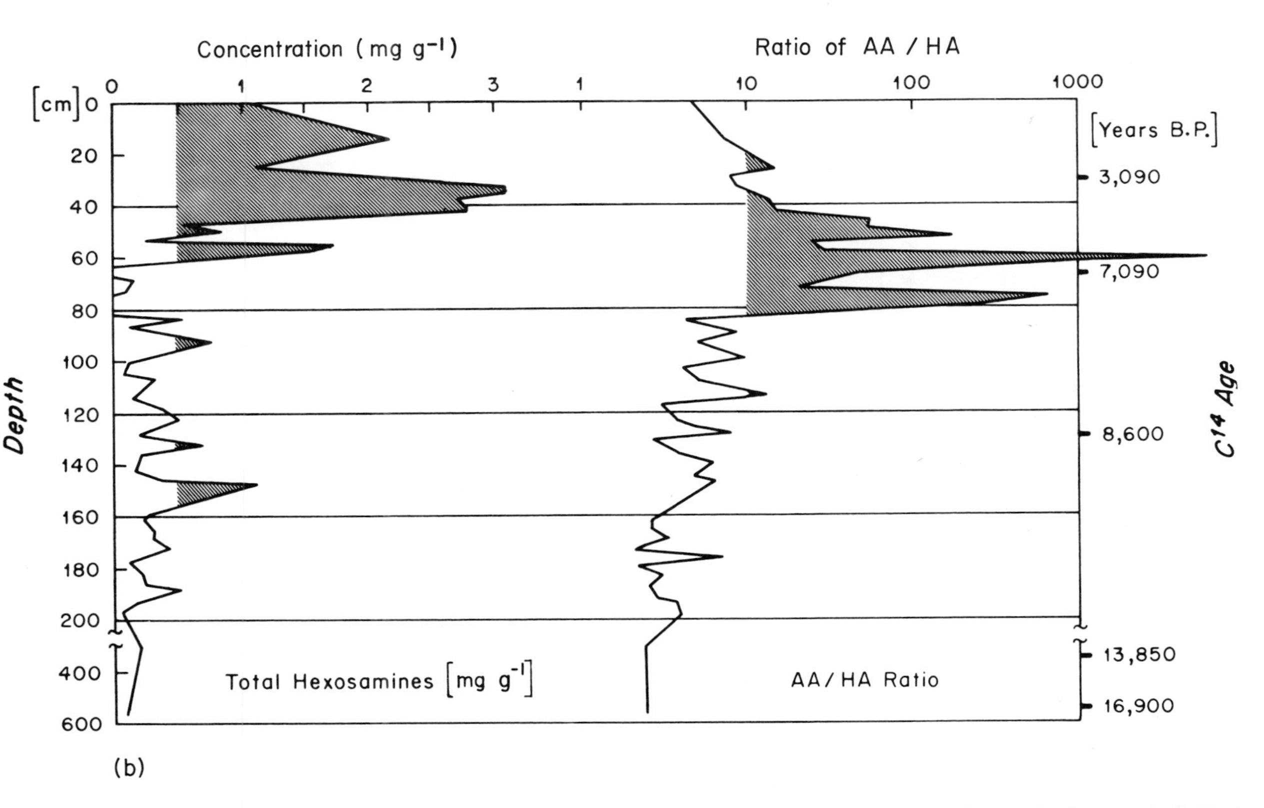

(b). Distribution of total hexosamines and amino acid/hexosamine ratios (AA/HA) in a Black Sea core. The spacing between individual samples is about 3 cm. The AA/HA values are plotted on a logarithmic scale. A zippeton pattern is used to emphasize concentration differences.

distinct temporal variations in the organic carbon content of the sediments of the Argentine Basin (Fig. 31.15) do not appear to correspond with sea level fluctuations or glacial periods. The temporal variations are probably attributable to two factors: (1) variations in surface primary productivity of the surface waters resulting from shifts in the Subtropical Convergence and, (2) variations in the amount of run-off from land due to eustatic changes in sea level during glacial and interglacial periods. It is probable, therefore, that fluctuations in the organic carbon content of the sediments of the Argentine Basin represent a resolution of two factors, the sedimentation rate and the primary productivity.

TABLE 31.9

*Total amino acids, hexosamines and ammonia in Black Sea sediments**

Years × 10^3	0–3 (4)†	3–7 (9)	7–8 (7)	8–9 (9)	9–10 (6)	10–11 (9)	13–17 (4)
Aspartic acid	109	105	145	177	159	166	151
Threonine	64	48	42	56	69	63	67
Serine	61	44	40	56	74	71	70
Glutamic acid	86	86	103	92	99	117	110
Proline	60	52	40	37	40	39	47
Glycine	138	127	160	155	151	157	150
Alanine	115	121	124	110	113	112	104
Cystine (half)	7	6	7	7	10	9	6
Valine	67	78	76	68	64	66	59
Methionine	12	11	5	3	5	4	3
Alloisoleucine	1·2	3·1	3·7	1·9	1·3	2·4	2·9
Isoleucine	45	57	46	36	35	31	28
Leucine	76	91	71	55	58	54	52
Tyrosine	26	25	12	8	6	9	9
Phenylalanine	42	54	36	30	20	23	36
Beta-Alanine	9	4	13	13	14	13	16
Lysine	61	53	58	69	61	45	67
Histidine	1	1	1	1	1	1	1
Arginine	21	32	19	25	21	18	21
Total amino acids (mg g^{-1} dry wt.)	16·7	36·8	2·99	1·64	1·73	0·66	0·40
Hexosamines (mg g^{-1})	1·81	1·13	0·30	0·40	0·39	0·25	0·16
Ammonia (mg g^{-1})	0·22	0·49	0·20	0·13	0·12	0·08	0·11
Organic carbon (%)	7·1	16·4	1·6	1·1	0·70	0·67	0·54
Organic nitrogen (%)	0·62	1·23	0·15	0·09	0·07	0·06	0·05
C/N ratio	11·5	13·3	11	12	10	11	11
AA/HA ratio	9	33	10	4	4	3	3
N recovered (%)	38	44	42	41	36	29	29

* Residues per 1000.
† Number of samples.

31.5.2. AMINO ACIDS

Because the Black Sea sediments were deposited under varying environmental conditions fluctuations in the amino acid composition, which are attributable to these environmental changes, can be clearly discerned. In comparison to the anoxic marine sediments, the fresh water oxic sediments are slightly enriched in aspartic and glutamic acids, and β-alanine, and are slightly impoverished in isoleucine, leucine and aromatic amino acids (Table 31.9). These differences, however, must be considered minor when compared to the 100-fold reduction in the total amino acid concentrations between the brackish anoxic sediments and the fresh water oxic sediments. The distribution of total (HCl hydrolysis) and soluble (hydrolyzed water extract) amino acids (Fig. 31.16(a)) reveals abrupt concentration changes which do not involve a fractionation of the individual amino acids. Environmental changes do not therefore appear to significantly affect the relative proportions of the extracted amino acids. It was formerly believed (Degens *et al.*, 1963) that separation of organic molecules by chromatography along clay surfaces during compaction was a significant diagenetic factor; in light of the present data this now seems unlikely.

The distribution of total hexosamines and the amino acid/hexosamine ratios in the sediments of the Black Sea are shown in Fig. 31.16(b)). During the period between 3500 and 7500 years B.P. the hexosamine content was negligible. Before and after this, hexosamine was a substantial fraction of the total nitrogenous organic matter. It appears that this sudden drop in hexosamine concentration is related to the onset of conditions of extremly high productivity, perhaps eutrophication, in the surface waters. During eutrophication the algal biomass, which contains little or no chitinous material (hexosamines), overwhelms the zooplankton biomass, which generally contains significant quantities of chitinous material, with the result that the sediment has a high amino acid/hexosamine ratio.

31.5.3. SUGARS

The degree of metal complexation of sedimentary carbohydrates appears to be dependent on the redox potential at the sediment-water interface (Fig. 31.9) and it seems to remain constant after burial (Fig. 31.13). For these reasons changes in environmental conditions at the sediment-water interface from high to low redox potentials may be discerned. For example, in reducing environments the percentage of the total extractable sugars present in a metal-bound form ranges from 55–70 %, whereas in oxidizing environments the range is 25–40 %. However, because the amino acids in reducing sediments become increasingly metal-complexed with time, (see Fig. 31.13) the degree of complexation of amino acids may not be as useful as that of sugars for

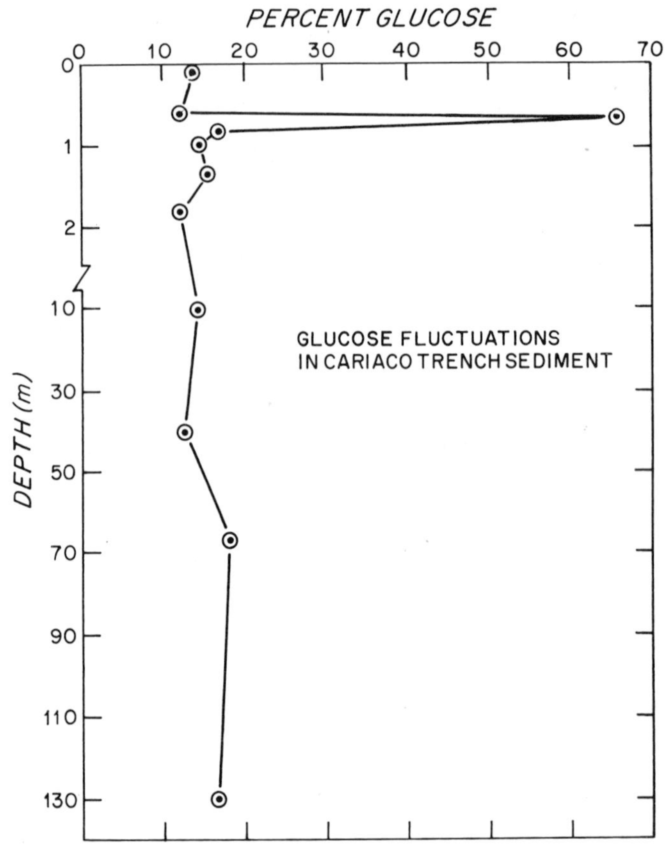

FIG. 31.17. Glucose fluctuations in Cariaco Trench sediments. The spike represents land-derived material (wood).

determining whether oxidizing or reducing conditions prevailed at the time of deposition.

The glucose content of sediments may be used to estimate the input of terrigenous material (see for example Fig. 31.17). Thus, the organic matter of typical marine sediments generally contains between 15–25% glucose (Table 31.7), whereas that of sediments of a purely terrigenous origin (i.e. wood, peat etc.) contains about 70% glucose.

31.5.4. MARINE SEWAGE DISPOSAL

The New York Bight sediment sample (see Tables 31.4 and 31.7) was collected at the periphery of a sewage disposal area at a water depth of about 30 m. Raw and both primary and secondary treated sewage are dumped in this area. The water in this area is oxic down to the sediment-water interface.

The carbohydrate composition of this sample is practically identical to that of adjacent non-contaminated sediment and also to that of deep-sea oxic sediments, such as those of the Argentine Basin (Mopper, 1973). For example, the relative proportions of the sugars in the acid extracts are nearly identical: the degree of metal-carbohydrate interaction is 60–70%; 20–25% of the total acid extractable sugars are present as metal-bound monomers, and the total sugar content relative to the total organic carbon is 4–5%. Furthermore, the amino acid distribution in the New York Bight sample is also very similar to that found in the other oxic surface sediment samples examined. Thus, despite the differences in the organic input into shallow and deep-sea oxygenated marine environments the end-product of the biological degradation occurring at the sediment-water interface is remarkably uniform.

These results imply that the treatment of raw sewage on land may be unnecessary because the rate of biological degradation of this material at the sediment-water interface in *shallow* oxic waters is comparable to, and perhaps more efficient than, primary and secondary sewage treatment. However, care should be taken to maintain a dumping rate which is lower than the maximum possible consumption rate in order to avoid overloading the system, which may result in the formation of periodic or permanent anoxic conditions.

31.6. Summary and Conclusions

Organic molecules in a living cell have a high degree of structural order and are, therefore, far removed from thermodynamic equilibrium. As a consequence, diagenesis of organic matter in sediments would be expected to proceed so as to increase thermodynamic stability. Initially, the organic material brought to sediments may be characterized as an ill-defined conglomeration of many biomolecules at various stages of decomposition. This organic conglomerate then proceeds through several stages of diagenetic meta-stability which progressively enhance the structural order through cross-linking. The final product of diagenesis –graphite– has again achieved a high degree of structural order. This order is achieved by the elimination of functional groups through condensation, deamination, decarboxylation etc.; CO_2, CH_4, NH_3, H_2O, and small organic molecules are released as by-products.

The major factors affecting the early stages of the diagenesis of organic matter in aquatic and sedimentary environments are (1) biological degradation; (2) organic–organic condensations; (3) organic–metal ion interactions; (4) organic–mineral interactions; and (5) increasing temperature. Early stages in the diagenesis of carbohydrates and proteins in the water column and in the surface sediments probably result in the formation of metal–organic associations. The degree of association appears to be related to the intensity and duration of biological degradation. Thus, in an anoxic environment where the biological activity is low, an organic residue is produced in which carbohydrate and amino acid–metal complexes are only weakly bound. Functional groups on the organic matter buried in this environment may be largely available for further reaction during early diagenesis ($<10^6$ years).

In an oxic environment intensive biological degradation appears to produce a strongly metal-bound, carbohydrate and amino acid-containing organic residue. The end-product of biological degradation in this environment may be resistant to abiotic alteration during early diagenesis since functional groups are probably largely unavailable for reaction as a result of intermolecular cross-linking and metal binding. Sewage material dumped into a shallow-water, oxic environment is degraded rapidly despite the high content of potentially toxic metals. It seems likely that these metals are tied-up in the ultimate metal bound carbohydrate and amino acid residue. If sewage were dumped into a reducing or deep-sea oxic environment degradation would probably proceed very slowly.

The alkali and alkali earth metals which probably occupied most of the coordination and complexation sites in living cells (i.e. plankton) are replaced to some degree by metals present in the sedimentary environment. The degree of replacement is probably related to the degree of biological degradation of this material at or near the sediment–water interface.

Metal binding appears to fix carbohydrate and amino acid material *in situ*, thereby inhibiting diffusion of these compounds. This finding conflicts with the previous belief that chromatographic separation of organic molecules along mineral surfaces is a significant diagenetic process. However, organic matter such as hydrocarbons, containing few, if any, functional groups, would not be immobilized by such metal binding.

The effects of mineral-catalyzed fragmentation, synthesis and polymerization of organic matter are either negligible or are indistinguishable from those of biological degradation and metal binding during early diagenesis; however, these processes probably become more significant during later stages of diagenesis.

Carbohydrates and proteins in sediments may be used to interpret palaeoenvironmental fluctuations. For example, the degree of metal–organic interaction appears to reflect the redox potential at the sediment–water interface. Likewise, changes in glucose contents of sediments may be used to estimate the relative terrigenous and marine organic inputs. The interpretation of palaeo-eutrophic conditions in the surface waters may be aided by an examination of the amino acid–hexosamine ratio of the underlying sediment.

REFERENCES

Allen, J. A. and Sanders, H. L. (1966). *Deep-Sea Res.* **13**, 1175.

Barber, R. T. (1966). *Nature, Lond.* **211**, 257.

Barber, R. T. (1968). *Nature, Lond.* **220**, 274.

Barber, R. T. and Ryther, J. H. (1969). *J. Exp. Mar. Biol. Ecol.* 3, 191.

Blumer, M. (1973). *Pure Appl. Chem.* (in press).

Bordovskiy, O. K. (1965). *Mar. Geol.* **3**, 5–31.

Breger, I. A. (1964). "Organic Geochemistry". Macmillan, New York.

Brown, F. S., Baedecker, M. J., Nissenbaum, A. and Kaplan, I. R. (1972). *Geochim. Cosmochim. Acta* **36**, 1185.

Burton, F. G. and Neumann, W. F. (1971). *Curr. Mod. Biol.* **4**, 47.

Craig, H. (1953). *Geochim. Cosmochim. Acta*,**3**, 53.

Craig, H. (1971). *J. Geophys. Res.* **76**, 5078.

Deacon, G. E. R. (1963). *In* "The Sea" (M. N. Hill, ed.), Vol. 2, pp. 281ff. Interscience, New York.

Degens, E. T. (1965). "Geochemistry of Sediments". Prentice-Hall, Englewood Cliffs, New Jersey.

Degens, E. T. (1967). *In* "Diagenesis in Sediments" (G. Larsen and G. V. Chilingar, eds.), pp. 343ff. Elsevier, Amsterdam-London-New York.

Degens, E. T. (1969). *In* "Organic Geochemistry" (G. Eglinton and M. T. J. Murphy, eds.), pp. 304ff. Springer-Verlag, Berlin-Heidelberg-New York.

Degens, E. T. (1970). *In* "Organic Matter in Natural Waters" (D. W. Hood, ed.), pp. 77ff. Inst. Mar. Sci., Univ. Alaska, Fairbanks.

Degens, E. T. (1976). "Molecular Mechanisms in Carbonate, Phosphate and Silica Deposition in the Living Cell". Springer-Verlag, Berlin-Heidelberg-New York.

Degens, E. T. and Matheja, J. (1968). *J. Brit. Interplanet. Soc.* **21**, 52.

Degens, E. T. and Ross, D. A. (1972). *Chem. Geol.* **10**, 1.

Degens, E. T., Emery, K. O. and Reuter, J. H. (1963). *Neues Jahrb. Geol. Palaeontol. Monatsh.*, 231.

Degens, E. T., Watson, S. W. and Remsen, C. C. (1970). *Science, N.Y.* **168**, 1207.

Degens, E. T., Von Herzen, R. P., Wong, H. K., Deuser, W. G. and Jannasch, H. W. (1973). *Geol. Rundsch.* **62**, 245.

Deuser, W. G. (1971). *Deep-Sea Res.* **18**, 995.

Duursma, E. K. (1965). *In* "Chemical Oceanography" (J. P. Riley and G. Skirrow, eds.), Vol. 1, pp. 433ff. Academic Press, London and New York.

Ebeling, A. W. (1962). *In* "Dana Report of the Carlsberg Foundation's Oceanographical Expedition". Vol. XI, p. 164.

Eglinton, G. and Murphy, M. T. J. (eds) (1969. "Organic Geochemistry". Springer-Verlag, Berlin-Heidelberg-New York.

Foree, E. G. and McCarty, P. L. (1970). *Environ. Sci. Technol.* **4**, 842.

Ghiselin, M. T., Degens, E. T., Spencer, D. W. and Parker, R. H. (1967). *Breviora, Harvard Mus. Comp. Zool.* **262**, 1.

Gordon, D. C. (1970). *Deep-Sea Res.* **17**, 233.

Handa, N. and Yanagi, K. (1969). *Mar. Biol.* **4**, 197.

Hare, P. E. (1969). *In* "Organic Geochemistry" (G. Eglinton and M. T. J. Murphy, eds.), pp. 438ff. Springer-Verlag, Berlin-Heidelberg- New York.

Harvey, G. R., Degens, E. T. and Mopper, K. (1971). *Naturwissenschaften* **58**, 624.

Harvey, G. R., Mopper, K. and Degens, E. T. (1972). *Chem. Geol.* **9**, 79.

Hecky, R. E., Mopper, K., Kilham, P. and Degens, E. T. (1973). *Mar. Biol.* **19**, 323.

Hellebust, J. A. (1965). *Limnol. Oceanogr.* **10**, 192.
Hunt, J. M. (1968). *In* "Advances in Organic Geochemistry" (G. D. Hobson, ed.), pp. 27ff. Pergamon Press, Oxford.
Jannasch, H. W. and Wirsen, C. O. (1973). *Science*, **180**, 641.
Jannasch, H. W., Eimhjellen, K., Wirsen, C. O. and Farmanfarmaian, A. (1971). *Science, N.Y.* **171**, 672.
Kanwisher, J. (1962). *Occas. Publ. Narragansett Mar. Lab.* **1**, 13.
King, K. and Hare, P. E. (1972). *Science, N.Y.* **175**, 1461.
Koblentz-Mishke, O. I., Volkovinsky, V. V. and Kabanova, J. G. (1968). SCOR South Pacific Symposium. Scripps Inst. Oceanogr., La Jolla, California.
Manheim, F. T. and Sayles, F. L. (1974). *In* "The Sea, Ideas and Observations on Progress in the Study of the Seas" (E.D. Goldberg, ed.), Vol. 5. Interscience, New York.
Manheim, F. T., Meade, R. H. and Bond, G. C. (1970). *Science, N.Y.* **167**, 371.
Martell, A. E. (1971). *In* "Organic Compounds in Aquatic Environments" (S. J. Faust and J. V. Hunter, eds.), pp. 239ff. Marcel Dekker, New York.
Matheja, J. and Degens, E. T. (1971). "Structural Molecular Biology of Phosphates". Gustav Fischer Verlag, Stuttgart.
Meade, R. H. (1973). *In* "Continental Shelf Sediment Transport" (D. J. P. Swift, D. B. Duane and O. H. Pilkey, eds.). Dowden, Hutchinson and Ross, Stroudsburg, Pennsylvania.
Meeuse, B. J. D. (1962). *In* "Physiology and Biochemistry of Algae" (R. A. Lewin, ed.), pp. 289ff. Academic Press, New York and London.
Menard, H. W. and Smith, S. M. (1966). *J. Geophys. Res.* **71**, 4305.
Menzel, D. W. (1966). *Deep-Sea Res.* **13**, 963.
Menzel, D. W. and Ryther, J. H. (1964). *Limnol. Oceanogr.* **9**, 179.
Menzel, D. W. and Ryther, J. H. (1970). *In* "Organic Matter in Natural Waters" (D. W. Hood, ed.), pp. 31ff. Inst. Mar. Sci., Univ. Alaska, Fairbanks.
Mitterer, R. M. (1972). *In* "Advances in Organic Geochemistry" (H. R. von Gaertner and H. Wehner, eds.), pp. 441ff. Pergamon Press, Oxford.
Mopper, K. (1973). "Aspects of the Biogeochemistry of Carbohydrates in Aquatic Environments". Ph.D. Thesis, Joint Program Mass. Inst. Technol. and Woods Hole Oceanogr. Inst.
Mopper, K. and Degens, E. T. (1972). *Tech. Rep. Woods Hole Oceanogr. Inst.* **72–68**, 1–117.
Mopper, K. and Degens, E. T. (1976). *Geochim. Cosmochim, Acta*, (in press).
Mulik, J. D. and Erdman, J. G. (1963). *Science, N.Y.*, **141**, 806.
Nissenbaum, A., Baedecker, M. J. and Kaplan, I. R. (1972). *In* "Advances in Organic Geochemistry" (H. R. von Gaertner and H. Wehner, eds.), pp. 427ff. Pergamon Press, Oxford.
Ogner, G. and Schnitzer, M. (1970). *Science, N.Y.* **170**, 311.
Ogura, N. (1970). *Deep-Sea Res.* **17**, 221.
Presley, B. J., Kolodny, Y., Nissenbaum, A. and Kaplan, I. R. (1972). *Geochim. Cosmochim. Acta*, **36**, 1073.
Rittenberg, S. C., Emery, K. O., Hülsemann, J., Degens, E. T., Fay, R. C., Reuter, J. H., Grady, J. R., Richardson, S. H. and Bray, E. E. (1963). *J. Sediment. Petrology*, **33**, 140.
Rowe, G. T. (1971). *In* "Fertility of the Sea" (J. D. Costlow, ed.), Vol. 2, pp. 441ff. Gordon and Breach, New York-London-Paris.

Ryther, J. H. (1963). *In* "The Sea" (M. N. Hill, ed.) Vol. 2, pp. 347ff. Interscience, New York.
Ryther, J. H., Menzel, D. W., Hulbert, E. M., Lorenzen, C. J. and Corwin, N. (1970). *Anton Bruun Rep.* No. **4**, 3.
Sackett, W. M. and Thompson, R. R. (1963). *Bull. Amer. Assoc. Petrol. Geol.* **47**, 525.
Sanders, H. L. and Hessler, R. R. (1969). *Science, N.Y.* **163**, 1419.
Sanders, H. L., Hessler, R. R. and Hampson, G. R. (1965). *Deep-Sea Res.* **12**, 845.
Siegel, A. (1971). *In* "Organic Compounds in Aquatic Environments" (S. J. Faust and J. V. Hunter, eds.), pp. 265ff. Marcel Dekker, New York.
Smith, K. L. and Teal, J. M. (1973). *Science, N.Y.* **179**, 282.
Sørensen, L. H. (1972). *Soil Sci.* **114**, 5.
Stevenson, F. J. and Cheng, C.-N. (1972). *Geochim. Cosmochim, Acta*, **36**, 653.
Tan, K. H., King, L. D. and Morris, H. D. (1971). *Soil Sci. Soc. Amer., Proc.* **35**, 748.
Wagner, F. S., Jr. (1969). *Contrib. Mar. Sci.* **14**, 115.
Wangersky, P. J. (1965). *Amer. Sci.* **53**, 358.
Weiss, A. (1969). *In* "Organic Geochemistry" (G. Eglinton and M. T. J. Murphy, eds.), pp. 737ff. Springer-Verlag, Berlin-Heidelberg-New York.
Welte, D. H. (1969). *Bull. Amer. Assoc. Petrol. Geol.* **49**, 2246.
Whitehead, W. L. and Breger, I. A. (1950). *Science, N.Y.* **111**, 335–337.
Whittaker, R. H. (1970). "Communities and Ecosystems". Collier-Macmillan Canada, Toronto.
Williams, P. M. (1971). *In* "Organic Compounds in Aquatic Environments" (S. J. Faust and J. V. Hunter, eds.), pp. 145ff. Marcel Dekker, New York.
Williams, P. M., Oeschger, H. and Kinney, P. (1969). *Nature, Lond.* **224**, 556.

Chapter 32

Interstitial Waters of Marine Sediments*

F. T. MANHEIM

U.S. Geological Survey, Woods Hole, Massachusetts 02543, U.S.A.†

32.1. INTRODUCTION AND HISTORICAL BACKGROUND

Interstitial waters are aqueous solutions that occupy the pore spaces between particles in rocks and sediments. Their composition reflects the nature of the original fluids buried with the sediments, particle-fluid reactions, and migration of fluids and dissolved components by convection and diffusion.

* Publication authorized by Director, U.S. Geological Survey, Woods Hole Oceanographic Institution.

† Present address: Department of Marine Science, University of South Florida, St. Petersburg, Florida 33701, U.S.A.

Pore fluids of deep sedimentary basins are often referred to as formation fluids, especially in petroleum geological literature. Most of the voluminous data for the composition of oil field brines and other formation waters (e.g. see, White *et al.*, 1963; Manheim, 1972) are for waters from reservoir rocks or permeable strata, obtained during flow or pumping tests performed in the course of exploratory drilling. Although non-reservoir rocks, such as shales, silts or poorly permeable carbonate rocks, may comprise as much as 50 per cent of sedimentary basins, the nature and origin of their fluids have only just begun to be investigated. Most of the work which has been done, has been carried out by Soviet workers under the leadership of P. A. Kriukov (Kriukov, 1971).

The geochemistry of interstitial fluids forms the subject of the present Chapter. However, these fluids are also significant from the geophysical, sedimentological and geotechnical points of view. In aquatic sediments, the composition and arrangement of pore fluids are critical factors in determining their strength and bearing capacity (Trask and Rolston, 1950; Richards, 1967) and their ability to conduct heat, sound (Nafe and Drake, 1963) and electricity (Kermabon *et al.*, 1969). The upward movement of pore fluids in relatively unconsolidated marine sediments under the influence of excess pressure has been shown to exert a strong erosive action when that pressure is relieved at the sediment-water interface. Under these conditions, sediment grains can "float" and be carried away by the slightest fluid movement. Indeed, submarine sapping through submarine discharge of fluid (see later) was held by Johnson (1940) to be a major causative factor in the formation of submarine canyons on the continental slope, and has probably been the agent that formed the depressions up to 150 m in depth on the sea floor off Florida (Kohout *et al.*, in prep.).

The first investigation of marine interstitial waters was published by Sir John Murray, the pioneer British oceanographer in conjunction with R. Irvine in 1895. Murray and Irvine squeezed fluids from a shallow "blue mud" from off the Scottish coast and found that, although the major components of the interstitial water were similar in their relative proportions to those of sea water, oxygen had been depleted, sulphate had been lost, and the bicarbonate content had increased. The relative proportions of the constituents indicated that the (largely bacterial) oxidation of organic carbon to CO_2 balanced most of the reduction of sulphate to form sulphide. In addition, when Murray realized that there was an appreciable amount of dissolved manganese in the pore water he revised his earlier concept of a largely volcanic origin for deep-sea manganese concretions, and suggested that, together with river-borne and volcanic supplies of Mn, a major source of the element might be diffusion into sea water from terrigenous marine sediments undergoing syngenetic reduction. Table 32.1 shows the results which Murray and Irvine obtained for the major elements in the interstitial waters of the blue mud.

The chlorinity of the interstitial waters from the coastal, and especially intertidal sediments, was studied by British biologists from the late 1920's (see references in Smith, 1955). The interstitial waters of many coastal sediments have chlorinities lower than those of the corresponding bottom waters because of the influence of fresh-brackish ground water seepages (Kudelin *et al.*, 1971). However, in estuaries, the chlorinity of the interstitial water is usually higher than the mean of that of the overlying water (Reid, 1932; Bruevich, 1957; Chernovskaya, 1958; Laevastu and Fleming, 1959; Callame, 1963; Sanders *et al.*, 1965). This phenomenon may arise in part through the salt wedge effect, and in part through cyclic gravitational convection of more saline (denser) solutions downwards through surficial sediments (Callame, 1960; Scholl, 1965). From an extrapolation of the results of laboratory experiments, Valyashko (1963) deduced that gravitational downward movement may widely affect the salinity distribution in the interstitial waters of fossil sedimentary basins, as well as that in unconsolidated surficial sediments. Such a mechanism was incorporated in the dolomitization concept of Adams and Rhodes (1960). However, Hsü and Siegenthaler (1969) suggested that the converse phenomenon, evaporative pumping, caused movement of Mg-rich solutions from the sea to evaporating lagoons (Fig. 32.1), thus facilitating the formation of dolomite.

Where beach ground water tables are not connected with fresh water aquifers from the land they may contain water which is more saline than sea water as a result of evaporation from the partly air-filled pore spaces. In this context, Emery and Foster (1948) have reported that they found crusts of salt cementing the top ~ 6 mm of some beach sands off southern California.

The chemical properties of interstitial waters of Recent sediments aroused considerable interest among Soviet geochemists, limnologists and oceanographers in the late 1930's (see for example Vinogradov, 1939; Bruevich *et al.*, 1938; Bruevich and Vinogradova, 1940a, b). The impetus for these investigations came partially from balneological (medicinal mud and spa) investigations of Black Sea Limans (e.g. Verigo, 1881; Shchukarev, 1937), and from soil studies. The Russian literature also frequently credits pioneer work to the noted Russian geochemist, V. I. Vernadskii. The main Russian advances in pore fluid studies occurred after World War II. Soviet oceanographers and geochemists (again led by S. V. Bruevich) including O. V. Shishkina, E. D. Zaitseva, T. I. Gorskhkova, N. G. Tageeva, M. M. Tikhomirova and N. I. Starikova, took advantage of the effective sediment squeezers developed by P. A. Kriukov (Kriukov, 1947, 1971) and extended interstitial water studies to the Russian northern seas, the Black, Caspian and Baltic Seas, and the Atlantic, Pacific and Indian Oceans.

Development of the piston corer by Kullenberg in 1944–1947 greatly expanded the scope of interstitial water studies. Kullenberg's own work on interstitial water revealed a sharp decrease in interstitial chlorinity with depth in the Baltic Sea. He attributed this pattern to the influence of underlying Pleistocene ice lake sediments (Kullenberg, 1952). Kullenberg, and

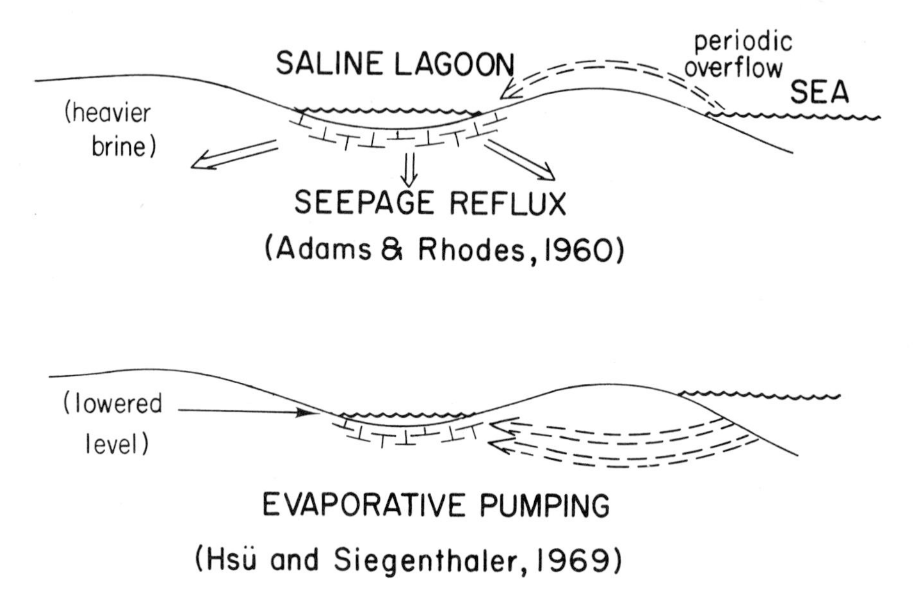

FIG. 32.1. Proposed mechanisms for interstitial brine movements promoting dolomitization of lagoonal sediments.

similar types of corers constructed by Russian workers, ultimately achieved the record sediment penetration of 34 m in the Bering Sea, and chemical studies of interstitial waters were extended to 27 m below the sea floor by Bruevich and Zaitseva (1958).

Meanwhile, Emery and Rittenberg (1952) and Rittenberg *et al.* (1955) in the United States were studying biogenic cycles and the regeneration of nutrients from the sediments of basins off southern California. These studies were followed by one of the few pre-1970 investigations of gases in marine sediments (Emery and Hoggan, 1958). Ammonia, carbon dioxide and methane were the principal gases detected and determined by mass spectrometry and gas chromatography. Earlier, Koczy (1958) had shown that nearly all the radon in sea water probably emanated from the radium in sea floor sediments.

Knowledge of the dissolved gases other than CO_2 in interstitial waters has advanced recently (Werner and Hyslop, 1968; Reeburgh, 1969; Hammond *et al.*, 1973). The lack of *in situ*, or pressurized, samples has limited the availability of data on absolute levels and concentration gradients of gases. Nevertheless, it has been shown that the dominant gases are CH_4,

TABLE 32.1

The composition of sea water and the interstitial water of terrigenous blue mud from the Scottish coast. (Expressed as weight per cent of total salt). (Murray and Irvine, 1895).

Component	Sea water	Interstitial water[a]
Cl^-	55·3	56·2
Br^-	0·188	0·191
SO_4^{2-}	7·69	4·10
CO_2	0·152	1·66
Ca^{2+}	1·19	1·07[b]
Mg^{2+}	3·76	3·82
K^+	1·11	1·12
Na^+	30·6	31·0
NH_3	—	0·053

[a] Chloride has increased owing to an appreciable loss of sulphate. For this reason Mg^{2+}, K^+, and Na^+ in terms of their ratio to chloride are slightly depleted in the pore fluids.
[b] The water may have become depleted in Ca^{2+} as a result of carbonate precipitation.

CO_2, N_2, O_2 and the noble gases. Nitrogen is depleted below the expected values in organic-rich strata, suggesting that it is converted to ammonia by bacteria (Hammond *et al.*, 1973). The methane is typified by its low $^{13}C:^{12}C$ ratio ($\delta^{13}C$ −60 to −90‰ on the PDB scale), and is probably formed mainly by methane bacteria. Claypool *et al.* (1973) found that the proportions of ethane (a minor component) increased in areas of high heat flow, suggesting that the initial stages in the formation of petroleum hydrocarbons by thermal processes occur at relatively low temperatures. Studies of noble gases by Clarke, *et al.* (1973) showed that ^3He and ^4He were present in amounts in excess of the values to be expected from solubility considerations.

Studies by Powers (1954, 1957, 1959) and Shishkina (1958, 1959) had already shown that there is a relative loss of magnesium and potassium from the pore fluids of terrigenous sediments; this has been ascribed to uptake of these elements by clays. The distributions of sulphur species, including sulphate, thiosulphate, native sulphur, organic sulphur, monosulphides and pyrite were investigated in Black Sea and Pacific Ocean sediments by Ostroumov *et al* (1961). However, the suggestion by Sillén (1961) that the ultimate composition of sea water may be controlled by the equilibria existing when silicate minerals react with sea water gave a powerful impetus to studies of sediment-water reactions in recent sediments.

Sulphur isotopes were studied in sediments from southern California and the Baltic Sea by Kaplan *et al.* (1963) and by Hartmann and Nielsen (1964), and the formation of pyrite and the migration of sulphur species has been documented by Berner (1969). Pore fluid–carbonate interactions have been studied by, among others, Berner (1966, 1971), Berner *et al.* (1970), Presley and Kaplan (1968), Thorstenson (1969), and Thorstenson and MacKenzie (1971). Recently, interest in carbonate equilibria has led to the initiation of the "Geochemical Leg" (Leg 15) of the Deep Sea Drilling Project (DSDP) in the Caribbean Sea. This has provided the opportunity for an intensive study of pore fluids and the equilibria controlling their composition as well as for interlaboratory checks on techniques. This work was carried out on cores having lengths of up to 370 m (see geochemical section in Initial Reports of the Deep Sea Drilling Expedition, Vol. 15).

Equilibria between interstitial waters and silicate rock systems have been dealt with by Mackenzie and Garrels (1965) and by DeSegonzac (1970), (see also Chapter 5). General diagenetic models incorporating carbonate, sulphate reduction and silicate reactions have been elaborated by Berner (1971) and by Manheim and Sayles (1974).

Although the greatest bulk of the data referring to the distribution of nutrient salts is to be found in the reports by Russian investigators (see below), recent interest has focused on the equilibria governing the reactions that proceed in near-shore organic-rich sediments, including those associated with sewer outfall sites. These will be briefly reviewed below.

Another development which has added greatly to our knowledge of pore fluids is the availability of long cores of sediment. Data from these extended the range which could be attained from the approximate 10 m limits imposed by coring to over 1000 m. Studies of long core drill sequences include those from the Caspian Sea (Tageeva and Tikhomirova, 1964; Strakhov, 1965); the Experimental Mohole off Baja California (Rittenberg *et al.*, 1963; Siever *et al.*, 1965); the JOIDES (Joint Oceanographic Institutions Deep Earth Sampling Program) drilling off Florida (Manheim, 1967) and the Deep Sea Drilling Project (DSDP) (Sayles and Manheim, 1975).

Considerable interest has been aroused about the role of pore fluids in the formation of authigenic phases, such as manganese nodules (Manheim,

1965; Lynn and Bonatti, 1965; Michard, 1971; Calvert and Price 1972; Bender, 1971) and the iron-rich deposits on, and around, the East Pacific Rise, which are possibly of hydrothermal origin (Bischoff and Sayles, 1972; Dymond *et al.*, 1973). Interstitial waters may also be involved in the synthesis of marine phosphorite (Bushinskii, 1963; Baturin, 1972), the origin of "excess" oceanic constituents such as boron (Horn and Adams, 1965), and in the diagenetic reactions leading to the production of petroleum and natural gas.

The phenomenon of interstitial brines discharging at the sea floor in zones of rifting or tectonic disturbance is now considered to be one probable mechanism involved in the formation of heavy metal deposits of major economic significance. For example, these discharges have probably led to the formation of the hot brines and copper, zinc, silver and gold-rich deposits in the central rift area of the Red Sea (Degens and Ross, 1969), and they may be of significance elsewhere. Heavy metal enrichments, including occurrences of sphalerite and native copper, though of lesser extent, have also been found in sediments from the African Rift Valley Lakes (Degens *et al.*, 1971), and in strata off the southern Atlantic coast of the United States (Lancelot *et al.*, 1970).

Several problems connected with interstitial brines are still unresolved. These include the specific pathways by which fluids are lost from sedimentary accumulations (especially geosynclinal or rapidly accumulated pockets of sediments), and the details of solid-liquid equilibria in deep evaporite deposits. One school of thought stresses the importance of membrane filtration in the formation of concentrated brines in sedimentary basins on land (Hanshaw and Coplen, 1973). However, no evidence for this mechanism has been obtained from marine studies to date, i.e. the salt contents of interstitial waters have not been found to be abnormally high, except where there has been contamination by percolation or diffusion from adjacent evaporite deposits. There is considerable doubt about the occurrence of membrane filtration or reverse osmosis in geological systems generally (see Manheim and Horn, 1968; Kriukov, 1971), but membrane processes may be important in sediments. For example, under suitable conditions, gradients may set up osmotic pressures and give rise to convection and fluid motion (Mokady and Low, 1968).

32.2. Sampling of Interstitial Waters

There are a number of problems involved in the collection and storage of interstitial waters, and some of these will be discussed below. However, analytical techniques are beyond the scope of this chapter.

The primary requirement for valid work on cores that are recovered from the sea floor and brought to the laboratory is that they should be in a state which is as close as possible to their *in situ* condition; thus, they should be sampled as soon as possible, to avoid evaporation of the pore water and exposure to the atmosphere. Even with piston cores the outer parts are frequently contaminated because of smearing along the core tube. With drill cores the outer parts of cores may be more contaminated because sea water is generally used as the circulating medium. It is important, therefore, to select and sequester those interior portions of cores which show the least signs of disturbance and deformation. However, it should be noted that where the core barrel is rotated into the sediments without pumping, to minimize the disruptive effect of the mud stream, even distorted sediment sections may have pore fluid compositions similar to those of interior "undisturbed" sections.

Stored whole cores may be affected both by evaporation and by selective condensation of pure water in gas pockets, as well as by the subsequent diffusion and mixing of waters of differing salinity. Comparative tests between those samples sequestered immediately and those taken after 4 months storage in sealed plastic sleeves at 4°C have shown that the interstitial salinity of the stored samples had significantly changed in both directions (Manheim, 1967). Other discrepancies may be induced by the use of cellulose acetate butyrate core liners which may initiate redox changes when used for storage.

There are several grounds for thinking that relatively uncontaminated pore fluid can be obtained from wire-line cores in drilling operations. In the area sampled during the JOIDES drillings off Florida (Manheim, 1967), submarine discharge of waters of low salinity from continental aquifers affected sub-seafloor sediments as far as 120 km from shore. Lack of contamination is indicated by the fact that as little as 60 ppm of chloride was found in their pore fluids, even though the holes were drilled with sea water as the circulating medium. The pore fluids of organic-rich sediments are frequently almost completely devoid of sulphate as a result of bacterial reduction. Since the sea water circulation medium has about 2·8 g SO_4^{2-} kg^{-1}, substantial freedom from contamination is implied. Similarly, pore fluids trapped in shales in evaporite strata from the Red Sea floor, i.e. between halite (NaCl) beds, have yielded fluids essentially saturated with respect to sodium chloride (Manheim *et al.*, 1974).

In contrast, rather poor results were obtained by efforts to introduce tracers into the drilling fluid during the Geochemical Leg of the DSDP. Dyes such as rhodamine B were so strongly adsorbed by the sediments that they could not be used as safe indicators of lack of contamination of the pore fluids. Furthermore, fresh water, temporarily employed as the drilling medium, exercised such strong disruptive effects on the core (probably partly as a result of osmotic gradients) that the results were inconclusive. However, the evidence cited previously, together with the generally smooth vertical concentration gradients found for reactive pore fluid components, suggests that contamination with sea water is not normally a serious problem, when appropriate precautions are taken.

In contrast to the lack of change resulting during the drilling, movement of sediment cores from a sea floor environment to a different one at the surface inevitably affects the distribution of some constituents because of the differences in pressure and temperature. The pressure used in extraction of the interstitial water has little effect on most constituents (Manheim, 1974). However, when the cores are raised to the surface, the change in pressure affects the carbonate equilibria, and may lead to exsolution of methane gas which will result in the formation of gas pockets in the sediments; it may even cause other gases to be swept out in the process. Mangelsdorf *et al.* (1969) have shown that temperature changes alter the distribution of cations in pore fluids, and this has been subsequently confirmed by other workers. Their results were in qualitative agreement with earlier data obtained independently by Kazintsev (1968) and by Krasintseva and Korunova (1968). Thus, movement of sea floor sediments at a temperature of 1–2°C to room temperature leads to increases in the concentrations of monovalent cations in the interstitial waters, and to decreases in the concentrations of the divalent cations. For potassium, the erroneous differences for surficial, unconsolidated clayey sediments are large enough to create a diffusion gradient that would supply more potassium to the oceans than would all the world's streams and rivers. The concentration of silica has also been shown to increase in interstitial waters when the sediments are warmed (Fanning and Pilson, 1971). Table 32.2 gives a compilation of the changes in the concentrations of interstitial constituents which occurred in four lithologically different sediment types from the Caribbean Sea when the pore fluids were extracted at different temperatures on board ship (Sayles *et al.*, 1973).

The data strongly support the concept that for major cations the variations observed are due to changes in the cation exchange capacity of the clay minerals. These processes are largely reversible and probably occur almost instantaneously with respect to individual clay micelles. The actual time required to achieve equilibrium (several hours) is probably related to diffusive migration of ionic constituents through the tortuous pore systems in response to the differences in the position of equilibrium resulting from the change of temperature. In contrast, the concentrations of anions apparently do not change significantly with temperature. The mechanisms which cause the variations of silicon and boron concentrations are still poorly understood, and the release of boron from sediments with increasing ambient temperature is sufficient to account for many of the boron anomalies found for Recent and sub-Recent marine sediments (Krasintseva and Shishkina, 1959; Kitano *et al.*, 1969; Initial Reports of DSDP, legs 1–15).

Fanning and Pilson (1971) have reported that the use of acid-washed filter paper led to neutralization of some of the alkalinity when pore water was passed through it. In this context, the author of this chapter has noted that the use of micropore filters composed of cellulose acetate-nitrate mixtures can result in nitrate contamination of filter effluents.

TABLE 32.2

Changes in composition of interstitial water resulting from raising the temperature from 4°C to 22°C during extraction. (Expressed as the percentage increase (plus), or decrease (minus), at 22°C relative to 4°C).*

Component	Site 147 Cariaco Trench (clayey-org.-rich)	Site 148 (clayey—carbonate)	Site 149 (carbonate)	Site 149 (siliceous)
B	+ 30	+ 61	+ 7	nd
Si	+ 26	+ 41	+ 27	0
K	+ 18	+ 24	+ 16	+ 12
Na	+ 0·9	+ 1·3	+ 1·0	+ 1·0
Li	− 3 (?)	0·0	0·0	nd
Ca	variable	− 6·5	− 3·1	− 1·1
Mg	− 7·3	− 7·3	− 3·1	− 1·4
Sr	− 19	− 7	0·9	nd
Cl	< 0·5	< 0·5	< 0·5	< 0·5
SO_4	< 0·5	< 0·5	< 0·5	<0·5

nd = no data.
* Data for sediments recovered during Leg 15 of the Deep Sea Drilling Project (Sayles *et al.*, 1973).

32.3. DISTRIBUTION AND MAJOR RESULTS OF INTERSTITIAL WATER STUDIES

32.3.1. PISTON CORES AND SHALLOW SAMPLING

32.3.1.1. *General distribution and properties*

A comprehensive, though not exhaustive, summary of the geographical distribution of interstitial water studies is given in Table 32.3; and the locations of the stations concerned are plotted in Fig. 32.2. As the results of the

TABLE 32.3

Summary of marine interstitial water studies from piston core, or shallow sediment, samplings. Symbols refer to the geographical locations indicated in Fig. 32.2.

Abbreviations: incr.–increased; alk.–alkalinity; exch.–exchange; brn–brown; gry–gray; refs–references. Where stations are too numerous, or otherwise awkward to plot, they have been designated by dashed boxes in Fig. 2.

Symbol	Location	Reference	Notes
		ANTARCTIC AND ADJOINING SEAS	
LKMW	SE Pacific sector	Li *et al.* (1973)	Ba concentrations of interest.
		ARCTIC AND NORTHERN SEAS	
	Barents and Kara seas	Bruevich (1958)	Oxidation-reduction phenomena occur, giving brn-gry interface in the sediment column.
Ch	Murmansk Bay	Chernovskaya (1958)	Incr. alk. with depth: P 0·15–0·3 mg l^{-1}; Si, 3–10 mg l^{-1} i.e. higher than overlying water.
TTK	Arctic and Russian Northern seas	Tageeva *et al.* (1961)	Only major constituents determined.
TT	Greenland Sea and ice stations, N. Pole	Tageeva and Tikhomirova (1961)	Major and minor constituents, exch. capacity and sediment parameters determined.
	Arctic, Kara, Laptev and Chukotsk seas	Tageeva (1965)	Cl 9–19‰; Zn, Fe and B determined; possibly some erratic and systematic errors
LBM	Ice island N. of Alaska	Li *et al.* (1969)	Mn concentrations 0·1–9 mg l^{-1} and diffusion calculations.
		Sea of Azov	
		Gorshkova (1955, 1961)	Nutrient salts, pH, and sediment parameters determined; marked variations with season.
		Shishkina (1961)	Major constituent analyses given for 6·5m piston cores; little change in chlorinity (ave. 208 meq kg^{-1}); alkalinity increases with depth.
		Starikova (1961)	NH_4^+–N 14–50 mg. l^{-1}; N_{org} 5 mg. l^{-1}; C_{org} 26–54 mg. l^{-1}; C/N 5–8 in pore fluid.

Symbol	Location	Reference	Notes
		BALTIC SEA	
		Kullenberg (1952)	Strong decrease in salinity with depth (from 15–16‰ at surface to < 10‰ at 10 m).
	(Praestö Fjord)	Mikkelsen (1956)	Showed diffusive gradient
	(incl. Bothnian)	Jerbo (1965)	Cl, Mg, Na, and geotechnical measurements; decreasing salinity with depth in cores, and northward in Bothnian Gulf (to < 1‰ S)
		Hartmann (1964)	Up to 10 mg l^{-1} interstitial Fe, Mn in stagnant sediments, contrasted with 10–100 μg Mn l^{-1} in bottom waters and 5 μg l^{-1} in intermediate and surface waters.
		Hartmann and Nielsen (1964)	Sulphur isotopes studied; lighter sulphur (H_2S) precipitated and remaining sulphate-sulphur enriched to as much as $\delta^{34}S$ = 60‰.
		Gorshkova (1957, 1970)	Nutrient constituents, pH and sediment parameters measured P 2–3 mg l^{-1}; NH_4^+–N 12–145 mg l^{-1}; NO_2^-–N 0–18 mg l^{-1}; Alk. 10–16 meq l^{-1}; Si 25–45 mg l^{-1}; pH 8·0–8·2.
		Bojanowski and Paslawska (1970)	Iodine 3·5–19 mg l^{-1}, poorly related to that in sediments.
		Bering Sea	
		Zaitseva (1954)	P, Si, NH_4^+–N and Alk determined between depths of 1–16 m in piston cores from water depths of 150–300 m; Alk and NH_4^+ increase with depth; P, Si, irregular.
		Shishkina (1955)	Experimental squeezing of clays showed little influence of pressure to 3000 kg cm^{-2}.
		Bruevich and Zaitseva (1958)	Nutrient and sediment parameter data.
		Bruevich (1958)	Salinity and chlorinity

Reference	Notes
Black Sea	
Verigo (1881)	Analysis of aqueous extracts of liman mud.
Arkhangel'skii and Zalmanzon (1931)	Chemistry of aqueous extracts of Black Sea gravity cores; exchange capacity data.
Shchukarev (1937)	Extensive chemical and physiochemical studies of liman muds.
Ostroumov (1953); Ostroumov *et al.* (1961)	Data on forms of sulphur in sediments and interstitial waters.
Bruevich (1952); Bruevich and Shishkina (1959)	Studies revealed decrease in chlorinity with depth in piston cores.
Shishkina (1957, 1959, 1962, 1969)	General interstitial water chemistry and sediment-water diagenesis.
Shishkina (1961)	Redox and pH relationships
Mokievskaya (1960)	Iron ranged from 1–1080 $\mu g\ l^{-1}$ most of which was shown to be organically bound.
Tageeva and Tikhomirova (1961)	Data on pore fluids of shallow cores from north-western Black Sea shelf.
Starikova (1959)	Studies of N_{org}, NH_4–N and C_{org} in pore fluids.
Zaitseva (1959, 1962)	Investigations of nutrient salts, exchange capacity and equilibria.
Manheim and Chan (1974)	Review and new data on submarine discharges, diffusion and palaeohydrology of Black Sea.
Caspian Sea	
Bruevich and Vinogradova (1940, 1946, 1947)	Data on salinity, nutrients and water content of gravity cores.
Tageeva (1958)	Studies on water content and exchange capacity.
Shoikhet and Sakhnovskaya (1956)	Deals with Eh, pH and bitumens.
Tageeva and Tikhomirova (1962)	Extensive monograph on sediment properties and interstitial water composition of gravity cores and drill cores down to a depth of 2000 m in eastern Caspian near shore region. Some dubious data, especially for trace elements.

Symbol	Location	Reference	Notes
		Gulf of California	
		Byrne and Emery (1960)	Data on SO_4^{2-}, S^{2-}, Cl, pH and gas generation.
		Berner (1964)	Deals with pS, sulphur species and diagenesis.
		Siever *et al.* (1965)	Studies of Cl, Na, Mg, Ca, K from piston cores.
		Gulf of Mexico	
		Grim and Johns (1954)	Chemical analyses of < 1 μm fraction of sediment, plus interstitial ion ratios obtained by leaching salts from sediment.
		Shepard and Moore (1955)	Sedimentological analyses, Cl^-, SO_4^{2-} in pore water from Texas coastal bay sediments.
		Powers (1959)	Studies of diagenesis.
		Mel'nik (1970)	Investigations of pH, Eh, albedo and colour of sediments off Cuba.
		See also interstitial water studies on drill cores: Manheim and Bischoff (1969); Kaplan and Presley (1969); Manheim *et al.* (1972), and publications cited therein.	
		Indian Ocean	
		Bordovskii (1960)	Alk. 1·9–5·4 meq. l^{-1}; NH_4–N, 0·03–1·1 mg. l^{-1}; Si 5–20 mg. l^{-1}; pH 6·5–8·0.
Numbered Stations (circles)		Shishkina and Zheleznova (1964)	Data on chlorinity of water from piston cores.
		Zheleznova and Shishkina (1964)	Data on Eh and pH.
	Waltair, India	Ganopati and Chandrasekhara (1962)	Studies on salinity and oxygen content of pore water of beach sediments.
M	Off W. Pakistan	Marchig (1970)	Data on Cl, Mg, and K all of which may be unreliable.

		Mediterranean	
BF	Off Monaco	Brouardel and Fage (1953)	Decreasing O_2 as sediment is approached.
		Gripenberg *et al.* (1948)	Titrated chloride in dried sediment to obtain original water content.
K	Eastern Med.	Kullenberg (1952)	Determined interstitial chloride in core leachates.
EC	Eastern Med.	Emelyanov and Chumakov (1962)	Relatively small increases in pore water Cl and moderate increases in Alk. with depth.
Ga	Banyuls (French–Spanish border)	Gadel (1968)	Studies on lagoonal interstitial waters.
BP	Rhone delta	Bellaiche *et al.* (1966)	Salinity in pore waters of cores = 33–36‰, compared with 38‰ in the overlying water.
	Tyrrhenian Sea and Ionian Sea	Shishkina (1972)	Major constituents determined, only small changes found.
	Eastern Med.	Milliman and Müller (1973)	Mg/Cl increased with depth in core, presumed due to breakdown of Mg – calcite.
		North Atlantic	
O	Wide distrib.	Shishkina and Bykova (1964); Shishkina *et al.* (1969)	Demonstrated a high degree of constancy for the major constituents of 5 piston core pore waters.
HP	Bermuda Rise	Harris and Pilkey (1966)	Showed that in the pore waters Cl, Mg, Ca and Si were relatively constant.
FS	Bermuda Rise	Fanning and Schink (1969); Fanning and Pilson (1971);	Found that silicate increased by 50% on bringing core from 0–3°C to ambient temperature; phosphate did not change.
SBB		Siever *et al.* (1965)	Major constituents plus Si.
BK 70	Mid-Atlantic Ridge	Bischoff and Ku (1970)	Found few changes in the major ions of the pore water except for B and Si; Mn less than 1 mg l^{-1}.
BK 71	N. Azores	Bischoff and Ku (1971)	Found sulphate content of pore waters decreases in reducing sediments, whereas Alk. and Mn increases.
H	S. of Canary Is.	Hartmann *et al.* (1973)	Extensive study of organic breakdown. Decomposition products in interstitial water account for 2% of organic matter, whereas total breakdown, estimated from sulphide-sulphur, is 15–20% of organic matter.

Symbol	Location	Reference	Notes
		NORTH AMERICAN COASTAL WATERS	
SMH	Cape Cod	Sanders *et al.* (1965)	Constant interstitial chlorinity in sediments of an estuary with fluctuating salinity.
SGKB	Cape Cod	Siever *et al.* (1963)	Studied pH and Cl.
	Long Island Connecticut	Friedman and Gavish (1970)	Investigated interstitial components of lagoonal deltaic and estuarine muds.
FS	Long Island	Friedman *et al.* (1968)	Data on major constituents and some trace elements of continental shelf sediment pore waters.
BST	Connecticut Shore, Shore, Maine	Berner *et al.* (1970)	Described alkalinity models and reduction processes in stagnant sediments; data on NH_3, SO_4^{2-}, Ca, Mg, pH and Alk.
P	Chesapeake Bay	Powers (1954, 1957, 1959)	Data on Mg and K uptake on clays; diagenesis.
BB	Chesapeake Bay	Burbanck and Burbanck (1967)	Gives data on the oxygen content in the upper cm of sediment.
Re	Chesapeake Bay	Reeburgh (1969)	Discusses gases in sediments: Ar, N, CO_2, H_2S, CH_4, and the apparent selective removal of N.
Ne	Rappahannock R. estuary, Chesapeake Bay	Nelson (1972)	Found that nutrient salts and pH increase in pore waters from an upper estuary towards mouth.
Sch	Florida	Scholl (1965)	Discusses salt exchange in mangrove swamp sediments.
Be66	Florida Bay	Berner (1966)	Gives a discussion on carbonate diagenesis.
Gi	Florida Bay	Ginsburg (1966)	Notes that an interstitial salinity 5–6 times greater than normal promotes dolomitization.
Se	Bahamas	Seibold (1962)	Data on Eh and pH in oolite pore waters.
D	Rio Ameca basin (Mexican coast)	Drever (1971a, b)	Discussion on diagenesis of interstitial waters; Mg replaces Fe in clays in stagnant deposits.
LF	Puget Sound Area (Lake Wash.)	Laevastu and Fleming (1959)	Demonstrated that fresh water replaces saline water in interstitial spaces less easily than vice-versa.
Sh	Southern Alaska	Sharma (1970)	Discussed the adjustment of glacial pore water to the marine environment.

K8	Juneau glacier area	Kitano *et al.* (1969)	Studies of glacial sediment pore water; borate higher in interstitial than bottom water.
AFA	Baffin Bay	Ali *et al.* (1972)	Data on major and some trace constituents, e.g. Li, Rb and Sr in piston cores.
SI	Saanich Inlet, British Columbia	Nissenbaum (1971); Nissenbaum *et al.* (1972); Presley, *et al.* (1973); Brown *et al.* (1972).	Comprehensive studies of interstitial constituents in reducing fjord sediments; very high sulphide and bicarbonate; isotopic and organic constituents studied.
		NORTHERN EUROPE	
De	La Rochelle Bay, French Coast	Debyser and Rouge (1956)	Studies of "soluble" and colloidal iron in pore waters.
Ca	French Atlantic coast	Callame (1963)	Role of gravitational convection and diffusion in interstitial waters.
Vo	Zuider Zee (Ijsselmeer)	Volker (1961)	Salinity history of Zuider Zee and its relationship to ground water salinity.
KM	Cuxhaven	Kühl and Mann (1966)	Studies of salinity, oxygen, nutrients and Alk. in interstitial water.
Gp	Vilaine Bay	Goni and Parent (1966)	Data on pH, SO_4^{2-}, H_2S, Cl, Ca and K.
NL	Slack estuary	Nooter and Liebregt (1971)	Study of interstitial water salinity in an estuary.
WD	Scheldt	Wollast and De Broeu (1971)	Studies on silica and mechanisms for its removal in an estuary. The silica concentration (14–28 mg $SiO_2\,l^{-1}$) inadequate for sepiolite stability. showed nearly all silica removed by organisms, rather than by inorganic mechanisms.
		NORWEGIAN SEA	
		Gorshkova (1960)	Data on Si, P, Alk and pH.
		Shishkina and Bykova (1966)	Studies of salinity variations.

Symbol	Location	Reference	Notes
		PACIFIC OCEAN	
P	Off Chile	Pamatmat (1971)	Studies of oxygen uptake by sediments, and of NO_3^- and NO_2^- content of interstitial waters.
		Arrhenius and Rotschi (1953)	Studies on Si in interstitial waters.
	NW Pacific	Zaitseva (1956)	Investigations of exchange capacity of sediments.
	General Pacific	Zaitseva (1966)	Studies of cation exchange capacity and interstitial ions. Determination of exchange equilibria (see text).
	NW and W Pacific	Shishkina (1958, 1959, 1966, 1969, 1972)	Most precise existing data on major constituents in interstitial waters from piston cores; demonstrated coherence of both Alk. increase and sulphate decrease. K, Ca and Na variations relatively small but significant. No change in chloride, and bromide changes minor.
	Deep trenches; Mariana, Yap, Philippino, Kurile-Kamchatka	Shishkina (1964)	Studies of major constituents in pore waters of reducing and oxidizing sediments.
	SW and NW Pacific	Ostroumov and Fomina (1960); Ostroumov and Volkov (1960)	Investigations of forms of sulphur in sediments and interstitial waters.
	Okhotsk Sea	Starikova (1956)	Studies of C_{org}, N_{org} and NH_4^+ in pore waters.
	NW Pacific	Starikova (1959, 1961)	Data on C_{org} and N_{org}, ammonia and bacterial counts in pore waters.
	Okhotsk Sea	Bruevich (1956, 1957)	Investigations of biogenic components and related parameters, and of chlorinity in pore waters.
Ok	Off Honshu, Matsushima Bay	Okuda (1960)	Data on pH, Cl and nutrient salts in shallow bay sediment pore waters.
	Central Pacific	Bordovskii (1960, 1961)	Studies on alkalinity and ammonia in pore waters.
(squares)	W. Pacific	Bruevich and Zaitseva (1960, 1964); see also Bruevich (1966)	Study on biogenic constituents, Si, NH_4^+–N, PO_4^{3-}–P, NO_2^-, NO_3^-, Alk. and pH (see text).

RU	Peru–Chile Trench	Romankevich and Urbanovich (1971)	Investigation of Org. matter, nitrogen and carbohydrates in interstitial water.
BS	E. Pacific Rise	Bischoff and Sayles (1972)	Studies of major and some minor constituents in pore waters; no appreciable anomalies found for metalliferous sediments.
		Red Sea	
SZ		Shishkina and Zheleznova (1964)	Study of chlorinity in pore water. Chloride content of 21·5 g kg^{-1} at surface of sediment; 22·6 g kg^{-1} at 17·5 m.
HS		Hendricks *et al.* (1969)	Data on interstitial water composition of metalliferous sediments from the central (hot brine) rift zone of Red Sea. Extensive chemical and spectrochemical analyses, many on fluids extracted on board ship.
BKP		Brooks *et al.* (1969)	Study on above area, on fluids extracted from stored box cores.
		Milliman and Manheim (unpublished)	Studies on piston cores from central southern Red Sea showed an increase of salinity with depth.
		(See also data by Manheim *et al.* (1974) for DSDP cores).	
		South Atlantic	
OC	Near Recife, Brazil	Okuda and Cavalcanti (1963)	Studies of pH, Cl and nutrient salts in shallow water
circles	Western S. Atlantic	Shishkina (1969)	Investigation of major constituents in pore waters.
SBB	Western S. Atlantic	Siever *et al.* (1965)	Data on Na, Ca, Mg, K, Cl and Si in pore water.
Ba	Off S. Africa	Baturin (1971)	Information on uranium in interstitial waters of phosphatic sediments.
		Southern California Offshore Area	
	Beach water tables, La Jolla	Emery and Foster (1948)	Studies of possible influences of ground waters on beach pore waters.
	Tomales Bay	Johnson (1967)	

Symbol	Location	Reference	Notes
	Offshore basins	Emery and Rittenberg (1952); see also Emery (1960)	Data on sediment parameters, interstitial waters and their nutrient components; studies on mechanism for expulsion and advection of pore fluid (vertically) from sediment.
	Offshore basins	Kaplan *et al.* (1963)	Information on sulphate depletion and sulphur isotope distribution in piston cores.
	San Pedro and Santa Catalina Basins	Presley and Kaplan (1968)	Investigation of effect of reducing conditions on carbonate equilibria and on Ca and trace elements.
	Offshore basins	Brooks *et al.* (1968)	Data on phosphate and trace elements.
	Offshore basins	Rittenberg *et al.* (1955)	Studies on nutrient cycles in pore waters.
	Offshore basin and Pacific	Presley *et al.* (1967)	Studies on manganese distribution.
		United Kingdom Coastal Waters	
MI	Firth of Forth	Murray and Irvine (1895)	Pioneering study of interstitial water in blue mud.
Br	Isle of Man	Bruce (1928)	Data on CO_2 equilibria; and sulphide in estuarine pore waters.
ASB	Tees Estuary	Alexander *et al.* (1932)	Salinity data for estuarine pore waters.
		Reid (1932)	Information on effect of varying salinity on marine animals in estuarine sediment.
Sm	Millport, Clyde	Smith (1955)	Data on brackish pore water from beach and shore sediments.
B	Whitstable, Kent	Brafield (1964)	Studies on oxygen in interstitial waters.
CP	Loch Fyne	Calvert and Price (1972)	Studies of depth distribution of Mn in pore waters.

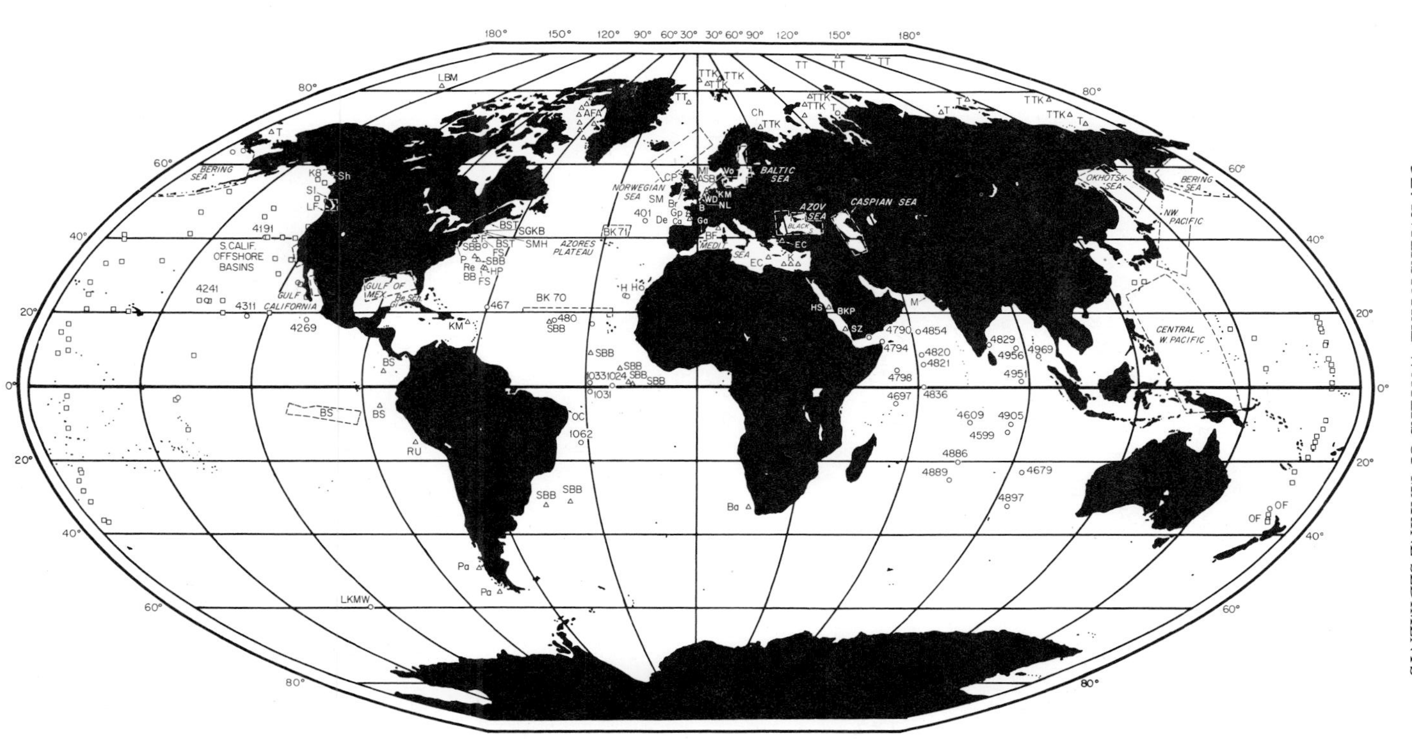

FIG. 32.2. Map showing locations of stations sampled for interstitial waters using piston and gravity corers. Symbols, letters and numbers refer to references listed in Table 32.2. Boxes designate areas subjected to more extensive sampling.

TABLE 32.4

Analyses of pore waters from deep-sea hemipelagic sediments. Data for piston core, northeast Pacific Ocean, approximately 51°N 161°E. Depth 5570 m. Data from Shishkina (1959). (All units except total solids are mg-equiv. kg^{-1}).*

Depth (cm)	Br^-	Cl^-	SO_4^{2-}	Alk	Total anions	Na^+	K^+	Ca^{2+}	Mg^{2+}	Total cations	Total solids g kg^{-1}
Btm. water	0·83	543	55·6	2·4	602	469	9·7	20·5	104	603	—
25–51	0·87	547	57·6	3·2	609	478	13·0	19·9	99	610	35·2
75–100	0·90	547	56·5	4·4	608	478	12·7	19·8	98	609	35·2
160–184	0·89	546	54·8	4·4	606	477	13·1	20·1	96	606	35·2
223–243	0·83	551	56·1	4·5	612	482	13·3	20·6	97	613	35·6

* The sediment is a grey-brown, fine-grained, clay with less than 0·1% carbonate and with organic carbon decreasing from 0·50% (at 25–51 cm depth) to 0·22% (at 223–248 cm depth). The water content decreases from 58% at 25–51 cm to 48·5% at 223–248 cm.

TABLE 32.5

*Analyses of pore waters from terrigenous-biogenic sediments. Data for piston core, Okhotsk Sea, approximately 56° N, 154° E. Depth 964 m. Data from Shishkina (1959).** (*All units except total solids are mg-equiv kg*$^{-1}$.).

Depth (cm)	Br^-	Cl^-	SO_4^{2-}	Alk	Total anions	Na^+	K^+	NH^+	Ca^{2+}	Mg^{2+}	Total cations	Total solids g kg^{-1}
Btm. water	0·79	526	54·3	2·3	583	451	9·5	—	21·0	102·5	584	—
10–60	0·84	523	49·8	10·3	584	454	11·3	0·7	20·5	97	584	33·0
60–117	0·86	526	37·4	17·3	582	—	11·0	1·5	19·0	95	—	34·0
117–172	0·88	524	28·5	24·8	578	456	10·7	2·4	17·8	93	580	33·3
172–226	0·90	532	13·4	32·8	579	—	—	3·1	15·6	95	—	33·2
409–420	0·93	532	5·5	40·5	579	456	10·7	2·8	14·2	95	579	33·1
470–480	0·95	535	2·9	42·8	582	461	10·4	2·9	12·9	94	581	33·5

* The sediment is a grey-green, diatomaceous clay with black sulphide staining in part. Organic carbon content is 1·5 %, amorphous silica 40 %, and carbonate less than 2 %.

Russian studies tend to be duplicated in summaries and reviews only the more comprehensive articles are cited in the table.

Shishkina has carried out some of the most accurate work on interstitial waters from a wide variety of geographical environments. For these reasons some of her data are included in Tables 32.4 and 32.5 as representative of typical interstitial waters of pelagic and terrigenous sediments. It is evident from the data in the tables that the chloride and bromide contents of interstitial waters of both pelagic and terrigenous deposits are relatively constant. Further, interstitial waters of pelagic deposits (Table 32.4) vary little with depth, with the exception of a moderate increase in alkalinity.

The interstitial waters of both the pelagic and terrigenous cores demonstrate the typical artificially produced differences in cations between bottom water and surficial sediments which result from raising the cores to a higher temperature before extraction (see above).

The principal changes in the composition of the interstitial water of the terrigenous sediment core are the progressive depletion of sulphate with depth and the increase in bicarbonate. At a depth of about 150 cm below the sediment-water interface, the sulphate and bicarbonate are present at almost equimolar concentrations, but at greater depth the bicarbonate is less than

TABLE 32.6

Variations in interstitial water parameters with depth in the sediment. The data from Table 5, are expressed as the amount (in mg equiv kg^{-1}) by which the pore water at the given depth, differs from the overlying sea water.

Depth (cm)	SO_4^{2-}	Alk	Ca^{2+}
10–60	− 4·5	+ 8·0	− 0·5
60–117	− 16·9	+ 15·0	− 2·0
117–172	− 25·8	+ 22·5	− 3·2
172–226	− 40·9	+ 32·3	− 5·4
409–420	− 48·8	+ 38·2	− 6·8
470–480	− 51·4	+ 40·5	− 8·1

that required to counterbalance sulphate lost (Table 32.6). The increasing loss of interstitial calcium suggests that calcium carbonate precipitation has removed bicarbonate from solution (see below).

Interstitial waters from sediments of Arctic and other far northern seas (Table 32.7) may be anomalous since the sediments are generally reducing below the upper brownish oxidized zone which is a few centimetres to some tens of centimetres in thickness. The manganese concentration of the inter-

TABLE 32.7

Sediment and interstitial water parameters for samples from the Greenland Sea and the Polar Arctic Ocean (Tageeva and Tikhomirova, 1961).

Depth (cm)	Median diam (μm)	<10 μm size fraction %	$CaCO_3$ (%)	C_{org} (%)	Water content (% wt)	Parameters of sediment: Hydroscopic water (% wt)	Molecular water (% wt)	Exch. capac. m eq/100 g	Components of interstitial water (g kg^{-1}): Cl^-	SO_4^{2-}	HCO_3^-	Ca^{2+}	Mg^{2+}	$Na^+ + K^+$
Station 23, 3326 m. Greenland Sea (see Fig. 1)														
0– 16	7·6	57·5	21·4	1·4	82·3	1·1	2·2	8·3	19·6	2·70	0·17	0·36	1·26	11·3
16– 46	3·8	92·8	25·2	1·1	62·7	1·5	2·9	10·9	19·5	2·49	0·17	0·39	1·20	11·2
46– 63	4·0	89·4	25·7	1·4	55·5	1·1	2·3	10·7	19·9	2·68	0·13	0·32	1·24	11·5
63– 88	4·9	91·2	24·9	1·4	58·0	1·1	2·2	9·5	20·6	2·88	0·30	0·30	1·14	12·3
88–113	3·8	92·8	25·3	1·3	55·4	1·2	2·4	9·0	20·2	2·61	0·099	0·35	1·28	11·6
113–147	3·9	91·6	25·5	1·4	60·3	0·9	1·9	11·2	19·9	2·65	0·11	0·38	1·22	11·5
147–167	5·5	70·1	21·3	1·8	43·4	1·1	2·1	8·8	20·2	2·59	0·12	0·32	1·19	11·8
STATION 16, 1574 m. LOMONOSOV RIDGE, CENTRAL ARCTIC OCEAN														
0– 19	4·3	83·8	2·4	1·4	84·3	2·1	4·2	18·4	21·0*	2·88*	0·16	0·41	1·45	11·8
19– 33	4·8	77·0	11·8	0·8	47·7	1·6	3·2	16·9	21·9*	2·85	0·19	0·39	1·49	12·4

Ranges of trace constituents in the pore water at Station 23 are (ppm) B 1·2–1·7; Si 4–10; I 0·3–1·6; Cu 0·2–3·0; Al 1–3; Ti 0·07–0·7; Mn 0·07–3·0; Mo 0·2–0·3; Zn 0·1–0·9; Fe 0·07–0·9.

* The fact that the chloride concentration of the pore water at Station 16 exceeds that of the overlying sea water suggest that some evaporation may have occurred. However, lack of detail about recovery and storage of the cores and the expressed water prevents this being definitely established.

stitial water within the oxidized zone is usually about 0·1 ppm or less, whereas within the reducing zone Mn concentrations may reach several ppm. In most areas, although the organic-carbon content of the sediments often lies between 1 and 2 %, sulphate deficiency and excess alkalinity are not appreciable in the reducing zone. In contrast, in sediments from other localities having similar organic-carbon concentrations there is often evidence for bacterial sulphate reduction. This phenomenon is probably not related to the absence of sulphate reduction, since despite the high latitude, bottom water temperatures are only a little colder than are those in temperate surface areas. It is more likely that the frequent, or semi-permanent, ice cover reduces plankton production in the surface waters and that this reduces the amount of freshly decomposed organic remains deposited so that the organic matter probably represents the more refractory organic residues. In addition, except in a few locations, deposition rates may be lower than those in corresponding temperate areas, and this will allow potential anomalies to be smoothed out by molecular diffusion.

In contrast, organic decomposition and inorganic oxidation proceed much more rapidly in sediments of tropical areas, and within the pore fluids of such sediments there may be much less mobility of labile reduced species, such as those of manganese (II) and iron (II). Thus, manganese concretions are almost unknown in tropical lake sediments, and in other lake sediments they are virtually limited to those of formerly glaciated terrains.

32.3.1.2. *The Black Sea: a natural laboratory for interstitial studies*

Relationships may be more complex in the interstitial waters of those sediments close to land or in those of restricted bodies of water which have unusual hydrological characteristics. A striking example of the latter environment is the Black Sea which contains a total of more than 540 000 km^3 of water having a salinity between 1/2 and 2/3 of that of normal sea water and which is largely anaerobic (see Chapter 15). It is further differentiated from most other stagnant marine (especially brackish) environments, by its great depth (over 2200 m). The interstitial waters of the sediments of the Black Sea have been studied more intensively than have those of any other area (Fig. 32.3). The density of sampling in the Black Sea has allowed interstitial chlorinity maps to be drawn (e.g. Fig. 32.4). These reveal not only the general decrease in chlorinity with depth in the sediments, but also regions of anomalously high or low chlorinity. The latter are interpreted in part, as being the result of submarine discharge of less saline water from land into the sea floor. A particularly clear example of this is to be found in the north-western slope of the Black Sea basin, where relatively flat-lying shelf strata have been eroded by slumping. Aquifers from the Crimean karst to the northeast appear to be the ultimate source of the interstitial waters which have chlorinities as low as 2·9 ‰; these

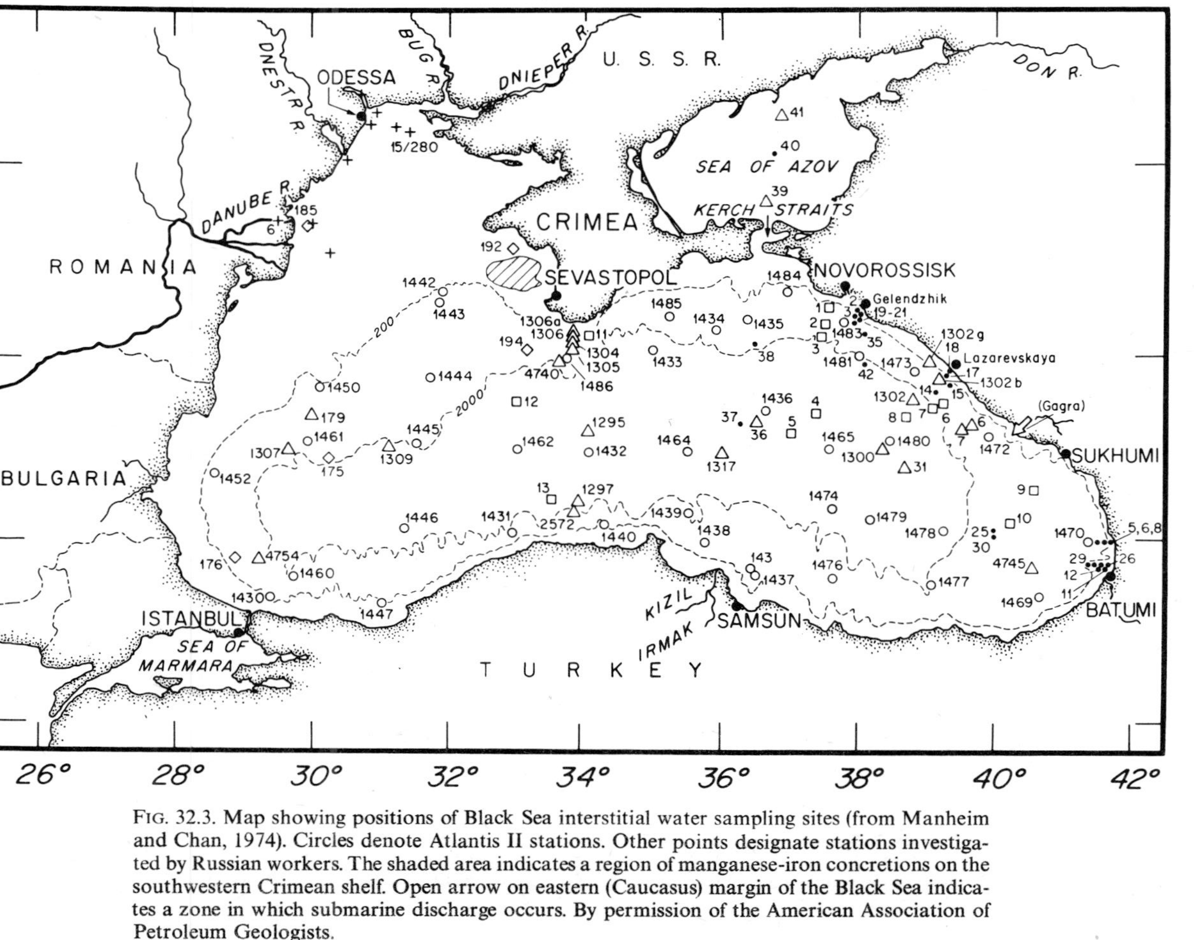

FIG. 32.3. Map showing positions of Black Sea interstitial water sampling sites (from Manheim and Chan, 1974). Circles denote Atlantis II stations. Other points designate stations investigated by Russian workers. The shaded area indicates a region of manganese-iron concretions on the southwestern Crimean shelf. Open arrow on eastern (Caucasus) margin of the Black Sea indicates a zone in which submarine discharge occurs. By permission of the American Association of Petroleum Geologists.

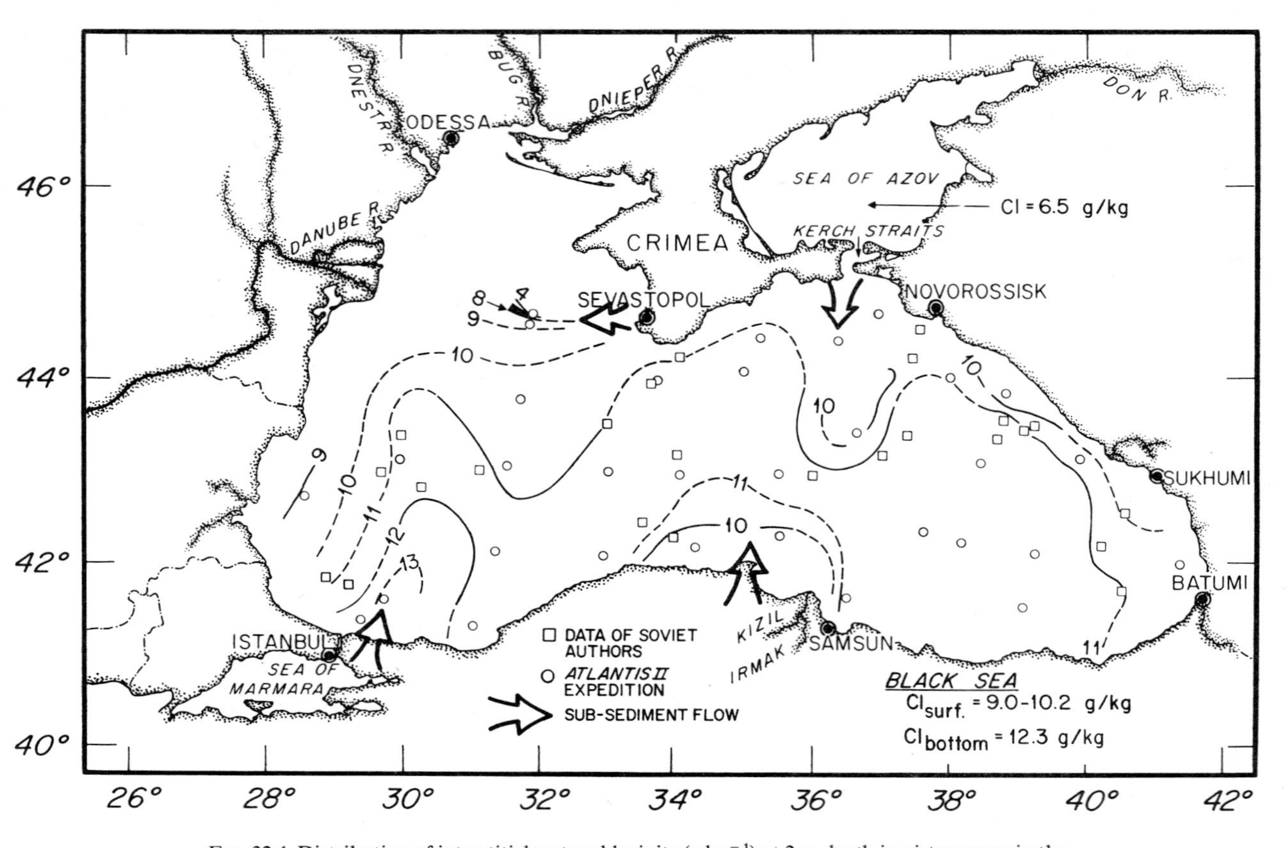

FIG. 32.4. Distribution of interstitial water chlorinity (g kg^{-1}) at 2 m depth in piston cores in the Black Sea. By permission of the American Association of Petroleum Geologists.

should be contrasted with the bottom waters at this location which have a chlorinity of 12‰. Calculations incorporating the mean decrease in chlorinity with depth for all available cores, the rate of sedimentation, and the diffusive migration of chloride suggest that during the period of isolation, between about 20 000 and 10 000 B.P., the Black Sea had a constant chlorinity of about 3·5‰.* After the eustatic rise in sea level and the breaching of the sill separating the Bosphorus and the Sea of Marmara, the Black Sea gradually increased in chlorinity to its present level of about 10‰ at the surface, and a maximum of 12·3‰ at depth.

The ionic constituents of the interstitial waters of the Black Sea sediments show interesting variations with increase in depth. Sodium and magnesium decrease by half or more and potassium nearly disappears. In contrast, calcium is strongly enriched in the interstitial solutions. These characteristics were exploited by Zaitseva to provide a unique natural experiment on the interrelationship between the exchange capacity of clayey sediments and the compositions of their associated interstitial waters. Careful analysis for interstitial components and exchangeable ions (including corrections for the leaching of calcium carbonate and for bound water in the sediments) yielded the data in Table 32.8 part a (Zaitseva, 1962, cited in Manheim and Chan, 1974).

When exchange coefficients based on mass balance equations of the type

$$Na^{+}_{exch} + K^{+}_{aq} \rightleftharpoons K^{+}_{exch} + Na^{+}_{aq} \quad K = \frac{(K_{exch})(Na_{aq})}{(Na_{exch})(K_{aq})}$$

$$Na^{+}_{exch} + \tfrac{1}{2}Ca^{+2}_{aq} \rightleftharpoons \tfrac{1}{2}Ca^{+2}_{exch} + Na^{+}_{aq} \quad K = \frac{(Ca_{exch})^{\frac{1}{2}}(Na_{aq})}{(Na_{exch})(Ca_{aq})^{\frac{1}{2}}}$$

were calculated the data in Table 32.6 were obtained. These show that in spite of gross changes in the ionic ratios, the exchange coefficients change very little with depth; this is remarkable when it is borne in mind that the ionic strength has decreased simultaneously. The data indicate that the ions in the exchangeable positions in the clays are responding predictably to changes in pore fluid chemistry. Moreover, the fact that respectively lesser and greater amounts of K and Ca are in exchangeable positions in sediments at depth than they are in surface muds indicates that the anomalies in the interstitial waters cannot be explained by ions moving into, or from, exchangeable positions. Evidently K is lost to an insoluble, probably silicate phase, in the sediments. Calcium in the pore fluids must be gained from a solid

* This figure is based on samples collected from deep parts of the Black Sea, i.e. below the shelf and upper slope regions. Note added in proof: recent data for DSDPLeg42b show that there is hypersaline pore fluid below the low salinity zone.

phase; perhaps by substitution of sodium for calcium in silicate minerals (Shishkina, 1972). The fact that alkalinity does not continue to increase, and in fact decreases, with depth in Black Sea sediments deeper than few metres is attributable to the bacterial formation of methane by synthesis from hydrogen and CO_2 (see below). During a cruise of *Atlantis II* in 1969 methane was found to be relatively abundant in Black Sea sediments, and sections of piston cores recovered lost their taped ends owing to ex-solution of dissolved gas.

Fluorine and phosphorus are other elements which are influenced by the increase in interstitial calcium with depth below the sediment surface. Shishkina *et al.* (1969) have shown that fluorine decreases from about 1·5 –

TABLE 32.8

Interstitial cations of the pore waters, and exchange capacity of sediments from the Black Sea (44°05′N 33°10′E; depth 1880 m; Zaitseva, 1962; Sta. 194).

a. Depth (cm)	Pore waters (meq kg^{-1})				Exchange capacity (meq/100 g dry sed.)				
	Na^+	K^+	Mg^{2+}	Ca^{2+}	Na^+	K^+	Mg^{2+}	Ca^{2+}	Total exch. cap.
Bottom water	300	6·4	68·4	13·9	—	—	—	—	—
0– 27	292	7·2	63·9	13·0	8·1	3·3	8·1	5·2	24·7
131–157	282	6·6	63·1	12·3	9·2	3·1	8·5	6·1	27·0
209–225	282	6·4	58·8	14·5	9·7	3·2	8·5	6·8	27·7
316–325	256	5·5	51·2	17·3	9·5	3·4	8·2	7·6	28·8
377–397	252	5·4	46·1	19·5	8·2	2·9	7·6	8·3	27·0
502–529	222	4·8	32·6	27·5	6·8	2·2	6·9	9·6	25·5

b. Exchange coefficients (see text) for K, Ca, and Mg relative to Na in the above Black Sea sediments and the averages for 5 typical red clay samples from a 7·5 m South Pacific piston core for which the total exchange capacity is 58 meq/100 g dry sediment (from Zaitseva; in Bruevich, 1966).

Location and description	K_K	K_{Ca}	K_{Mg}
Black Sea sediment			
0–27 cm	16·5	23	12·9
131–151 cm	14·5	22	11·2
209–225 cm	14·3	20	11·0
316–325 cm	16·8	18	10·8
377–397 cm	16·6	20	12·5
502–529 cm	15·0	19	15·3
"Red clay", mean of 5 South Pacific samples			
	15·0	15	8·2

1·8 mg kg^{-1} in near surface interstitial waters (F/Cl ratio of 1·2 – 1·5 × 10^{-5}) to ⩽ 1·0 mg kg^{-1} (F/Cl 1·1 × 10^{-5} or less). This relationship and the fact that interstitial phosphate does not increase, in spite of evidence for the breakdown of organic matter (e.g. sulphate depletion and methane formation), indicates that a carbonate fluorapatite phase is probably being precipitated. However, the amount of this mineral phase is far too small (⩽25 ppm) to be detectable in the solid phase. Collophane, or carbonate fluorapatite, typical of marine phosphorites, can only become a significant component of the

TABLE 32.9

Distribution of manganese in the pore waters of some terrigeneous marine sediments. Crimean Shelf. (Sevastyanov and Volkov, 1967, cited in Manheim and Chan, 1974).

Station	Depth (cm)	Sediment type	Mn (mg l^{-1})
15A	0–1	Mud, clayey, yellow brown; concretions	0·09
A	2–10	Mud, clayey, gray, sulphide streaks	0·56
B	18–24	Mud, clayey, gray-yellow	0·59
B	30–50	Mud, clayey, gray-yellow; H_2S below 48 cm	0·56
B	58–75	Mud, clayey, gray-yellow	0·43
C	90–110	Mud, clayey, light gray-bluish; shells	0·56
C	140–160	Mud, clayey, light gray-bluish; shells	0·90
C	175–200	Mud, silty-clayey, bluish-gray; shells	1·5
C	210–250	Mud, clayey, bluish gray, black sulphide	1·8
C	270–290	Mud, clayey, bluish gray, black sulphide	1·4

Arctic Basin, 82° N, 156° W; 20 cm brown oxidized sediment overlying a dark green reducing zone (from Li et al.*, 1969).*

Depth (cm)	pH	ΣCO_2(mmol l^{-1})	Mn (mg l^{-1})
0–5	7·99	2·53	<0·05
5–10	8·07	2·70	<0·05
10–20	8·06	2·71	<0·05
20–30	8·12	(2·85)	0·60
30–40	8·11	2·98	2·38
40–50	7·90	3·10	3·56
50–60	8·30	3·12	4·36
60–70	8·19	3·30	5·34
70–80	8·19	3·42	6·84
80–90	—	—	6·40
100–110	8·32	3·90	7·80
150–160	8·22	3·88	4·60
200–210	8·35	—	4·60

sediment when such precipitation takes place at the surface of sediments under conditions where the rate of deposition of mineral detritus is low and calcium carbonate is simultaneously leached. In general, the F/Cl ratios of interstitial waters of open-ocean sediments are about $13{\cdot}8 \times 10^{-5}$, i.e. a factor of two higher than that of sea water (Shishkina, 1964).

The sedimentary conditions prevailing in the Black Sea have influenced the distribution of manganese and have led to the formation of manganese-iron concretions on the Crimean shelf where the bottom waters are aerobic. Table 32.9 shows interstitial water data for the sediments immediately below this surficial nodule layer (Sevastyanov and Volkov, 1967) and, for comparison, the interstitial manganese concentrations in a core from the Arctic Ocean (Li *et al.*, 1969). At the redox boundary the manganese concentration increases sharply to values of 1 mg l^{-1}, or more, within the reducing zone. Similar phenomena have also been observed by Hartmann (1964) and by Gorshkova (1970) for other sediments which show transitions from oxidizing to reducing environments. Using the rate of diffusion of manganese (calculated from the gradients observed by Sevastyanov and Volkov (1967) for Crimean nodule-bearing sediments) in a steady state model it can be shown that the rate of supply of manganese corresponds to a nodule accretion rate of about 0·5 cm per 1000 years. This is slower than that which has been inferred for some manganese nodules from oxic near-shore environments (Manheim, 1965; Ku and Glasby, 1972), but far more rapid than that for deep-ocean nodules (Ku and Broecker, 1969). In contrast to iron, the concentration of manganese in interstitial waters seems to be stable for periods of time of the order of days or even weeks, even at concentrations which thermodynamically would be supersaturated at the prevailing P_{O_2} (Bischoff and Ku, 1971; Bischoff, personal communication). The concentrations do not necessarily remain constant, however, if the interstitial fluid remains in contact with the sediments.

32.3.1.3. *Organic reactions*

The organic reactions occurring in sediments are complex and still incompletely understood. However, a number of processes have been identified. The following major reactions occur in approximately the following sequence, many of them being mediated by bacterial or enzymatic agencies.

1. *Oxidation of organic matter by molecular oxygen.* One of the ways in which oxidation of organic matter in the interstitial water can be studied is by observation of changes in the dissolved oxygen concentrations. Typical overall oxidation processes are indicated in Equations 1–3.

$$(CH_2O) + O_2 \rightarrow CO_2 + H_2O \qquad (1)$$

$$\underset{\displaystyle NH_2}{CH_3\underset{|}{C}HCOOH} \text{ (alanine)} + 2O_2 \rightarrow 2CO_2 + NH_3 + 2H_2O \quad (2)$$

$$2NH_3 + 4O_2 + 2e^- \xrightarrow{\text{autotrophic bacteria}} NO_2^- + NO_3^- + 3H_2O \text{ (nitrification)} \quad (3)$$

Direct measurements of oxygen concentrations (Brafield, 1964), and the presence of oxidized zones in sediments (see for example, Lynn and Bonatti, 1965), suggest that the depth of oxygen penetration in most shallow water marine sediments is probably only a few centimetres. Even for sediments deposited several hundreds of kilometres from the shore piston coring has shown that the oxidized zones may have depths of only 20 cm to 1 m. As yet, there is no direct evidence as to what extent, if at all, oxygen has penetrated the interstitial waters within those deep-sea sediments which appear to be oxidized since much of the oxidation undoubtedly occurred while the sediment was at, or close to, the sediment-water interface. Intriguing data were provided by Pamatmat (1971) who found an oxygen uptake rate of 1·6 to 2 ml m^{-2} h^{-1} at the surface of Southeast Pacific sediments at a water depth of 2900 m; although these values seem very high. However, these data alone are inadequate to permit the depth of penetration of oxygen into the sediments to be calculated.

2. *Anaerobic oxidation of organic matter using NO_3^-, NO_2^- or metal oxides as oxygen donors.* Once all the molecular oxygen has been consumed nitrate and nitrite ions serve as oxygen donors for further breakdown of organic compounds, being themselves reduced to molecular nitrogen or nitrous oxide.

e.g. $$(CH_2O)(NH_3) + 4NO_3^- \rightarrow 6CO_2 + 6H_2O + 2N_2 + NH_3 + 4e^- \quad (4)$$

Nitrite ion may accumulate as an intermediate product of nitrate reduction (Emery and Rittenberg, 1952; Rittenberg, *et al.*, 1955). After this has been reduced, it appears that oxides such as MnO_2 and Fe_2O_3 may serve as electron acceptors in microbially mediated reactions, being themselves consequently reduced.

3. *Anaerobic oxidation of organic matter using SO_4^{2-} as an oxygen donor.* The anaerobic oxidation of organic matter having a C:N:P ratio typical of that of marine organisms (106:16:1) using sulphate as the oxygen donor can be represented as follows (Nissenbaum, (1971):

$$1/53(CH_2O)_{106}(NH_3)_{16}H_3PO_4 + SO_4^{2-} \rightarrow$$
$$2CO_2 + 2H_2O + 16/53NH_3 + S^{2-} + 1/53H_3PO_4 \quad (5)$$

From this it can be seen that complete reduction of the sulphate content of normal seawater (27 mmol l^{-1}) will yield 54 mmol l^{-1} of CO_2, 27 mmol l^{-1} of S^{2-}, 8 mmol l^{-1} of NH_3 and 0·5 mmol l^{-1} of PO_4^{-3}. An equilibrium pH of 7·0 was calculated by Nissenbaum for the reaction; this pH would remain constant as long as the reactants remained in the same proportions. Uptake of sulphide by iron oxides to form monosulphides or pyrite, equilibrium with solid calcium carbonate, weathering of silicates, or methane production (see below) could raise pH values to those frequently found for pore waters. Berner *et al.* (1970) found that the interstitial reactions in a number of organic rich sediments from off the northeast coast of the U.S. could be expressed in terms of a simple alkalinity model. Thus,

$$\text{Alk} = \text{Alk}_{\text{res}} + \Delta c_{NH_4} + \Delta c_{Ca} + \Delta c_{Mg} - 2\Delta c_{SO_4} \quad (6)$$

where Alk_{res} refers to the alkalinity expected for sea water of the given chlorinity. The units Δc refer to the differences (in mmol) between the observed concentration of the specified ion and that of the ion in sea water after correction for any dilution indicated by the chlorinity values. Ammonia may be important in these reactions as it reaches values in excess of 10 mmol l^{-1} (see for example, Table 32.10). However, in other situations it may be necessary to employ much more complex reaction schemes involving the solid materials and the organic matter present. (Thorstenson and Mackenzie, 1971; Gardner, 1973). In this context, the importance of silicates has been amply demonstrated (see below).

4. *Fermentation and methane synthesis in the absence of sulphate.* After the sulphate has been consumed the breakdown of carbohydrates, proteinaceous matter and fatty acids continues by fermentation. Some methane and CO_2 may be produced by decarboxylation of organic substrates such as acetic acid:

$$CH_3COOH \xrightarrow{(\textit{Methanosarsi barkeri})} CO_2 + CH_4 \quad (7)$$

Most of the methane, together with trace quantities of ethane, is produced from the fermentation products, CO_2 and H_2, by bacterial synthesis. This "anaerobic respiration", is known to occur in sewage sludge digestors, the rumens of cows and in many other anaerobic systems (Stainer *et al.*, 1970; Wolfe, 1971). The production of H_2, and methane synthesis, may be represented schematically as follows:

$$\text{Sugars} \rightarrow \text{pyruvic acid} \rightarrow \text{formic acid (HCOOH)} \xrightarrow{(\textit{Escherica coli})} H_2 + CO_2 \quad (8)$$

$$4H_2 + CO_2 \xrightarrow{\text{(Methane bacteria)}} CH_4 + 2H_2O \quad (9)$$

TABLE 32.10 *Distribution of nutrient salts, and other properties, in interstitial waters of terrigenous sediments. Bering Sea, 59° N, 178·5° E water depth 3633 m (from Bruevich and Zaitseva, 1956)**

Depth (cm)	Description	H_2O	C_{org}	N	C/N	Alk (meq kg^{-1})	NH_4–N mg kg^{-1}	PO_4–P mg kg^{-1}	Si mg kg^{-1}	Cl (‰)
50–60	Grn-gry soft mud, sandy lenses	70·0	0·91	0·12	7·5	3·54	1·7	0·15	—	19·36
150–160	Grn-gry med, no H_2S	58·2	0·91	0·12	7·5	4·09	1·9	0·17	14·8	19·26
250–260	Dk grn-gry mud	57·2	—	—	—	7·49	5·3	0·37	14·8	19·30
350–360	Grn-gry mud	63·7	0·82	0·078	10·5	11·6	9·2	0·62	18·2	19·36
446–456	Dk gry med, faint tr H_2S	52·6	0·73	0·10	7·3	15·7	12·0	0·81	16·8	19·20
550–560	Dk gry med, more cons, H_2S	54·6	—	—	—	20·6	14·9	1·14	15·5	19·33
651–661	Dk gry mud, more cons, H_2S	54·2	—	—	—	23·7	17·3	1·94	—	19·37
760–770	Dk grn mud, more cons, H_2S	—	0·79	0·084	9·4	—	—	—	—	—
840–860	Dk gry mud with v. sl. tr H_2S	43·2	0·77	—	—	32·1	23·0	1·92	—	19·40
895–905	Dk blue-gry clay, tr H_2S	42·5	—	—	—	32·9	26·3	2·35	22·8	19·55
943–955	Dk gry, more cons clay, tr H_2S	42·1	0·66	0·085	7·7	31·2	—	0·56	22·0	19·73
1108–	Dk gry clay, tr H_2S	36·7	—	—	—	36·7	30·4	1·94	—	19·54
1216–1260	Dk gry dense clay, tr H_2S	39·2	1·05	0·11	9·5	35·8	—	2·20	17·0	19·74
1310–1320	Dk gry clay, tr H_2S	48·6	—	—	—	40·4	30·6	3·53	21·8	19·60
1415–1460	As above	45·7	0·86	0·072	11·9	40·5	33·0	2·40	—	19·57
1510–1550	As above	54·0	0·72	0·086	8·4	40·6	31·0	3·30	16·0	19·36
1630–1636	As above	47·5	—	—	—	42·9	31·8	3·25	18·0	19·74

* H_2O given as a percentage of wet sediment; C, N as a percentage of dry sediment; other parameters in terms of interstitial waters. Abbreviations: v-very, sl-slight, grn-green gry-gray, dk-dark, tr-trace, cons-consolidated.

Saanich Inlet, British Columbia Core 1, (from Nissenbaum et al., *1972). All constituents (except Org. C) are expressed as mmol kg^{-1} of pore water. C:N:P ratios are on an atomic basis, and refer only to CO_2, NH_4, and PO_4.*

Depth (cm)	ΣCO_2	NH_4	PO_4	SO_4	C:N:P	C:N	dissolved Org.C (mg/kg)	$\delta^{13}C$ (‰) Org.C	$\delta^{13}C$ (‰) CO_2
0–15	2·7	0·15	0·042	22·9	64:3·6:1	17·8	60	−21·4	−11·2
40–50	13·6	0·53	0·035	11·5	390:15·2:1	25·6	58	—	−37·1
85–100	23·7	2·05	0·155	0·3	152:13·2:1	11·7	78	—	−11·3
135–150	29·7	3·50	0·073	0·7	407:48:1	8·4	116	—	+3·3
175–185	36·2	4·26	0·122	0·3	297;35:1	8·5	120	−21·7	+9·4

or

$$C_2H_5OH + H_2O \xrightarrow{(Methanobacillus\ omelianskii)} CH_3COOH + 2H_2 \quad (10)$$

$$4H_2 + CO_2 \xrightarrow{(Methanobacillus\ omelianskii)} CH_4 + 2H_2O \quad (11)$$

Although the details of methane synthesis are apparently not well known it is recognized that it requires strictly anaerobic conditions and a well-buffered system. It would thus be favoured by the presence of solid calcium carbonate. The occurrence of this reaction may provide an alternative explanation for the loss of alkalinity (bicarbonate) in highly reducing, gas-rich sediments. Hydrogen used in the synthesis is quickly utilized by methane synthesizing organisms and may therefore, be difficult to detect. Many of the reports of the presence of hydrogen in entrained gases of oilfield waters may be erroneous because the gas may have been generated by the reaction of water with the bare steel pipes in the well. However, hydrogen has been detected in marine sediments from areas subject to pollution (Werner and Hyslop, 1968).

A selection of data on the nutrient elements of interstitial waters of recent terrigenous and organic-rich sediments is given in Table 32.10. These data demonstrate the considerable variability of the concentration of these elements. Oxidizing sediments contain far lower concentrations of these nutrients.

32.3.2. MARINE DRILLING

32.3.2.1 *The significance of deep-drilling samples*

On land, millions of holes have been drilled into fluid-bearing strata and a vast number of analyses have been performed on water samples recovered from them. However, there is still considerable controversy about the factors influencing the compositions of these waters. Two principal factors contribute to this uncertainty. Firstly, holes are usually drilled to tap permeable horizons on the continent, and for this reason intervening poorly permeable sediments which are unpromising for petroleum or potable water have largely been ignored; thus, sampling tends to be selective, and does not cover waters from all parts of the sedimentary column. Secondly, in sedimentary basins buried beneath the continents the fluids are strongly affected by the deep penetration of meteoric water and also by other factors such as variations in erosional and tectonic history. These give rise to complex and irregular fluid mixing and migration phenomena.

In contrast, the absence of artesian pressures from most marine sediments (with the exception of some from continental margin areas) results in slow diffusive processes being the chief force acting on the fluids. The importance of diffusion relative to advection in the escape of fluids during sediment consolidation can be evaluated by the "scale height"; this is the ratio of the diffusion coefficient (D, $cm^2 s^{-1}$) to the sedimentation rate (U, $cm s^{-1}$) (Lerman, 1971 and Hurd, 1973). For ocean sediments with average D values of $1 \times 10^{-6} cm^2 s^{-1}$ (or perhaps somewhat greater) and sedimentation rates ranging from 0·1 to 10 cm per 1000 years, scale heights of 30–3000 m would be obtained. This means that except for the most rapidly accumulated sediments (or those deposited under certain unusual conditions), the movement of the bulk fluids expelled from the sediment during consolidation will have a much smaller influence on the composition of interstitial waters than will ionic and molecular diffusion.

Thus, marine drilling programmes provide an opportunity to trace continuously changes in the composition of interstitial waters occurring down the sedimentary column from recently-buried back to fossil sediments. It has been found that the marine sedimentary column includes some "oilfield brines" analogous to those found on the continent. Such drilling programmes throw light on some important geochemical questions, such as the reactions occurring at the basement rock-sediment interface, the long-distance migration of fluids and ions, palaeosalinities, and the organic reactions which take place in deep-buried sediments. The distribution of the sites which have been investigated during the Deep Sea Drilling Project is shown in Fig. 32.5. Some of the chief geochemical results of marine drilling programmes have been reviewed by Manheim and Sayles (1974), and the more important of these are summarized below.

32.3.2.2. *Caspian boring*

One of the first marine borings to be extensively investigated was a 1100 m drillhole sunk in very shallow water near Baku, off the central southwest coast of the Caspian Sea. The core obtained was studied for a variety of parameters, including interstitial water composition, physical properties, petrology, sediment composition, exchange capacity and diffusional permeability (see for example, Strakhov, 1965). Curiously, this investigation has been almost ignored in the Western literature and for this reason some of the interesting results obtained are given below. Pore solutions at a depth of 3 m in the sediment resemble present Caspian Sea water, with the exception that the alkalinity is higher. Below this depth, the bicarbonate concentration is more or less constant at 1 meq l^{-1}, whereas all other major constituents, excluding K, increase with depth. Thus, the chloride concentration increases from about 5 g l^{-1} to a maximum of 85 g l^{-1} (total salinity $\sim$139 g l^{-1}) at a depth of about 600 m (Pushkina, 1963, 1965). This increased salt content is probably the result of contamination from evaporites. Below this depth, salinity and chlorinity decrease sharply at the upper boundary of the strata corresponding to the Pliocene "Productive" beds. Oil exploration in adjoining parts of Azerbaidzhan (USSR) has shown that the less saline waters found in the

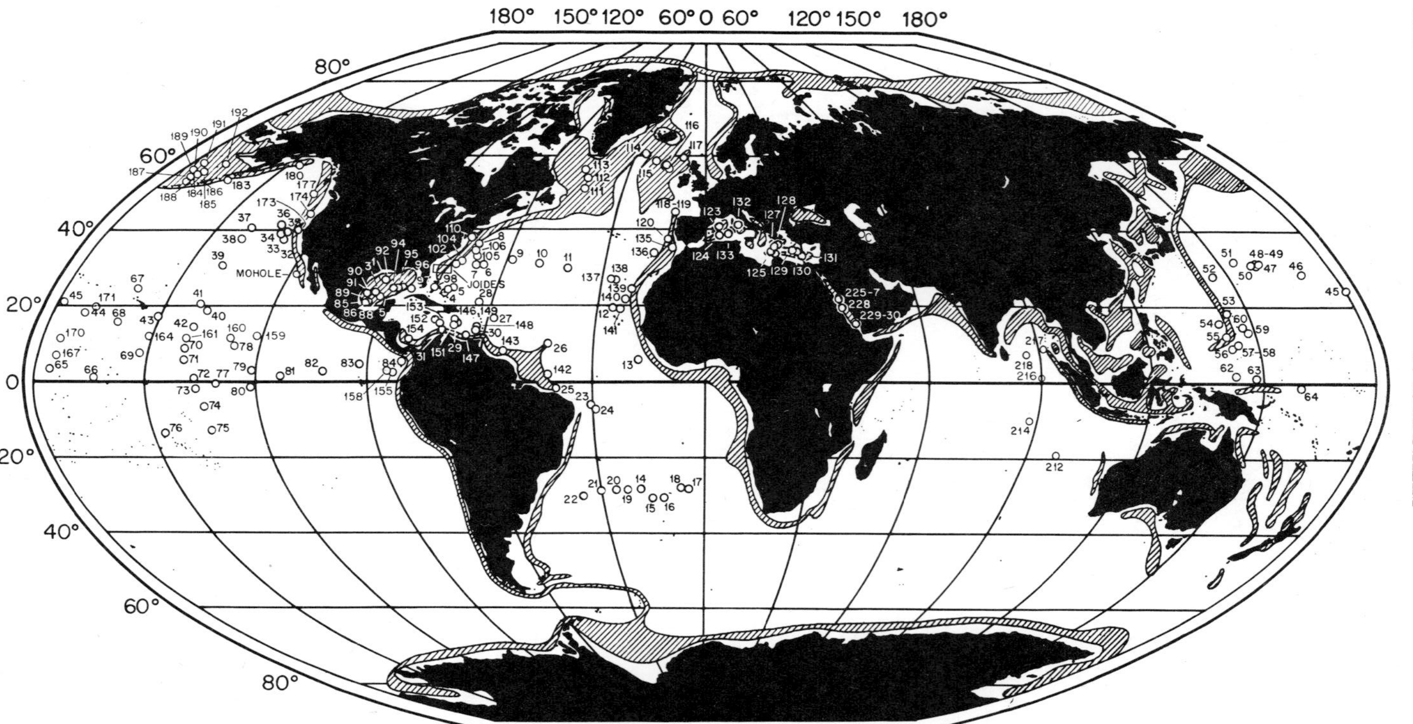

FIG. 32.5. Distribution of DSDP sites at which studies of interstitial waters have been carried out (modified from Manheim and Sayles, 1974). Shaded areas are those subject to rapid geosynclinal accumulation ($\geq$ 3 cm/1000 years) of terrigeneous, or partly terrigenous, sediments. Further stations along the S.E. Atlantic coastal margin and in the northern Gulf of Mexico are indicated in Figs. 32.6 and 32.7 respectively.

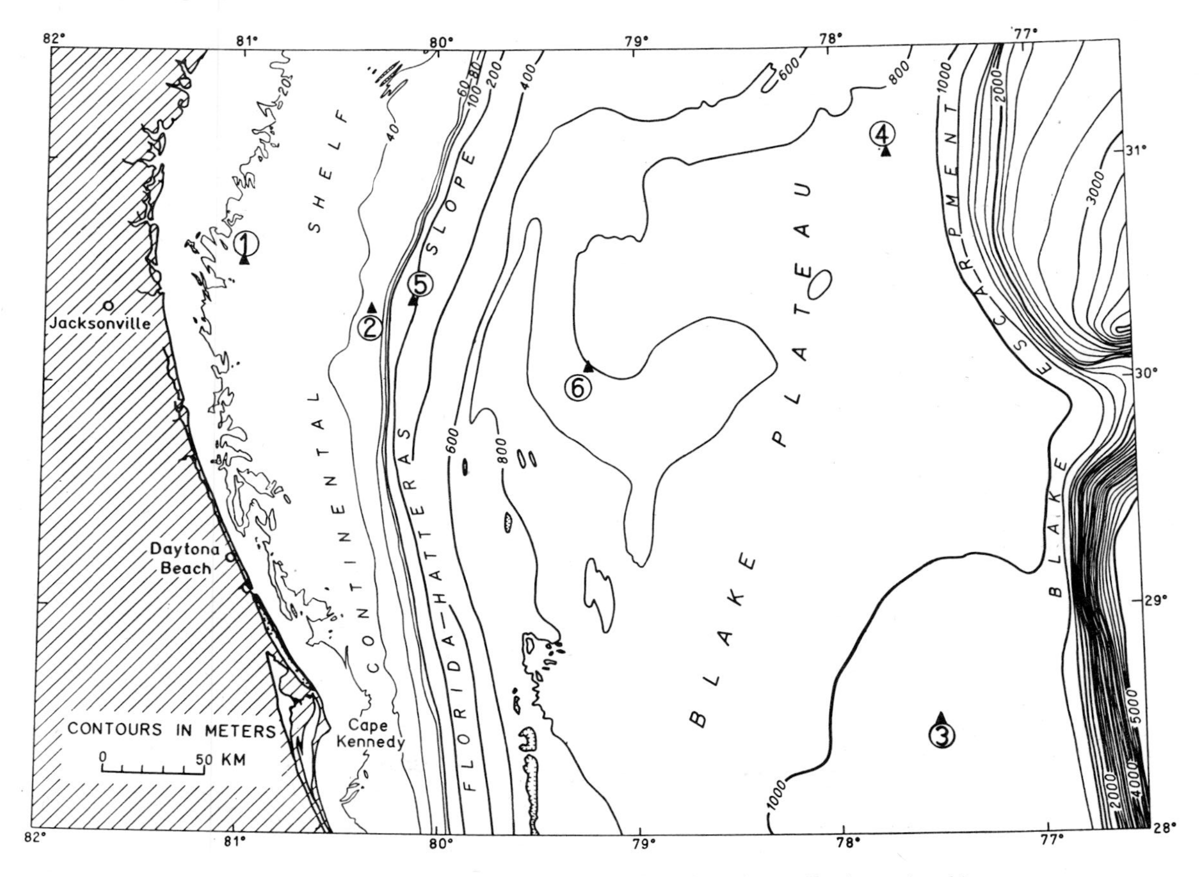

FIG. 32.6. Distribution of JOIDES sites off Florida and South Carolina (reproduced by courtesy of J. S. Schlee).

sedimentary column in this area are associated with the permeable lenses found with highly clayey, undercompacted sediments in which they occur at high pressures (70–80 atm). Sukhrev and Krumboldt (1962) consider that this decreased salinity results from the influence of meteoric ground water. However, analogy with similar phenomena in deep strata along the northern Gulf of Mexico (Schmidt, 1973) would suggest that the less saline water in the Caspian strata deeper than 1000 m may be partly the result of de-watering of clays and the transfer of electrolyte-deficient waters to mobile pore fluids.

32.3.2.3. *Joint Oceanographic Institutions' Deep Earth Sampling Program (JOIDES)*

Analyses of material from the first drilling on the deep ocean floor, the Experimental Mohole near Guadalupe Island, off Baja California, Mexico (Fig. 32.5), showed rather surprisingly that there were no significant changes in interstitial chlorinity or salinity down to a depth of 138 m within the sediment (Rittenberg *et al.*, 1963; Siever *et al.*, 1965). The super-deep drilling program was abandoned mainly because of burgeoning costs. The next initiative which was taken by the JOIDES Committee was the drilling of relatively shallow holes from dynamically positioned (anchorless) drilling ships. Six sites were drilled in 1965 on the shelf and continental slope off Jacksonville, Florida (see Fig. 32.6) by the dynamically positioned vessel, *Caldrill.* The Experimental Mohole was drilled by a dynamically positioned barge which maintained its position using four outboard motors responding to positioning signals from a sonic beacon at depth. However, all subsequent work has been carried out by fully mobile vessels. The *Caldrill* and the Shell Oil Co. vessel, *Eureka* can only drill where the water is less than 2000 m deep, whereas the larger DSDP vessel, *Glomar Challenger*, can work in 8000 m of water.

The JOIDES interstitial water studies gave results very different from the Experimental Mohole. At JOIDES site No. 1, more than 40 km seaward of Jacksonville, Florida, fresh water started to flow from the drill pipe from strata at about 130 m below the sea floor. The interstitial waters (Manheim, 1967) at both J–1 and J–2 (the latter more than 100 km seaward of the coast) were either fresh or brackish throughout the full extent of the drilling (i.e. to a core depth of ca. 300 m). This startling observation was attributed in part to the influence of submarine aquifers which carry fresh water seaward from the land. In contrast, on the Blake Plateau at water depths between 300 m and 1000 m increased chlorinities and salinities were found in the interstitial waters, possibly as a result of diffusion from deeper Lower Cretaceous evaporitic strata analogous to those penetrated in a drilling in Florida (Manheim and Horn, 1968).

Similarly, brackish interstitial waters have also been found adjacent to continental areas in the Caspian Sea (Kudelin *et al.*, 1971), off Panama (Presley *et al.*, 1973), off the coast of the Gulf of Mexico (especially along Mississippi and Alabama (Lang and Newcome, 1964)), off Southern California (Greene, 1970) and in other coastal areas. It is evident that submarine discharge was far more prevalent during the Pleistocene glacial maximum, when hydraulic gradients were intensified by the drop in sea level, and fresh water may have discharged at the ocean floor at depths of several thousand metres.

32.3.2.4. *Eureka drillings in the Gulf of Mexico*

A large number of core holes were drilled on the continental slope of the Gulf of Mexico by the Shell Oil Co. over the period 1965–1967 (Lehner, 1968); maximum penetration into the sediment column was about 300 m. One of the major results of the analyses of the interstitial waters from them was the discovery that the pore waters from holes drilled near diapiric structures showed systematic increases in chlorinity and salinity with depth (Manheim and Bischoff, 1969, Fig. 32.7). Some of the drillings at sites on diapirs actually

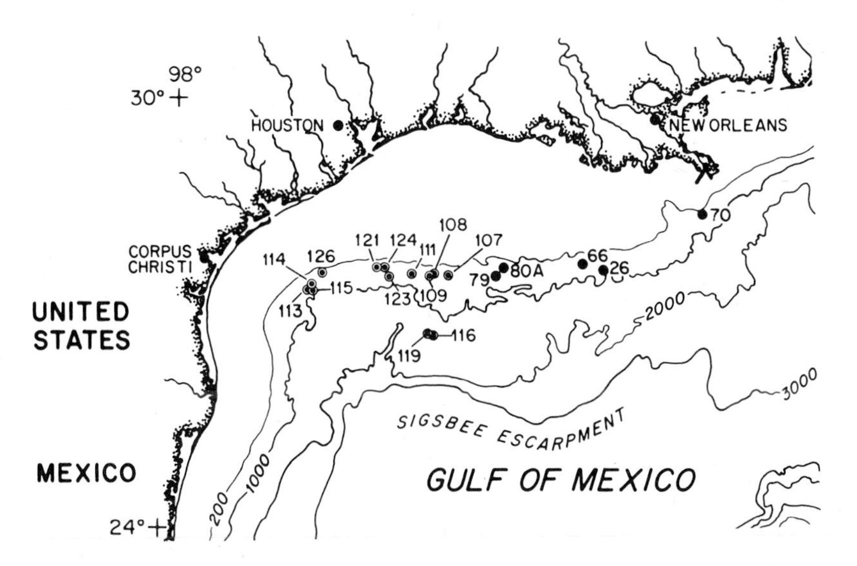

FIG. 32.7. Distribution of Eureka sites in the northern Gulf of Mexico (pore water investigations only). From Manheim and Bischoff (1969) and unpublished data.

penetrated salt deposits. Interstitial waters from holes away from the diapiric structures showed little change in most conservative properties (e.g. chloride and total salinity) with depth. However, loss of sulphate and other components arising from diagenetic reactions with the enclosing sediments did occur. Thus, it is likely that salt diffusion proceeded outwards and upwards through the clayey sediments from the saturated interstitial brine zone adjacent to the evaporite deposits. The rate of this diffusion which has been estimated for chloride ion to be 1 to 4 × 10^{-6} $cm^2\,s^{-1}$, could have added 10 times as much sodium and chloride to the pore fluids above the salt diapirs as was initially present. Thus, caprock can be formed not only by the leaching of salt plugs in the ground water zone, but also by pure sea water permeating through sediments.

The relationships between the chlorinities of interstitial waters and those of salt bodies are shown schematically in Fig. 32.8. In this example a salt stock

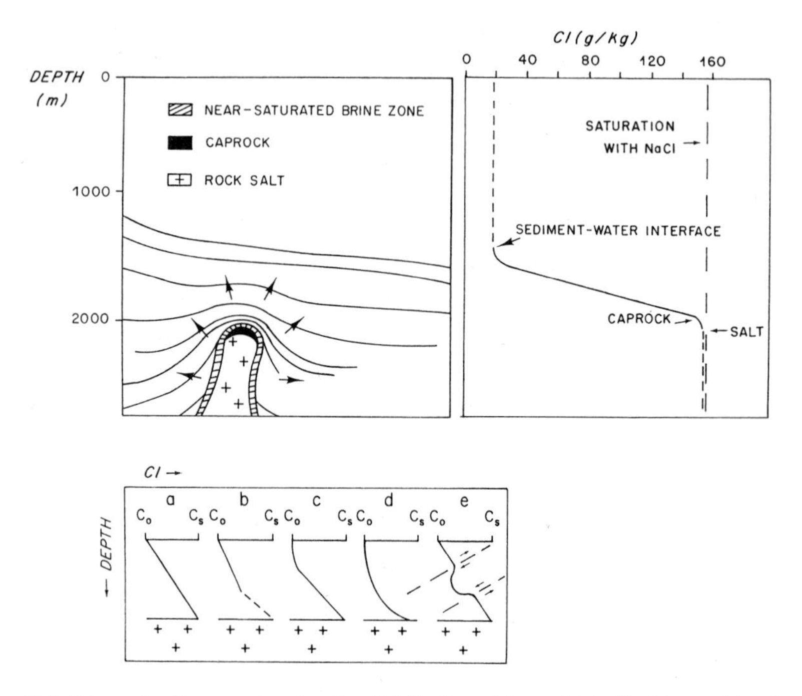

FIG. 32.8. Schematic diagram showing the distribution of interstitial chloride resulting from diffusion of salt from a diapir. C_0 and C_s denote the concentrations of chloride in the bottom water (normal sea water) and in the saturated NaCl brine respectively. The lower figure depicts interstitial chloride gradients which may be set up under a variety of different conditions (see text).

is surrounded by a zone of NaCl-saturated brine. The concentration gradient set up between the brine and the sea water-saturated pore fluids caused molecular diffusion of sodium and chloride ions in the direction of lower concentrations. Ultimately as shown in the diagram, a steady-state, or near steady-state distribution was obtained, in which the salt lost to the overlying water column was just balanced by salt which dissolved and was transported from the diapir. If the diffusion coefficient for chloride ion in the sediment remained constant with depth (case a) under steady-state conditions there will be a linear relationship between the interstitial chloride concentration and depth over the range from sea water concentration (19 g kg^{-1}) to saturation with respect to NaCl (about 160 g kg^{-1}). Case b shows the distribution which would have been attained if the diffusion coefficient decreased sharply in the lower section of the site. Case c illustrates the distribution which would have resulted either where there was a more rapid movement of Cl near the sediment-water interface, or when a marked increase in the sedimentation rate moved the interface upwards. Case d depicts the curve expected if the diffusion constant remained the same with depth, but if a steady-state distribution had not yet been attained. The influence of faulting and consequent disruption of the diffusion profile is illustrated in case e.

The fact that the concentrations of some constituents (e.g. Mg, K, Br and B) characteristic of later stages of evaporite formation increase in the interstitial waters of some bore holes suggests that later stage evaporites are being leached. Thus, potassium concentrations as high as 6·2 g kg^{-1} have been found in the interstitial waters from a drilling at a site in 580 m of water off the Louisiana coast.

32.3.2.5. *Deep Sea Drilling Project* (*DSDP*)

The DSDP ship, the *Glomar Challenger*, is capable of drilling in all of the world's ocean basins and from the outset pore waters were recovered from the cores collected. The interstitial water collections commenced with Leg 1 in Fig. 32.3, continued through Leg 19 and included Legs 22 and 23. Subsequent legs have included a limited programme of routine interstitial water sampling, full investigations being limited to special research topics. The data from the first fifteen legs is available in the Initial Reports of the DSDP and has been summarized by Manheim and Sayles (1974). The locations of drilling sites updated to include Legs 22 and 23, are plotted in Fig. 32.5; the Experimental Mohole and JOIDES drill sites are also indicated in this figure.

The first major finding to emerge from the DSDP was a confirmation of earlier Experimental Mohole observations that over most of the World Ocean, the chlorinity and salinity of the interstitial waters changed very little with depth and were similar to within ca. 1% to those of the overlying sea water. However, Palaeocene sediment trapped beneath a 30 m thick fine-grained basalt body (Site 214, Leg 22, Ninetyeast Ridge of the Indian Ocean) had a chlorinity of 21.3 g kg^{-1}, in contrast to that of 19·6 g kg^{-1} in the pore water of the sediments overlying the basalt. There is little evidence to suggest that the water of early Tertiary and later oceans differed markedly in salinity and chlorinity from that at the present day (see for example, Chapter 5). The above example is not an exception to this as the increased salinity is believed to have been the result of diagenetic changes which occurred beneath the impermeable layer of basalt. In slowly deposited sediments, such as the red clays of the Pacific Ocean, even the other ionic constituents hardly differ in proportion from those in sea water, provided that compensation is made for the effect of temperature on the extraction (see above). For the pore waters of such sediments this constancy extends to sediments as old as the Late Jurassic, and is to a large extent the result of diffusive mixing with surficial sea water.

The increase in salinity of the interstitial waters of sediments overlying the evaporites found during the Eureka Drillings in the Gulf of Mexico (see Section 32.3.2.4) has also been observed for the pore waters of sediments from the deep Gulf of Mexico (Manheim and Sayles, 1970), from off West Africa and Brazil (Waterman *et al.*, 1972), from the Mediterranean Sea (Sayles *et al.*, 1972), from the Red Sea (Manheim *et al.*, 1974), from the Gulf of Aden (E. Bune, and D. D. Kinsman, personal communications) and from the Timor Trough, north of Australia (J. Heirtzler, personal communication).

Interstitial chlorinity profiles from the Mediterranean Sea are depicted in in Fig. 32.9 (Sayles *et al.*, 1972), from which it can be seen that at each site in the Mediterranean this parameter increases with depth. This is presumably the result of the diffusion of NaCl from an evaporitic stratum of Late Miocene age. In contrast, the interstitial waters from Site 135, westwards of the Straits of Gibraltar show no significant change of chlorinity with depth. In addition to chloride and sodium, other chemical components also migrate outwards from evaporite bodies at rates which are controlled by several factors in addition to the rates of diffusion. For example, although anhydrite or gypsum nearly always accompany rock salt in evaporites, the dissolution of these minerals, and hence the transport of Ca^{2+} and SO_4^{2-}, is much slower than are those of Na^+ and Cl^-. Furthermore, the calcium and sulphate may both be subsequently removed from solution; the former by precipitation as $CaCO_3$, and the latter by reduction to sulphide in the presence of abundant organic matter. In the sediments of the Challenger Knoll in the central Gulf of Mexico (DSDP Site 2) petroleum evidently acted as an oxidizable substrate for the bacterial reduction of anhydrite-gypsum, and native sulphur was probably formed as an intermediate product. Feeley and Kulp, (1957) have postulated that sulphate and H_2S could react to form native sulphur directly. However, Manheim and Sayles (1970) consider this to be thermodynamically improbable except in the presence of large concentrations of H_2S, and this conclusion had been confirmed experimentally by Davis *et al.* (1970). If sulphate was not the oxidizing agent it would have been necessary to postulate that oxygen was able to penetrate through to the caprock area of the

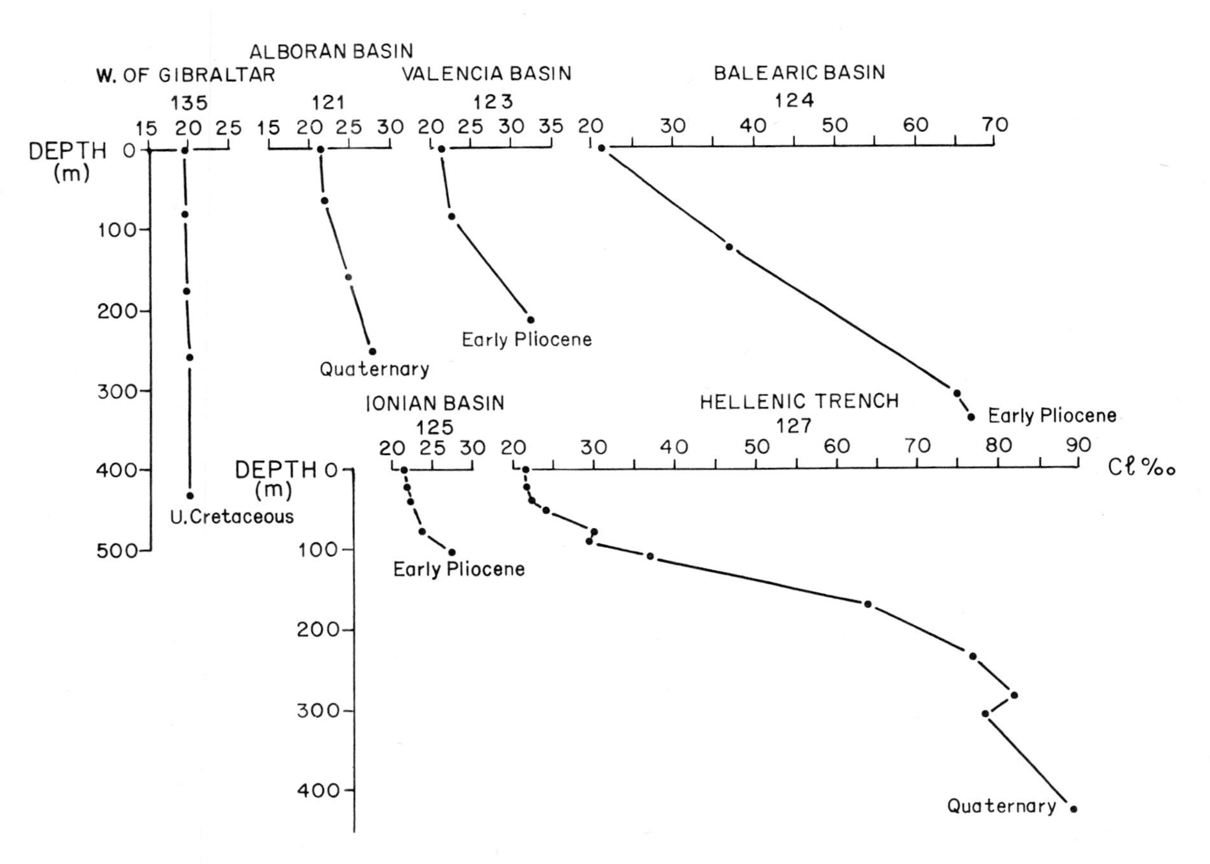

FIG. 32.9. Distribution of interstitial chloride concentrations (g kg^{-1}) with depth at sites in the Mediterranean (from Sayles *et al.*, 1972). Geological ages refer to the oldest strata sampled.

Challenger Dome and there react with H_2S originating from sulphate reduction to form native sulphur. The exact nature of the sulphur-forming reaction is uncertain, but since the Challenger Knoll lies in the centre of the Gulf of Mexico in 3500 m of water, the existence of sulphur in its sediments provides the first definite proof that native sulphur can be formed at depth in sea water-permeated sediments, without the necessity for oxygenated meteoric ground waters to act as an intermediary.

Constituents other than Na^+, Ca^{2+}, Cl^- and SO_4^{2-} which are sometimes enriched in the pore waters overlying evaporites include Mg^{2+}, K^+, B, Br^- and trace metals, and an example of the composition of such a pore fluid from the Red Sea is given in Table 32.11, (Manheim *et al.*, 1974). The sharp drop-off in magnesium content above the halite zone indicates that, although the leachates from halite, polyhalite and possibly tachyhydrite* – bearing rocks are rich in magnesium, this element is quickly removed from the solution during diffusion through the overlying evaporitic rocks, carbonates and shales. In this context, it is relevant to note that on Leg 23 (Red Sea) ordered dolomites were found to be prevalent immediately above the evaporitic strata, whereas calcium-rich dolomites occur higher in the section. This suggests that a considerable part of the magnesium has been removed in the form of dolomite. In a similar manner, other constituents are preferentially removed by reaction with the strata overlying evaporites. Only the concentrations of poorly reactive ions such as Cl^- and Br^-, and to a lesser extent Li^+ and Na^+, in the pore fluids give an undistorted picture of the effects of evaporite dissolution and the diffusion of their ionic components. However, these components may also reach overlying strata at different rates owing to differences in their diffusional mobilities (see below).

32.4. DIAGENESIS

The net gain or loss of components as a result of sediment-pore fluid reactions can be assessed from a consideration of the average composition of the interstitial waters of the various types of sediments from the World Oceans (Table 32.12A). The differences between these averages provide an indication of the direction of such reactions. However, the absence of experimentally significant differences does not necessarily imply that such reactions are not proceeding because diffusive movements in very slowly deposited sediments may be sufficient to make the anomalies insignificant. The values shown in the table are, in general, those for the interstitial waters from unconsolidated sediments. Much larger differences from present day sea water composition are exhibited by the interstitial waters of the more consolidated ones. How-

* $CaMg_2Cl_6 \cdot 12H_2O$

TABLE 32.11

Major constituents of interstitial water in Red Sea sediments at 21° 19·9′ N, 38° 08·0′ E, depth 1795 m. Data from Manheim et al. *(1974)** *(Concentrations are expressed as g kg^{-1}, H_2O refers to total water content of bulk sediment, determined by loss on heating at 110–120° C)*

Depth (m)	Description	Na^+	K^+	Ca^{2+}	Mg^{2+}	Sr^{2+}	Cl^-	SO_4^{2-} †	HCO_3^-	Sum	H_2O	pH
Surface	Red Sea water	11·8	0·43	0·46	1·42	0·009	21·4	3·06	0·15	38·8	—	8·3
46	Green-gray clayey chalk	14·7	0·40	0·88	1·28	0·018	25·8	3·39	0·20	46·5	29	7·2
82	Clayey-silty chalk	28·7	0·22	0·98	0·84	0·027	46·1	3·57	0·15	80·6	19	7·0
136	Gray, silty nanno chalk	61·3	0·38	2·32	0·92	0·046	99·7	2·43	0·04	167·2	15	6·5
167	Gray, silty nanno chalk	80·2	0·32	2·44	0·86	0·034	129·7	1·55	$<0·05$	215·1	—	6·5
186	Gray, silty nanno chalk	93·6	0·50	1·81	0·67	—	149·0	1·37	$<0·05$	247·0	19	—
282	Dark shale and anhydrite	71·0	1·29	4·90	12·6	0·099	157·1	1·90	$<0·05$	250·7	23	6·5
350	Dark shale between halite	67·4	0·93	6·70	16·8	—	164·8	1·20	$<0·05$	257·6	14	6·2

* The core penetrated the uppermost Miocene evaporites, including halite, at depths of 180 m and below.

† Sulphate was determined from equilibrium calculations using the remaining ionic constituents and the WATEQ computer program (Truesdell and Jones, 1973). The solubility of anhydrite provides the constraining condition.

TABLE 32.12A

*Average concentrations of major, and some minor components, of interstitial waters of sediments recovered during the Deep Sea Drilling Project (data from Initial Reports, Vol. 1–19); excluding evaporite-influenced sites)**

Component	Terrigenous† Mean	N	SD	Carbonate Mean	N	SD	Pelagic clay Mean	N	SD	Nannofossil ooze Mean	N	SD
	a			a		a	a		a	a		a
Na	10·8	9	0·6	10·9	20	0·2	10·8	15	0·20	10·8	12	0·1
K	0·30	9	0·07	0·39	20	0·03	0·38	15	0·02	0·38	12	0·02
Ca	0·40	9	0·20	0·48	20	0·16	0·42	15	0·05	0·41	12	0·06
Mg	1·08	9	0·09	1·20	20	0·09	1·25	15	0·06	1·24	12	0·11
Cl	19·4	7	0·3	19·5	19	0·2	19·5	15	0·20	19·5	12	0·2
SO_4	1·05	9	0·77	2·43	20	0·20	2·45	15	0·48	2·36	12	0·06
HCO_3	0·45	9	0·27	0·20	20	0·10	0·20	14	0·08	0·20	12	0·10
	b		b	b		b	b		b	b		b
Sr	8·7	8	3·0	15·7	19	6·0	11·1	15	5·6	15·9	10	5·9
Br	72	1	0	67	1	0	66	5	2	67	1	0
B	4·1	6	1·4	4·7	8	0·7	5·7	10	1·2	6·1	3	2·3
Ba	2·4	9	4·5	0·15	14	0·08	0·15	14	0·08	0·3	9	0· 4
Mn	—	—	—	—	—	—	2·4	3	0·7	0·5	1	0
Li	0·09	3	0·05	0·22	6	0·05	0·22	8	0·03	0·24	8	0·06

N is the number of drill sites
SD is the standard deviation.
a = g kg^{-1}
b = mg kg^{-1}

* Correction has been made for the "temperature of extraction effect" for potassium and magnesium (see Section 32.2).
† Clayey sediments having accumulation rates > 3 $cm/10^3$ years.

TABLE 32.12B

Average concentrations of trace elements in interstitial waters of sediments recovered during the Deep Sea Drilling Project (data from Initial Reports Vols. 5, 9, 15 and 20), excluding evaporite-influenced sites.[*]

Component	Terrigenous			Carbonate			Pelagic clay			Nannofossil ooze		
ppb	Mean	N	SD	Mean	N	SD	Mean	N	SD	Mean	N	SD
Fe	10	3	8	78	7	65	17	6	14	46	9	64
Co	—	—	—	1·7	7	0·8	1·7	3	0·8	2·2	9	2·1
Ni	26	3	15	25	7	14	19	6	9	18	9	9·6
Cu	11	3	7	26	7	23	12	6	9	12	9	5
Mn	608	4	708	748	5	915	2970	4	1776	931	7	1166
Zn	160	3	49	210	1	—	258	3	129	185	2	5
Li(ppm)	0·18	3	0·061	—	—	—	0·32	2	0·12	—	—	—
Si(ppm)	4·7	3	1·0	22	7	9·1	8·8	6	5·0	21·7	9	9·2
B(ppm)	3·5	3	0·6	—	—	—	5·7	2	0·3	—	—	—
PO_4(ppm)[†]	—	—	—	1·4	2	0·5	—	—	—	1·4	5	2·1

N is the number of drill sites
SD is the standard deviation.

* Lithological categories are somewhat different from those used in Table 12a as it was necessary to combine samples from several depths to obtain sufficient pore water for analysis.

† Excluding one very low value.

ever, the values given in the table are probably not truly representative of World Ocean interstitial waters, since the sites drilled during the DSDP were chosen in an attempt to reach basement, and hence are mainly those with a thin sedimentary cover.

Average concentrations of the major components of the interstitial waters of the upper 100 m of sediment at 56 sites (selected to represent "end member" sediment-environmental types are shown in Table 32.12a). Attempts have been made to correct for the effect of extraction temperature on the concentrations of K and Mg (see legend of Table 32.12a). However, no correction has been applied for other elements, such as Si, Sr and B which may be affected by temperature variations. It is probable that the average bicarbonate concentration shown for the interstitial waters of terrigenous sediments is too low as a result of loss of CO_2 during storage.

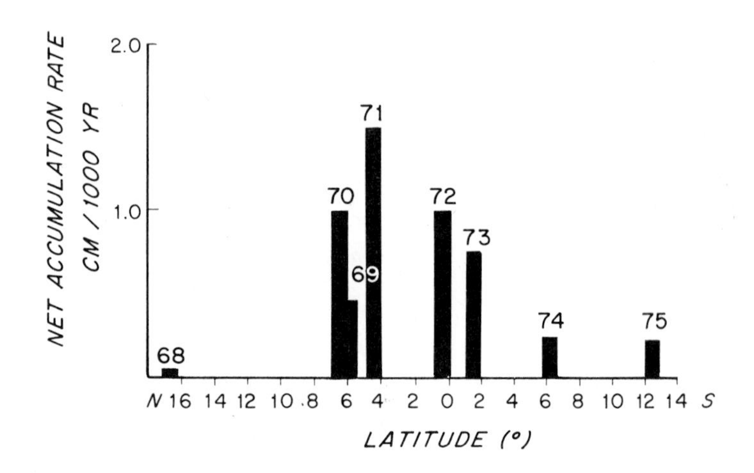

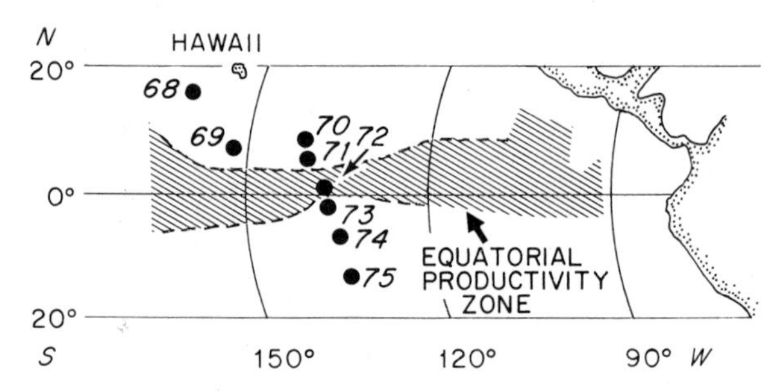

FIG. 32.10a Map showing the location of coring stations in, and around, the equatorial productivity zone in the Central eastern Pacific (DSDP Leg 8), (see Fig. 32.10b) together with the net sediment accumulation rates.

Care must be taken in interpreting the trace element data in Table 32.12b because such elements are particularly susceptible to change during core recovery (e.g. by admission of oxygen to the system and change in pressure and temperature, and there is also a risk of contamination during sample manipulation and fluid extraction). This problem is particularly severe for iron because on oxidation to Fe^{3+} it tends to precipitate, and in so doing may affect the concentrations of other trace elements by co-precipitating them.

In the interstitial waters of the terrigenous deposits Na^+ is depleted by ~1%, K^+ by ~23%, Mg^{2+} by ~16% and SO_4^{2-} by ~62% with respect to their concentrations in average sea water. Indeed, Mg^{2+} and K^+ are almost totally depleted in the pore waters of some clayey terrigenous deposits. Real enrichments with K^+ and Mg^{2+}, relative to their sea water concentrations, are found only where it is known, or suspected, that diffusion from evaporitic deposits has occurred, (see however, Milliman and Müller, 1973).

Calcium is frequently strongly enriched in pore waters of calcareous deposits, but this enrichment is not invariably caused by dissolution of calcium carbonate (see below). Strontium is often markedly enriched in rapidly deposited carbonate-rich sediments (Figs. 32.10a, b). Manheim *et al* (1971)

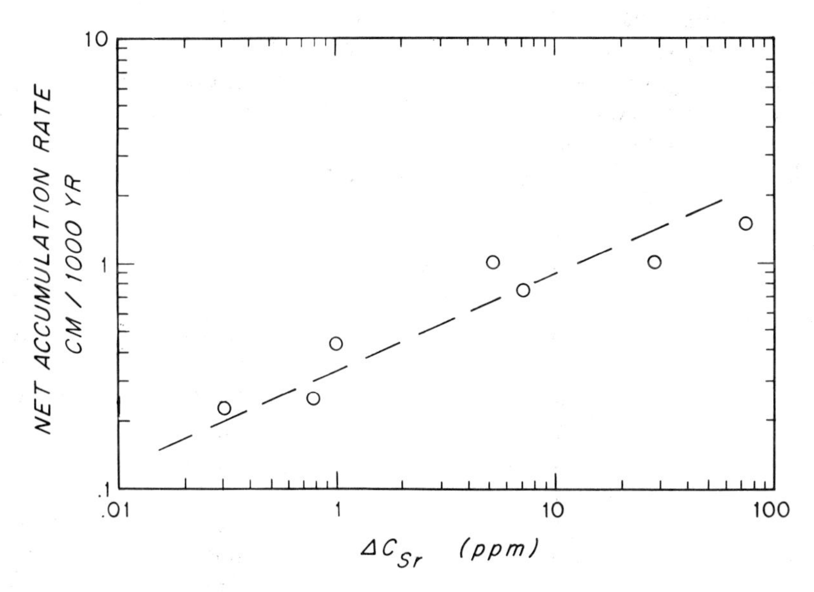

FIG. 32.10b Diagram illustrating the relationship between the excess strontium concentration (ΔC_{Sr}), in the pore waters at sites shown in Fig. 10a, and the net sediment accumulation rate. ΔC_{Sr} is the difference between the mean interstitial Sr concentration at each drill site and the concentration of the element in sea water (8 mg kg^{-1}).

have concluded that most strontium enrichments (other than those associated with evaporite leaching) occur when coccolith-foram-calcite (0·15% Sr) re-crystallizes to a calcite which is poorer in Sr (0·05%). The strontium enrichment resulting from this process can give concentrations 15 times greater than those in normal sea water. In contrast to the pore waters of rapidly-deposited carbonate sediments (in which re-crystallization and diagenetic effects can be correlated with marked changes in the concentrations of Ca and Sr in the pore waters) those of terrigenous deposits generally show little change in Sr. The pore waters of such deposits are much depleted in B, and its apparent enrichment in the pore waters of pelagic and nannofossil oozes is only the result of failure to correct for temperature-extraction effects. Silicon (4–5 mg kg^{-1}) shows a moderate enrichment relative to the bottom waters. The concentration of silicon in biogenic deposits (uncorrected for temperature of extraction effects) is fairly constant at about 30 mg kg^{-1} as a result of the presence of amorphous silica. Lithium is depleted in the pore waters of terrigenous deposits by ~50%, whereas it is slightly enriched in those of other sediment types. Barium is strongly enriched in the pore waters of terrigenous deposits owing mainly to the depletion of sulphate. Although the solubility of barium sulphate is the main factor controlling the concentration of barium in sea water, and in some interstitial waters, much higher barium concentrations can occur in the presence of organic complexing agents (Desai *et al.*, 1969; Li *et al.*, 1973; Michard *et al.*, 1974). The influence of complexing agents may indeed be part of the reason why the Ba concentration of some interstitial waters may exceed 10 ppm. The mutual interaction between the sulphate and barium concentration gradients in the interstitial waters of sediment cores may be the cause of the small barite crystals often found in oceanic sediments. However, discrete barite concretions have not yet been found in the oxidized surface layers of stagnant sediments even though barium concentrations in the sediments can exceed 1%. This contrasts with the behaviour of manganese, which occurs at a concentration in the pore water similar to that of barium, but which gives rise to discrete manganese-iron concretions in lake, near-shore and shallow water deposits (Manheim 1965). This apparent anomaly can be explained by the fact that $BaSO_4$ precipitation is rapid and is hardly influenced by kinetic factors. In contrast, oxidation and hence precipitation of Mn^{2+} occurs preferentially at surfaces at which substantial kinetic barriers are overcome by catalysis. For these reasons, $BaSO_4$ is intimately intermingled with sediments that are undergoing continuous deposition, whereas MnO_2 phases can (but do not invariably) occur discretely. The large barite concretions occurring on the sea floor (see for example, Revelle and Emery, 1951) are probably relicts formed under unusual environmental conditions, such as those of partly desiccated lagoons.

Silicate diagenetic processes at depth in oceanic sediments are well illustra-

TABLE 32.13

Variations in the concentrations of interstitial water components with depth at 3 sites on Ninetyeast ridge, Indian Ocean. Data for sites 214 and 217 of the DSDP from Manheim et al. *(1973); data for site 216 from Presley (1973). Major components as g kg*$^{-1}$

Site	Depth (m)	Age	Description of sediment	Ca	Na	Cl	Si(ppm)
214	8	Pleistocene	White foram-nannofossil ooze	0·44	10·8	19·4	9·8
	364	Palaeocene (?)	Gray silty sand	1·19	10·7	20·1	7·1
	384	Palaeocene (?)	Gray silty sand	1·33	10·7	20·3	6·5
	400	Palaeocene (?)	Grayish lignite	1·34	10·6	20·2	4·2
	424	Palaeocene (?)	Dk. gray clay pebble congl.	1·56	10·0	19·6(?)	1·9
	486	Palaeocene (?)	Grayish-blue silty clay, hard	3·19	9·6	21·3	2·4
216	9	Pleistocene	Yellow-gray nanno-foram ooze	0·46	11·2	19·6	15
	166	Early Plio.	Blue-white foram-nanno ooze	2·50	11·1	20·1	26
	250	Oligocene	White nannofossil chalk	3·47	9·6	21·5	26
	354	Maestricht. (U. Cretaceous)	Blue-green clayey, glauconite-carbonate rock	5·53	8·7	22·1	3·0
217	7	Pleistocene	Greenish foram-clay-nanno ooze	0·41	10·0	19·3	12
	73	Early Mio.	Clayey foram-nannofossil ooze	0·73	10·6	19·6	16
	155	M. Miocene	Clayey nannofossil ooze	1·15	10·5	19·9	15
	234	Oligocene	Clayey foram-nannofossil ooze	1·55	10·1	19·9	20
	348	Early Eocene	Radiolarian nannofossil chalk	2·20	9·7	20·8	20
	383	Palaeocene	Foram nannofossil chalk	2·46	9·4	21·1	12
	397	Palaeocene	Foram-nannofossil chalk	2·70	9·0	20·7	13
	454	Maestricht	Nannofossil chalk	3·32	8·6	21·2	3·3
	502	Maestricht	Clayey foram-nanno chalk	4·52	8·1	22·0	2·4

ted by cores from DSDP sites on the Ninetyeast Ridge, in the Indian Ocean (Table 32.13; Fig. 32.11). Not only does the calcium content of the pore waters increase by a factor of 10 over the normal oceanic value, but Na^+, K^+ and Mg^{2+} are markedly depleted. It seems likely that the depletion of the alkalis

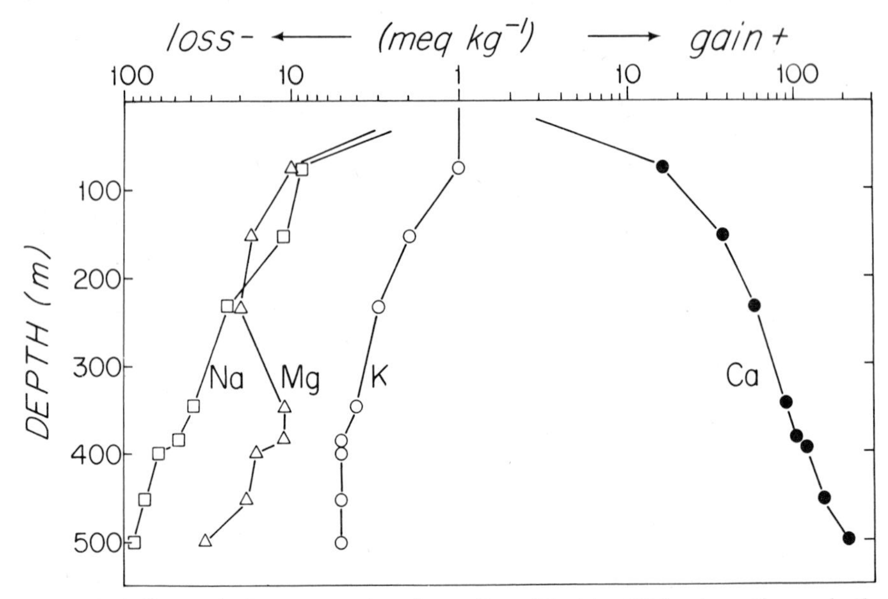

FIG. 32.11. Changes in the concentration of some ions of the interstitial waters with core depth at Site 217, Ninetyeast Ridge, Indian Ocean. Figures give gains or losses of ions relative to those in the pore water of the surficial sediments after correction for the chlorinity differences. The plot involves two "back-to-back" logarithmic scales, thus there is no zero; however changes smaller than 1 meq kg^{-1} are inconsequential and hence the 1 starting point is appropriate.

has occurred as a result of their uptake in authigenic silicates and alumino-silicates.

A study of alkalinity and CO_2 species suggests that the high calcium and low magnesium concentrations result from a dolomitization – magnesianization reaction namely:

$$2CaCO_3 + Mg^{2+} \rightarrow CaMg(CO_3)_2 + Ca^{2+} \quad (12)$$

Confirmation that this reaction does, in fact, occur is provided by the observation that small amounts of dolomite are widespread in DSDP sediment cores. However, the amount of calcium which can be provided by this source is limited by the amount of magnesium available in the pore fluids. Manheim and Sayles (1974) have pointed out that even if Mg had diffused continuously from the overlying sea water it would be unlikely that the sediments could accumulate more than ca. 1% of dolomite through interstitial water reactions, unless additional sources of Mg^{2+} were available.

Authigenic formation of silicates and alumino-silicates of the reverse weathering type is another reaction which may occur in the pore water-sediment system (MacKenzie and Garrels, 1966). In reactions of this type silicon is removed from solution (note the very low interstitial silicon values in the lowest part of the cores from the three sites in Table 32.13) and hydrogen ion is liberated (see equation 13).

Hydrogen ions liberated by exchange with sodium ions leads to an attack on $CaCO_3$, and an increase in the concentration of dissolved calcium (equation 14). Reaction (15), which is obtained by combining reactions (13) and (14), depicts the net formation of clinoptilolite (a zeolite widely found in marine sediments) and the release of dissolved Ca^{2+} and CO_2. Other authigenic silicates which could also exchange their hydrogen ions for sodium ions include palygorskite, illite, mixed-layer clays and montmorillonite. In these, as well as reactions to be discussed below, silica is presumed to be supplied as amorphous silica which has originated either biogenically or from volcanic glass.

$$CaCO_3 + 2Na^+ + 2Al_2O_3 + 14SiO_2 + 13H_2O \rightarrow \underset{\text{(clinoptilolite)}}{Na_2CaAl_4Si_{14}O_{36} \cdot 12H_2O} + CO_2 + 2H^+ \quad (13)$$

$$2H^+ + CaCO_3 \rightarrow Ca^{2+} + H_2O + CO_2 \quad (14)$$

$$2CaCO_3 + 2Na^+ + 2Al_2O_3 + 14SiO_2 + 12H_2O \rightarrow Na_2CaAl_4Si_{14}O_{36} \cdot 12H_2O + 2CO_2 + Ca^{2+} \quad (15)$$

One of the difficulties involved in accepting the above reactions as a source of calcium is the disposal of the CO_2 produced from the decomposition of calcium carbonate. In fact, total CO_2 and alkalinity have been found to decrease rather than to increase with depth in the interstitial waters of deeper strata in which diagenesis is prominent (Presley *et al.*, 1970; Takahashi *et al.*, 1973). It is very unlikely that CO_2 has been lost as a result of methane synthesis (eqn. 10 and 11) since this process does not take place in the presence of sulphate, which in fact occurs at concentrations of 10–14 mmol kg^{-1} in the deepest interstitial waters at all the sites listed in Table 32.11*, and also because corresponding quantities of organic matter are required as sources of free hydrogen.

Submarine weathering processes involving basaltic volcanic material yield the following net quantities of oxides, expressing the percentage change in

* Calculations show that the molar product of $CaSO_4$ is $1{\cdot}6 \times 10^{-3}$ quite similar to the apparent solubility product for gypsum, i.e. $K' = 1{\cdot}8 \times 10^{-3}$. This raises the possibility that either (a) a relatively small amount of gypsum could be formed in sediments without the necessity for a preceding evaporitic stage, or (b) sulphate could be removed from interstitial or connate, brines without invoking bacterial sulphate reduction by simply raising the calcium concentration to a sufficiently high level.

mass between weathered and fresh rocks (% wt) (Thompson, 1973):

Oxide	*Loss*
SiO_2	18–22
CaO	4–25
MgO	1–8
Al_2O_3	3–(9)
FeO (total Fe)	3–6
Na_2O	1–2
MnO	0·1
	Gain
K_2O	2–4
H_2O	10–24

However, it seems unlikely that this type of submarine weathering can be responsible for the trends observed for sediment pore waters. For example, pore fluids are depleted in sodium and rarely enriched in magnesium (except in the presence of evaporites), even where no carbonates are present to explain its uptake as dolomite. Further, the weathering reaction would probably be inhibited owing to the high pH if hydrogen ion is not continuously supplied to the system. pH values in the examples given lay in the normal 7–8 range. It is unlikely that the phenomena observed in Table 32.13 and Fig. 32.10b can be explained purely in terms of an exchange capacity shift in favour of sodium on the pre-existing clays (Shishkina, 1972). The amounts of exchangeable cations in the sediments are probably inadequate to account for the amounts of calcium added (Zaitseva, 1966), and a temperature increase with depth would favour the release, not the uptake, of monovalent cations (Sayles *et al.*, 1973).

The distributions shown in Table 32.13 are most readily accounted for by invoking silicate exchange reactions in which calcium-rich silicates are converted to sodium-containing minerals by the incorporation of Na from pore fluids, for example:

$$2Na^+ + \underset{\text{(anorthite)}}{CaAl_2Si_2O_8} + \underset{\text{(amorphous silica)}}{2SiO_2} + H_2O \rightleftharpoons \underset{\text{(analcite)}}{2NaAlSi_2O_6 \cdot H_2O} + Ca^{2+}$$

Although analcite has actually been found in oceanic sediment cores (presumably as an alteration product of basaltic glass), it is probable that the final alteration products will be more complex phases, for example montmorillonite, which, with nontronite, is often found in alteration residues of basalts and volcanic glass at the sediment-basement interface in many DSDP cores. In such clay minerals the sodium has been incorporated exclusively into the (newly formed) exchangeable layers in which it constitutes about 40% of the total ions (Zaitseva, 1966). Magnesium released by submarine weathering of basalt is probably incorporated in part in the montmorillonites and other silicates, such as chlorites, and in part is utilized in dolomitization – magnesianization reactions.

Initially, the addition of calcium to pore waters will tend to lead to the precipitation of calcium carbonate. However, once the carbonate ion has been removed in this way, calcium can continue to accumulate provided that the breakdown of calcium aluminosilicates continues.

The observed increase in the concentration of interstitial chloride with depth at the Ninetyeast sites is consistent with the abstraction of water by the formation of *hydrous* authigenic minerals (Table 32.13). These increases cannot be attributed to leaching of evaporites because the basaltic basement was penetrated at one of the sites (216) without discovery of an evaporite bed. Clearly, therefore, soluble salts can be concentrated in pore fluids by the removal of water molecules during diagenetic reactions. It can be predicted that at very great depths and temperatures dehydration and loss of interlayer water in clays will occur. This effect has been observed in deep cores from the Gulf Coast of the United States the interstitial waters of which had abnormally low salinities presumably resulting from the release of bound water to the pore fluids (Schmidt, 1973).

32.5. The Influence of Sea Floor Sediments on Ocean Waters

When an interstitial water deviates in ionic composition from that of the overlying sea water, diffusion gradients will be set up which will cause the sediment-interstitial water complexes to behave either as sources or sinks for dissolved constituents. Even before 1900 Murray and his H.M.S. *Challenger* colleagues recognized that a substantial proportion of the manganese in ocean waters might have emanated by diffusion from terrigenous muds as a result of post-depositional migration. Dittmar's discovery that the Ca/Cl ratio of ocean water increases with depth suggested to some workers that calcium may emanate from sea floor sediments, and Koczy (1953) has shown that most of the radon in deep ocean waters comes from the sediments. Recently, extensive data from the GEOSECS programs have revealed the presence of excess He^3 in deep ocean water; it is thought that this is primordial helium which has diffused through the sea floor (Jenkins *et al.*, 1972; see also Chapter 8). Nutrients such as nitrate, ammonia, phosphate, and especially silicate, are enriched in the lower parts of strongly stratified water bodies such as the Black Sea and the California offshore basins (Skopintsev, 1962; Bruevich, 1962; Rittenberg *et al.*, 1955; see also Chapter 15). Some workers consider that this enrichment may in part result from release of these nutrients from the sediments. Thus, Fanning and Pilson (1973) have suggested that the excess silicate in anoxic waters of the Cariaco Trench can only have originated from bottom sediments (Fig. 32.12), and Jones and Fanning (1975) have proposed that the silicate in those parts of the eastern Caribbean lying below sill depth may have a similar origin. However, because of the relative speed of mixing processes over the floors of most of the ocean basins recent work (including GEOSECS stations in the Atlantic) has failed to establish the

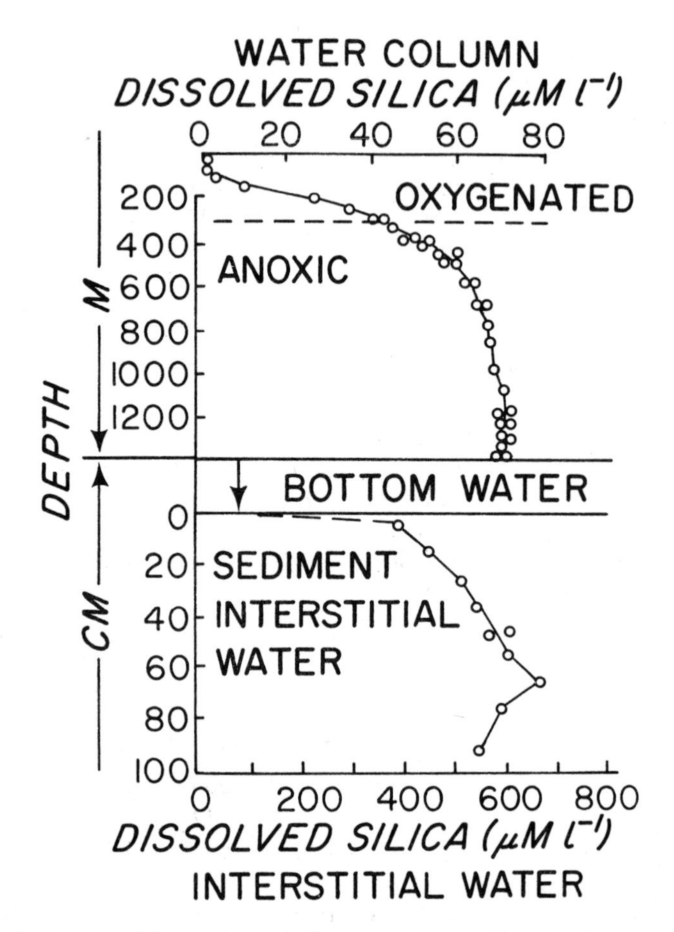

FIG. 32.12. Distribution of dissolved silica in the water column and interstitial waters of the sediments of the Cariaco Trench, Caribbean Sea. Bottom scale for silica refers to interstitial water and bottom water; top scale refers to water column (arrow refers to silica concentration in lowest 20 m of water column. (Modified from Fanning and Pilson, 1973; by permission of Pergamon Press)

existence of significant ion transport from bottom sediments (see for example, Brewer *et al.*, 1972; Bernat *et al.*, 1972; Spencer, 1973).

The study of possible sink and source relationships can be approached in another way by calculating the dissolved flux from the interstitial waters of ocean sediments to the overlying bottom waters. In such calculations a steady-state distribution of the dissolved ions is assumed. These calculated fluxes may be either positive or negative, i.e. out of, or into, the sediments respectively. The calculation can be made in two different ways. In one of these the surficial sediments are regarded merely as a channel through which elements from the deeper-lying sediments, or the igneous (volcanic) basement, are migrating. In the second of these it is considered that there is a short-range cycling of constituents between near interfacial sediments and bottom waters. Fluxes which have been computed using the first approach, have employed estimates of the mean pore fluid composition of the upper 100 m of all oceanic sediments (mainly based on data from the Deep Sea Drilling Project). Estimates made using the second approach have been based on piston core data published by both Russian (see for example, Bruevich, 1966, Shishkina, 1972) and Western workers. An attempt has been made to utilize both of these approaches to measure the fluxes of various ions to the oceans from the underlying sediment (Table 32.14). In doing this, for the first approach the average interstitial water composition for each sediment type at a depth of 50 m (Table 32.12) has been assumed to be the mean value for the upper 100 m of sediment and has been computed on an areal basis according to the following data: pelagic clays 102×10^6 km^2; carbonate-rich sediments 128×10^6 km^2; siliceous oozes 37×10^6 km^2; terrigenous sediments 32×10^6 km^2. In the second of these approaches the areal average composition of the interstitial water at a sediment depth of 1 m has been used in the calculation, because the pronounced concentration gradient often observed for the interstitial waters of the uppermost metre of sediments tends to flatten out appreciably below this depth. For several elements the change in the gradient occurs closer to the sediment-water interface, and thus for these elements the calculated fluxes are likely to be underestimates.

Diffusion coefficients for free solution (see Li and Gregory, 1974) have been corrected by application of the tortuosity factors, appropriate to the particular lithology and porosity data, given by Manheim (1970), Smirnov (1971) and Manheim and Waterman (1973). No correction has been made for the reactivity of the sediments with the dissolved constituents of the pore waters since this reactivity is, in fact, reflected in the concentration gradients themselves, and no account has been taken of the effect of mixed salt systems on the diffusion coefficients.

Gradients and diffusion coefficients are indicated in Table 32.14 together with data on the fluxes of dissolved components via interstititial waters and

TABLE 32.14

Fluxes of dissolved components to the oceans from sediments and from rivers (excluding those from the Antarctic Ocean, Arctic Seas and land-locked seas).

Fluxes used in this table have been compiled using the expression

$$F = D_0 \cdot \frac{\Delta c}{\Delta x} \cdot t.A.$$

Where:

F	is the flux from the total sea floor (g yr^{-1})
D_0	is the diffusion coefficient (cm^2 s^{-1}) for the particular ion. Tabulated values given in Table 32.14 have been taken mainly from Li and Gregory (1974).
Δc	is the amount (g cm^{-3}) by which the concentration of the ion in the pore water at a particular level differs from that in the immediately overlying sea water.
Δx	is the distance (cm) which the ion has to migrate from that level to the sediment surface.
t	is the number of seconds in a year ($3{\cdot}1 \times 10^7$)
A	is the area of the ocean floor $3{\cdot}2 \times 10^{18}$ cm^2

In the table:

M	is the estimated average overlying sea water composition (g kg^{-1}) (data mainly from Culkin, 1965, Bruevich, 1966 and Turekian, 1969)
C_1	and C_{50} are respectively the average compositions (g kg^{-1}) of the interstitial water at depths of 1 m and 50 m in the sediment.
ΔC_1 and ΔC_{50}	are the amounts (g cm^{-3}) by which these concentrations differ from that of the overlying sea water.
F_{c_1} and $F_{c_{50}}$	are the respective fluxes calculated from the above equation assuming a tortuosity factor of 2·5 (Manheim and Waterman, 1974).
F_R	is the yearly flux of the particular ion (g yr^{-1}) entering the oceans from rivers and streams (data mainly from Turekian, 1969).

TABLE 32.14—*continued*

Component	D_0 (4° C) $cm^2 s^{-1} . 10^6$	M $g\,kg^{-1}$	Cl $g\,kg^{-1}$	C_{50} $g\,kg^{-1}$	Δc_1 $g\,cm^{-3}$	Δc_{50} $g\,cm^{-3}$	F_{c_1} $g\,yr^{-1}$	$F_{c_{50}}$ $g\,yr^{-1}$	F_R $g\,yr^{-1}$	F_{c_1}/F_R
Na^+	7·2	10·76	<10·8?	<10·8?	($-50 . 10^{-6}$)	($50 . 10^{-6}$)	($-14.0 . 10^{13}$)	($-3·0 . 10^{12}$)	$12 . 10^{13}$	1·1
K^+	10·7	0·387	0·38	0·36	$-7 . 10^{-6}$	$-30 . 10^{-6}$	$-2·8 . 10^{13}$	$-2·4 . 10^{12}$	$5 . 10^{13}$	0·56
Ca^{2+}	4·3	0·413	0·415[a]	0·45	$-2 . 10^{-6}$	$-37 . 10^{-6}$	$+3·0 . 10^{13}$	$-2·4 . 10^{12}$	$5 . 10^{13}$	0·007
Mg^{2+}	3·9	1·29	1·25	1·21	$-40 . 10^{-6}$	$-80 . 10^{-6}$	$-6·0 . 10^{13}$	$-2·4 . 10^{12}$	$11 . 10^{13}$	0·55
Cl^-	11·1	19·35	19·4	19·47	$+20 . 10^{-6}$	$+120 . 10^{-6}$	$+8·0 . 10^{13}$	$+1·0 . 10^{13}$	$22 . 10^{13}$	0·36
SO_4^{2-}	5·2	2·71	2·50	2·28	$-21 . 10^{-5}$	$-43 . 10^{-5}$	$-41·0 . 10^{13}$	$-1·6 . 10^{13}$	$42 . 10^{13}$	1·0
HCO_3^-	6·4	0·14	0·19	0·27	$+50 . 10^{-6}$	$+130 . 10^{-6}$	$+12·0 . 10^{13}$	$+5·2 . 10^{12}$	$21 . 10^{14}$	0·057
Br^-	11·0	0·067	0·069	0·069	$+2 . 10^{-6}$	$+2 . 10^{-6}$	$+8·0 . 10^{12}$	$+1·6 . 10^{11}$	$7 . 10^{11}$	10·4
		$\mu g\,kg^{-1}$	$\mu g\,kg^{-1}$	$\mu g\,kg^{-1}$						
Sr^{+2}	4·3	7600	7800	13300	($+2 . 10^{-7}$)	$+53·0 . 10^{-7}$	$+3·0 . 10^{11}$	$+1·7 . 10^{11}$	$18 . 10^{11}$	0·27
SiO_4^{4-}—Si	5·4[b]	3500	9000	10000	$+5·5 . 10^{-6}$	$+6·5 . 10^{-6}$	$+1·1 . 10^{13}$	$+2·8 . 10^{11}$	$50 . 10^{13}$	0·022
O_2	14·2	4500	(100?)	0	$-4·4 . 10^{-6}$	$-4·5 . 10^{-6}$	$-2·6 . 10^{13}$	$-5·0 . 10^{11}$	$22 . 10^{13}$	0·12
PO_4^{3-}—P[c]	4·6	75	300	500	$+2·2 . 10^{-7}$	$+4·2 . 10^{-7}$	$+3·8 . 10^{11}$	$+1·3 . 10^{10}$	$7 . 10^{11}$	0·54
NH_4^+—N	10·8	10	2500	7000	$+2·5 . 10^{-6}$	$+7·0 . 10^{-6}$	$+1·0 . 10^{13}$	$+1·5 . 10^{12}$	$21 . 10^{13}$	0·048
NO_3^-—N	10·4	42	200	—	$+1·6 . 10^{-7}$	—	$+6·3 . 10^{11}$	—	$36 . 10^{12}$	0·018
$H_2BO_3^-$	4·6	4600	4200	4000	$-4·0 . 10^{-7}$	$-6·0 . 10^{-7}$	$-6·8 . 10^{11}$	$-2·0 . 10^{10}$	$70 . 10^{10}$	1·0
F^-	7·9	1300	2400	—	$+1·1 . 10^{-6}$	($1·0 . 10^{-6}$)	$+4.2 . 10^{12}$	$+8·0 . 10^{10}$	$17 . 10^{11}$	2·5
I^-	10·9	50	700	—	$+6·5 . 10^{-7}$	—	$+2·5 . 10^{12}$	—	$7 . 10^{10}$	0·36
Mn^{+2}	3·8	1	1000	1100	$+1·0 . 10^{-6}$	$+1·1 . 10^{-6}$	$+1·4 . 10^{11}$	$+3·1 . 10^{9}$	$35 . 10^{10}$	0·40
Fe^{+2}	3·9	3·4	(20)	(46)	($+17·0 . 10^{-9}$)	($-42·0 . 10^{-9}$)	($+2·5 . 10^{10}$)	($+1·4 . 10^{8}$)	($24 . 10^{12}$)	<0·01?
Co^{+2}	3·8	0·1[f]	(1)	1·6	($+0·9 . 10^{-9}$)	($+1·5 . 10^{-9}$)	($+1·0 . 10^{9}$)	($+3·0 . 10^{7}$)	$7 . 10^{9}$	0·14
Cu^{+2}	4·0	0·9	(5)	17	($+4·0 . 10^{-9}$)	$+16.0 . 10^{9}$	$+46·0\ 10^{9}$	($+5·0 . 10^{8}$)	$17 . 10^{10}$	0·35
Zn^{+2}	3·9	5	(25)	152	($+20·0 . 10^{-9}$)	$+15·0 . 10^{-8}$	($+3·0 . 10^{10}$)	$+4·0 . 10^{9}$	$70 . 10^{10}$	0·043
Ni^{+2}	3·7	4	(5)	22	($+1·0 . 10^{-9}$)	$+15·0 . 10^{-9}$	($+1·4 . 10^{9}$)	$+4·0 . 10^{8}$	$11 . 10^{9}$	0·13
Li^{+2}	5·7	170	(200)	210	($+30·0 . 10^{-9}$)	$+40·0 . 10^{-9}$	($+6·5 . 10^{10}$)	$+1·7 . 10^{9}$	$12 . 10^{10}$	0·54
Ba^{2+}	4·6	21	(100)	400[e]	($+80·0 . 10^{-9}$)	$+400·0 . 10^{-9}$	$+12·0 . 10^{10}$	$+1·2 . 10^{10}$	$36 . 10^{10}$	0·37
H^+	63·0	0·01	0·03	0·03	($+2·0 . 10^{-11}$)	$+2·0 . 10^{-11}$	($+5·0 . 10^{8}$)	$+1·0 . 10^{7}$	$3 . 10^{9}$	0·17
$(UO_2)^{+2}$	2·3	3	2·60	?	?	?	?	?	$35 . 10^{7}$	?

Values shown in parenthesis are estimated ones and subject to considerable uncertainty.

[a] Sayles *et al.* (1973) and Sayles (personal communication).
[b] Fanning and Pilson (1974) give a value $3·3 \times 10^{-6}\ cm^2\ s^{-1}$ by direct measurement in sediment.
[c] Data for $H_2PO_4^-$.
[d] Borate ion assumed to have a similar diffusional mobility to that of phosphate ion.
[e] Value influenced by high barium content of the sulphate-depleted interstitial water.
[f] Data from Brewer *et al.* (1972).

the ionic input from the world's rivers. It should be appreciated that for many ions it is not yet feasible to make a reliable estimate for the actual mean composition of either the pore fluids or the bottom water of the World's Oceans, and parentheses are used in the table to denote those values that are particularly suspect because of either factors such as the inadequacy of the data, or the uncertainty of the corrections to be applied for temperature of extraction (in fact only the data for K, Mg, B, Si and Sr have been corrected for this effect). In addition, there is considerable doubt about the validity of any published average river water compositions because they do not take full account of day to day, seasonal and anthropogenic changes. Furthermore, the effects of submarine weathering and ground water discharge on the trace element composition of sea water have also been ignored. A final factor leading to uncertainty is that the concentrations of many ions in the interstitial waters do not vary linearly with depth. Despite these limitations, the fluxes calculated on the basis of the average compositions of the interstitial waters of the uppermost 1 m of sediment (i.e. those nearest to the sediment sea water interface through which diffusion into sea water actually takes place) are likely to give an indication of those to be expected in nature.

It may be concluded from the data in Table 32.14 (see Fig. 32.13) that, of the elements having a positive flux, with a few exceptions (e.g. halogens), the steady state gradients of interstitial components inferred from the concentrations at the mean depth of 50 m below the sediment surface would contribute annual ion fluxes which are less than those from streams and rivers. However, it is difficult to predict the relative importance of the contributions of the river and the sea floor inputs to the oceanic geochemical balance. One reason for this is that, whereas bottom sediments are in intimate contact with ocean water over great areas, rivers discharge at the peripheries of the ocean, and much of their ionic contributions may be rapidly removed from the system and buried in near-shore sediments. It may, in addition, be noted that the fluxes calculated from the mean composition of the interstitial waters in the upper 1 m of sediments are generally quite similar to those from rivers. Exceptions to this include the halogens for which the sea floor flux exceeds that from rivers by a factor of 5–40, and iron for which the sea floor flux is less than 1/100 of that from rivers. In contrast, certain constituents, e.g. sulphate, boron, potassium, magnesium, and perhaps sodium have negative fluxes, that is, they are being removed from ocean bottom water by uptake into the sediments.

The constituents in the table may be divided into four general types according to the mechanism controlling their release from, or uptake by, the sediments. *Type 1* embraces those which are liberated as a result of the decomposition of the organic matter in the sediment. This may involve either direct decomposition and lead to the liberation of constituents such as I, Br, P,

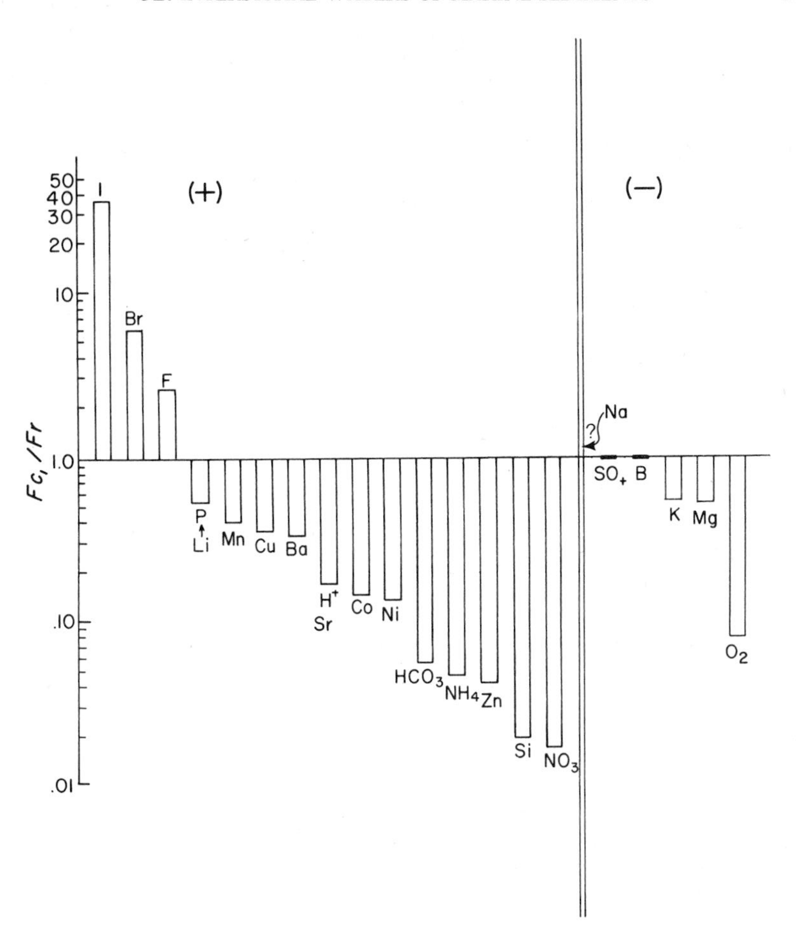

FIG. 32.13. Fluxes of dissolved constituents from sediments into ocean water (plus sign) and into sediments from bottom water (minus sign). F_{c1}/F_r refers to the ratio of the diffusive flux (based on interstitial water concentration gradients at a depth of 1 m in the sediment) to the total yearly river input to the major oceans. See Table 32.14 and text for data and calculations.

NH_3–N, HCO_3^-, or indirect decomposition (e.g. for Si and probably also Zn and Cu which are rendered more soluble by organic complexation). *Type 2* elements are those which are reduced to more soluble lower oxidation states under the anoxic condition prevailing in most pore fluid environments (e.g. iron (II), manganese (II) cerium (III) and perhaps lead (II), cobalt (II) and nickel (II)). *Type 3* constituents comprise those whose solubilities are controlled by carbonate and phosphate equilibria (e.g. calcium, bicarbonate, phosphate, fluoride and perhaps maganese (II) and iron (II). *Type 4* elements are those whose fluxes are controlled by equilibria and exchange reactions involving silicate minerals (e.g. sodium, potassium, calcium, magnesium, silicon and lithium).

In conclusion, it should be pointed out that ocean water composition may be more affected by interaction with bottom sediments than the flux calculations indicate. Two examples of this are given below. Calculation suggests that pore fluids contribute much less silica to the oceans than do rivers; however, most of the silica supplied by the latter appears to be removed and deposited in estuarine and coastal regions (Wollast and De Broeu, 1971; Fanning and Pilson, 1973). Elements (e.g., Cu, Co, Ni) which are accumulated in ferro-manganese nodules have ratios to Mn which more strongly resemble those of interstitial waters than those of sea water. Thus,

	Ferro-manganese nodules*	*Sea water**	*Interstitial water* (Table 32.12, 50 m)*
Mn	20·0	20	20·0
Cu	0·4	18	3·2
Ni	0·6	80	2·6
Co	0·3	8	0·3

* Values normalized to Mn = 20

This does not necessarily mean that interstitial ions from a given point dominate the metal supply for the nodules formed on the overlying sediments. Rather, it implies that bottom waters acquiring trace metals by upward flux from the sediments in one area of the ocean floor may lose them to the sediment in another. Whereas the former phenomenon is important in the formation of many lacustrine and shallow water marine nodules, the latter may influence deep-ocean nodules.

References*

Adams, J. E. and Rhodes, M. L. (1960). *Bull. Amer. Assoc. Petrol. Geol.* **44**, 1912.

Alexander, W. B., Southgate, B. A. and Bassindale, R. (1932). *J. Mar. Biol. Assoc. U.K.* **18**, 297.

Ali, S. Z., Friedman, G. M. and Amiel, A. J. (1972). *J. Sediment. Petrol.* **42**, 794.

Arkhangel'skii, A. D. and Zalmanzon, E. S. (1931). *Byull. Mosk. Obshchest. Ispyt. Prir., Otd. Geol.* **9**, 282.

Arrhenius, G. O. S. and Rotschi, H. (1953), cited in Arrhenius, G.O.S. (1963). *In* "The Sea, Ideas and Observations" (M. N. Hill, ed.), Vol. 3, pp. 655–727. Wiley, New York.

Baturin, G. N. (1971). *Dokl. Akad. Nauk SSSR,* **198**, 1186.

Baturin, G. N. (1972). *Oceanology,* **12**, 1020.

* Note added in proof; the main literature search was concluded in 1973.

Bellaiche, G., Leenhardt, O. and Pautot, G. (1966). *Compt. Rend. Acad. Sci.* **263**, 808.
Bender, M. L. (1971). *J. Geophys. Res.* **76**, 4212.
Bernat, M., Church, T. and Allegre, C. J. (1972). *Earth Planet. Sci. Lett.* **6**, 75.
Berner, R. A. (1964). *Geochim. Cosmochim. Acta,* **28**, 1497.
Berner, R. A. (1966). *Amer. J. Sci.* **264**, 1.
Berner, R. A. (1969). *Amer. J. Sci.* **267**, 19.
Berner, R. A. (1971). "Principles of Chemical Sedimentology," 240 pp. McGraw Hill, New York.
Berner, R. A., Scott, M. R. and Thomlinson, C. (1970). *Limnol. Oceanogr.* **15**, 544.
Bischoff, J. L. and Ku, T. L. (1970). *J. Sediment. Petrol.* **40**, p. 960.
Bischoff, J. L. and Ku, T. L. (1971). *J. Sediment. Petrol.* **41**, 1008.
Bischoff, J. L. and Sayles, F. L. (1972). *J. Sediment. Petrol.* **42**, 711.
Bojanowski, R. and Paslawska, S. (1970). *Acta Geophys. Pol.* **18**, 277.
Bordovskii, O. K. (1960), *Tr. Inst. Okeanol., Akad. Nauk SSSR,* **42**, 107.
Bordovskii, O. K. (1961). *Okeanolog. Issled.* 91.
Bordovskii, O. K. (1964). *Okeanolog. Issled.* 91.
Brafield, A. E. (1964). *J. Anim. Ecol.* **33**, 97.
Brewer, P. G., Spencer, D. W. and Robertson, D. E. (1972). *Earth Planet. Sci. Lett.* **16**, 111.
Brooks, R. R., Presley, B. J. and Kaplan, I. R. (1968). *Geochim. Cosmochim. Acta,* **32**, 397.
Brooks, R. R., Kaplan, I. R. and Peterson, M. N. A. (1969). *In* "Hot Brines and Recent Heavy Metal Deposits in the Red Sea," (E. T. Degens and D. A. Ross, eds), pp. 180–203. Springer-Verlag, New York.
Brouardel, J. and Fage, L. (1953). *Compt. Rend. Acad. Sci.,* **237**, 1605.
Brown, F. S., Baedecker, M. J., Nissenbaum, A. and Kaplan, I. R. (1972). *Geochim. Cosmochim. Acta,* **36**, 1185.
Bruce, J. R. (1928). *J. Mar. Biol. Assoc. U.K.* **15**, 535.
Bruevich, S. V. (1952). *Dokl. Akad. Nauk SSSR,* **84**, 575.
Bruevich, S. V. (1956). *Dokl. Akad. Nauk SSSR,* **111**, 391.
Bruevich, S. V. (1956). *Tr. Inst. Okeanol. Akad. Nauk SSSR,* **17**, 41.
Bruevich, S. V. (1957). *Dokl. Akad. Nauk SSSR,* **113**, 387.
Bruevich, S. V. (1958). *Dokl. Akad. Nauk SSSR,* **118**, 767.
Bruevich, S. V. (1962). *Tr. Inst. Okeanol. Akad. Nauk SSSR,* **54**, 31.
Bruevich, S. V. (1966). "Khimiya Tikhogo Okeana" (S. V. Bruevich, ed.), pp. 263–358. Izd. Akad. Nauk, Moscow.
Bruevich, S. V., Pevzniak, R. M., Ponizovskaya, V. L. and Sibiryakov, M. A. (1938). *Compt. Rend. Acad. Sci. URSS,* **21**, 282.
Bruevich, S. V. and Vinogradova, E. G. (1940a). *Dokl. Akad. Nauk SSSR,* **27**, 575.
Bruevich, S. V. and Vinogradova, E. G. (1940b), *Dokl. Akad. Nauk SSSR,* **27**, 579.
Bruevich, S. V. and Vinogradova, E. G. (1946). *Compt. Rend. Acad. Sci. URSS.* **54**, 419.
Bruevich, S. V. and Vinogradova, E. G. (1947). *Gidrokhim. Mater.* **13**.
Bruevich, S. V. and Zaitseva, E. D. (1958). *Tr. Inst. Okeanol. Akad. Nauk SSSR,* **26**, 8.
Bruevich, S. V. and Shishkina, O. V. (1959). *Dokl. Akad. Nauk SSSR,* **127**, 673.
Bruevich, S. V. and Zaitseva, E. G. (1960). *Tr. Inst. Okeanol., Akad. Nauk SSSR,* **42**, 4.
Bruevich, S. V. and Zaitseva, E. G. (1964). *Tr. Inst. Okeanol., Akad. Nauk SSSR,* **67**, 56.

Burbanck, W. D. and Burbanck, G. P. (1967). *Chesapeake Sci.* **8**, 14.
Bushinski, G. I. (1964). *Devop. Sedimentology,* **1**, 62.
Byrne, J. V. and Emery, K. O. (1960). *Bull. Geol. Soc. Amer.* **71**, 983.
Callame, B. (1960). *Bull. Inst. Oceanogr. Monaco,* **1181**, 4.
Calvert, S. F. and Price, N. B. (1972). *Earth Planet. Sci. Lett.* **16**, 245.
Chernovskaya, E. N. (1958). *Tr. Murmansk. Biol. Sta.* **4**, 7.
Clarke, W. B., Horowitz, R. M. and Broecker, W. (1973). *in* "Initial Reports of the Deep Sea Drilling Project", Vol. 20, p. 777. U.S. Gov. Printing Office, Washington, D.C.
Claypool, G. E., Presley, B. J. and Kaplan, I. R. (1973). "Initial Reports of the Deep Sea Drilling Project, Vol. 19, p. 879–884. U.S. Gov. Printing Office, Washington, D.C.
Culkin, F. (1965). *In* "Chemical Oceanography" (J. P. Riley and G. Skirrow, eds), Vol 1, pp. 121–161. Academic Press, London and New York.
Davis, J. B., Stanley, J. P. and Custard, H. C. (1970). *Bull. Amer. Assoc. Petrol. Geol.* **54**, 2444.
Debyser, J. and Rouge, P. E. (1956). *Compt. Rend. Acad. Sci.* **243**, 2111.
Degens, E. T. and Ross, D. A. (eds.) (1969). "Hot Brines and Recent Heavy Metal Deposits in the Red Sea." Springer–Verlag, New York.
Degens, E. T., Deuser, W. G., von Herzen, R. P., Wong, K. K., Wooding, F. B., Jannasch, H. W. and Kanwisher, J. W. (1971). *Tech. Rep. Woods Hole Oceanogr. Inst.* Ref. 71–52, pp. 1–20.
Desai, M. V. M., Kosly, E. and Gangully, A. K. (1969). *Curr. Sci.* **5**, 107.
Dunoyer de Segonzac, G. (1970). *Sedimentology,* **15**, 281.
Drever, J. I. (1971). *Science, N.Y.,* **172**, 1334.
Drever, J. I. (1971). *J. Sediment. Petrol.* **41**, 982.
Dymond, J., Corliss, J. B., Heath, J. R., Field, C. W., Dasch, E. J. and Veeh, H. H. (1973), *Bull. Geol. Soc. Amer.* **84**, 3355.
Emelyanov, E. M. and Chumakov, U. D. (1962). *Dokl. Akad. Nauk SSSR* (in transl.), **143**, 141.
Emery, K. O. (1960). "The Sea Off Southern California" Wiley & Sons, New York.
Emery, K. O. and Foster, J. F. (1948). *J. Mar. Res.* **1**, 644.
Emery, K. O. and Rittenberg, S. C. (1952). *Bull. Amer. Assoc. Petrol. Geol.* **36**, 735.
Emery, K. O. and Hoggan, D. (1958). *Bull. Amer. Assoc. Petrol. Geol.* **42**, 2174.
Fanning, K. A. (1973). Written communication.
Fanning, K. A. and Schink, D. R. (1969). *Limnol. Oceanogr.* **14**, 59.
Fanning, K. A. and Pilson, M. E. Q. (1971). *Science, N.Y.* **173**, 1228.
Fanning, K. A. and Pilson, M. E. Q. (1973). *Geochim. Cosmochim. Acta,* **37**, 2405.
Fanning, K. A. and Pilson, M. E. Q. (1974). *J. Geophys. Res.* **79**, 1293.
Feeley, H. W. and Kulp, J. L. (1957). *Bull. Amer. Assoc. Petrol. Geol.* **41**, 1802.
Friedman, G. M., Fabricand, B. P., Imbibo, E. S., Brey, M. E. and Sanders, J. E. (1968). *J. Sediment. Petrol.* **38**, 1313.
Friedman, G. M. and Gavish, E. (1970). *J. Sediment. Petrol.* **40**, 930.
Gadel, F. (1968). *Vie Milieu, Ser. B, Oceanogr.* **10**, 291.
Ganopati, P. N. and Chandrasekhara Rao, G. (1962). *J. Mar. Biol. Assoc. India,* **4**, 44.
Gardner, L. R. (1973). *Geochim. Cosmochim. Acta,* **37**, 53.
Ginsburg, R. N. (1964). Guidebook, S. Florida Carbonate Sediments. Geol. Soc. America Conv., 72 pp.
Glasby, G. P. (1973). *J. Roy. Soc. N.Z.* **3**, 43.

Goni, J. and Parent, C. (1966), *Bull. Bur. Rech. Geol. Min.* **1**, 19.
Gorshkova, T. I. (1955). *Tr. Vses. Nauch–Issled. Inst. Rybi Okeanogr.* **31**, 123.
Gorshkova, T. I. (1957). *Dokl. Akad. Nauk SSSR*, **113**, 863.
Gorshkova, T. I. (1960). *Okeanolog. Issled.* **2**, 113.
Gorshkova, T. I. (1961). *In* "Sovremennye osadki v morei i okeanakh" (N. M. Strakhov, ed.), pp. 477–503. Izdat. Acad. Nauk SSSR, 477.
Gorshkova, T. I. (1970). *In* "Khimicheskie resursy morei i okeanov" (S. V. Bruevich, ed.), pp. 67–78, Nauka, Moscow.
Greene, H. G. (1970). "Geology of S. Monterey Bay and its relationship to the ground water basin and salt water intrusion". U.S. Geological Survey Open File Report, 50 pp.
Gripenberg, S., Karlik, B. and Snellman, L. (1948). *Kgl. Vetensk. Vitterh. Samällets Hdlr. 6te Följd, Ser. B, Bd.* **5**, (*13*), 21.
Grim, R. E. and Johns, W. D. (1954). *Clays and Clay Minerals, Nat. Acad. Sci., Nat. Res. Counc. Publ.* **327**, 81.
Hammond, D. E., Horowitz, R. M., Broeker, W. S. and Bopp, R. (1973). "Initial Reports of the Deep Sea Drilling Project", Vol. 15, U.S. Gov. Printing Office, Washington, D.C.; *Spec. Geochem. Sec.* **20**, 765.
Hanshaw, B. B. and Coplen, T. B. (1973), *Geochim. Cosmochim. Acta*, **37**, 2311.
Harriss, R. C. and Pilkey, O. H. (1966). *Deep-Sea Res.*, **13**, 867.
Hartmann, M. (1964). *Meyniana*, **14**, 3.
Hartmann, M. and Nielsen, H. (1964). *Geol. Rundsch.* **58**, 621.
Hartmann, M., Müller, P., Suess, E. and Van der Weijden, C. H. (1973), "Meteor" *Forschungsergeb., Reihe C*, **12**, 74.
Hendricks, R. L., Reisbick, R. B., Mahaffey, E. O., Roberts, D. B. and Peterson, M. N. A. (1969). *In* "Hot Brines and Recent Heavy Metal Deposits in the Red Sea" (E. T. Degens and D. A. Ross, eds), pp. 407–429. Springer-Verlag, New York.
Horn, M. K. and Adams, J. A. S. (1965), *Geochim. Cosmochim. Acta*, **30**, 279.
Hurd, D. C. (1973), *Geochim. Cosmochim. Acta*, **37**, 2257.
Hsü, K. J. and Siegenthaler, C. (1969), *Sedimentology*, **12**, 11.
Jenkins, W. J., Beg, M. A., Clarke, W. B., Wangersky, P. J. and Craig, H. (1972), *Earth Planet. Sci. Lett.* **16**, 122.
Jerbo, A. (1965). *Medd. Statens Järnvägars Centralförv., Geotek. Kont.*, **11**, 159.
Johnson, D. W. (1940). *J. Geomorphology*, **2**, 134.
Johnson, R. G. (1967). *Limnol. Oceanogr.* **12**, 1.
Jones, S. L. and Fanning, K. A. (1975). *Eos*, 373.
Kaplan, I. R., Emery, K. O. and Rittenberg, S. C. (1963). *Geochim. Cosmochim. Acta*, **27**, 297.
Kazintsev, E. A. (1968). *In* "Porovye rastvory i metody ikh izucheniya" (Pore Fluids and Methods for Studying Them) (G. V. Bogomolov, ed.), pp. 178–190. Nauka i Teknika, Minsk.
Kermabon, A., Gehin, C. and Blavier, D. (1969). *Geophysics*, **34**, 554.
Kitano, Y., Danamori, S., Kato, K., Kanamori, N., Yoshioka, R., Knowles, L. I., Kunze, M. and Hood, D. W. (1969). *Geochem. J.*, **3**, pp. 99.
Koczy, F. F. (1958). *Proc. 2nd U.N. Int. Conf. Peaceful Uses of Atomic Energy*, **18**, 351.
Kohout, F. A. K., Manheim, F. T. and Leve, G. W. (1976). (In prep.)
Krasintseva, I. V. and Shishkina, O. V. (1959). *Dokl. Akad. Nauk SSSR*, **128**, 815.
Krasintseva, I. V. and Korunova, V. V. (1968). *In* "Porovye rastvory, i metody ikh izucheniya" (G. V. Bogomolov, ed.), pp. 191–204. Nauka i Teknika, Minsk.

Kriukov, P. A. (1947). *In* "Sovremennye metody issledovaniya fiziko-khimicheskikh svoistv pochv," Vol. 2, Moscow.
Kriukov, P. A. (1971), "Gornye pochvennye i ilovye rastvory (Pore Fluids of Rocks, Soils and Sediments)", Izdat Nauka, Sib. Otdel., Novosibirsk.
Ku, T. L. and Glasby, G. (1972), *Geochim. Cosmochim. Acta,* **36**, 699.
Ku, T. L. and Broecker, W. S. (1969), *Deep-Sea Res.* **16**, 625.
Kudelin, V. I., Zektser, I. S., Meskheteli, A. V. and Brusilovskii, S. A. (1971), *Sov. Geol.,* **1**, 72.
Kühl, H. and Mann, H. (1966), *Helgoländer, Wiss. Meeresunters.* **13**, 238.
Kullenberg, B. (1952), *Kgl. Vetensk. Vitterhets Sämhallets Hdlgr, 6te följd, Ser. B., Bd* **6** (**6**), 38.
Laevastu, T. and Fleming, R. H. (1959), *Preprints Int. Oceanogr. Congr. Wash. D.C.,* 967.
Lancelot, Y., Hathaway, J. C. and Hollister, C. D. (1970), "Initial Reports of the Deep Sea Drilling Project", Vol. 11, p. 901. U.S. Govt. Printing Office, Washington, D.C.
Lang, J. W. and Newcome, R. (1964), *State of Mississippi Bd. Water Commissioners, Bull.* **64–5**, 17.
Lehner, P. (1968). *Bull. Amer. Assoc. Petrol. Geol.* **53**, pp. 2431.
Lerman, A. (1971). *Adv. Chem. Ser. Amer. Chem. Soc.* **106**, 30.
Li, Y. H., Bischoff, J. L. and Mathieu, G. (1969). *Earth Planet. Sci. Lett.* **7**, 265.
Li, Y. H., Ku, T. L., Mathieu, G. and Wolgemuth, K. (1973). Unpubl. MS.
Li, Y. H. and Gregory, S. (1974). *Geochim. Cosmochim. Acta,* **38**, 703.
Lynn, D. C. and Bonatti, E. (1965). *Mar. Geol.* **3**, 457.
MacKenzie, F. T. and Garrels, R. M. (1965). *Science, N.Y.* **150**, 57.
MacKenzie, F. T. and Garrels, R. M. (1966). *Amer. J. Sci.* **264**, 407.
Mangelsdorf, P. C., Wilson, T. R. S. and Daniell, E. (1969). *Science, N. Y.* **165**, 171.
Manheim, F. T. (1965). *Narragansett Mar. Lab., Occ. Pap. 1965–3,* 217.
Manheim, F. T. (1967). *Trans. N.Y. Acad. Sci, Ser. II,* **29**, 839.
Manheim, F. T. (1970). *Earth Planet. Sci. Lett.* **9**, 307.
Manheim, F. T. (1972). "Encyclopedia of Geochemistry and Environmental Sciences" (R. W. Fairbridge, ed.). Reinhold, New York.
Manheim, F. T. and Horn, M. K. (1968). *Southeast. Geol.* **9**, 215.
Manheim, F. T. and Bischoff, J. L. (1969). *Chem. Geol.* **4**, 63.
Manheim, F. T., Sayles, F. L. and Friedman, I. (1969). "Initial Reports of the Deep Sea Drilling Project," Vol. 1. pp 403–410 U.S. Gov. Printing Office, Washington, D.C.
Manheim, F. T. and Sayles, F. L. (1970). *Science,* N.Y. **170**, 57.
Manheim, F. T., Sayles, F. L. and Waterman, L. S. (1971). "Initial Reports of the Deep Sea Drilling Project, Vol. 8, pp 857–872. U.S. Gov. Printing Office, Washington, D.C.
Manheim, F. T. and Sayles, F. L. (1974). *In "The Sea"* (Goldberg, E. D., ed.), Vol. 5, Marine Chemistry. John Wiley & Sons, New York.
Manheim, F. T. and Chan, K. M. (1974). Interstial waters of Black Sea sediments: new data and review. *In* "The Black Sea – Geology, Chemistry, and Biology" (E. T. Degens and D. A. Ross, eds). Amer. Assoc. Pet. Geol. Memoir 20. George Banta Co., Inc.
Manheim, F. T., Waterman, L. S., Woo, C. C. and Sayles, F. L. (1974). Interstitial water studies on small core samples, Leg 23 (Red Sea). "Initial Reports of the Deep

Sea Drilling Project," Vol. 23, pp. 955–967. U.S. Gov. Printing Office, Washington, D.C.

Manheim, F. T. and Waterman, L. S. (1974). Diffusimetry (diffusion constant estimation) on sediment cores by resistivity probe. "Initial Reports of the Deep Sea Drilling Project," Vol. 22, pp. 663–670. U.S. Gov. Printing Office, Washington, D.C.

Marchig, V. (1970). *Geol. Rundsch* **60**, 275.

Mel'nik, V. I. (1970). Okeanologicheski issledovanie, pp. 22–41. Nauka, Moscow.

Michard, G. (1971). *J. Geophys. Res.* **76**, 2179.

Michard, G., Church, T. M and Bernat, M. (1974). *J. Geophys. Res.* **79**, 817.

Mikkelsen, V. M. (1956). *Medd. Dansk. Geol. Foren. Bd. 13*, H. **2**, 104.

Milliman, J. D. and Manheim, F. T. (1970). Unpubl. MS.

Milliman, J. D. and Müller, J. (1973), *Sedimentology* **20**, 29.

Mokady, R. S. and Low, P. F. (1968). *Soil Sci.* **105**, 112.

Mokievskaya, V. V. (1960). *Tr. Okeanogr. Kom., Akad. Nauk SSSR*, **10**, 21.

Murray, J. and Irvine, R. (1895). *Trans. Roy. Soc. Edinburgh* **37**, 481.

Nafe, J. E. and Drake, C. L. (1963). *In* "The Sea, Ideas and Observations" (M. N. Hill, ed.), Vol. 3, pp. 523–552. Wiley, New York.

Nelson, B. W. (1972). *Geol. Soc. Amer. Mem.* **137**, 417.

Nissenbaum, A. (1971). *In* "Advances in Organic Geochemistry," pp. 427–440. Pergamon Press, Braunschweig.

Nissenbaum, A., Presley, B. J. and Kaplan, I. R. (1972). *Geochim. Cosmochim. Acta,* **36**, 1007.

Nooter, K. and Liebregt, F. (1971). *Bijdragen Tot Dierkunde*, **41**, 23.

Okuda, T. (1960). *Trab. Inst. Biol. Marit. Oceanogr. Univers. Recife*, **2**, 7.

Okuda, T. and Cavalcanti, L. B. (1963). *Trab. Inst. Biol. Marit Oceanogr. Univers. Recife*, **3**, 27.

Ostroumov, E. A. (1953), *Tr. Inst. Okeano. Akad. Nauk, SSSR,* **7**, 71.

Ostroumov, E. A. and Fomina, L. S. (1960). *Tr. Inst. Okeanol. Akad. Nauk SSSR,* **32**, 206.

Ostroumov, E. A. and Volkov, I. I. (1960). *Tr. Inst. Okeanol. Akad. Nauk SSSR*, **42**, 117.

Ostroumov, E. A., Volkov, I. I. and Fomina, L. S. (1961). *Tr. Inst. Okeanol. Akad. Nauk SSSR,* **59**, 93.

Pamatmat, M. M. (1971). *Limnol. Oceanogr.* **16**, pp. 536–550.

Powers, M. C. (1954). *Clays and Clay Minerals, Nat. Acad. Sci. Nat. Res. Counc. Publ.* **327**, 68.

Powers, M. C. (1957). *J. Sediment. Petrol.,* **27**, 355.

Powers, M. C. (1959). *Clays and Clay Minerals, 6th Nat. Conf.* **2**, 309.

Presley, B. J. (1973) written communication.

Presley, B. J., Brooks, R. T. and Kappel, H. M. (1967). *J. Mar. Res.,* **25**, 355.

Presley, B. J. and Kaplan, I. R. (1968). *Geochim. Cosmochim. Acta*, **32**, 1037.

Presley, B. J., Goldhaber, M. V. and Kaplan, I. R. (1970). "Initial Reports of the Deep Sea Drilling Project," Vol. 5, pp. 513–512. U.S. Gov. Printing Office, Washington, D.C.

Presley, B. J., Culp, J., Petrowski, C. and Kaplan, I. R. (1973), "Initial Reports of the Deep Sea Drilling Project," Vol. 15. U.S. Gov. Printing Office. Washington, D.C; *Spec. Geochim. Sec.* **20**, 785.

Presley, B. J. and Kaplan, I. R. (1972). "Initial Reports of the Deep Sea Drilling Project," Vol. 9, pp. 841–844. U.S. Gov. Printing Office, Washington, D.C.

Pushkina, Z. V. (1963). *Dok. Akad. Nauk SSSR* (in transl.), **148** (2), 433.

Pushkina, Z. V. (1965). *In* "Postsedimentatsionnye izmeneniya chetvertichnykh i pliotsenovykh glinistykh otlozhenii Bakinskogo arkhipelago (Postsedimentary changes in Quaternary and Pliocene clayey deposits of the Baku Archipelago)" (N. M. Strakhov, ed.), pp. 160–203. Nauka, Moscow.
Reeburgh, W. S. (1969), *Limnol. Oceanogr.* **14**, 368.
Reid, D. M. (1932). *J. Mar. Biol. Assoc. U.K.* **18**, 299.
Revelle, R. and Emery, K. O. (1951). *Bull. Geol. Soc. Amer.* **62**, 707.
Richards, F. A., ed. (1967), "Marine Geotechnique," 327 pp. Univ. Illinois Press, Urbana.
Rittenberg, S. C., Emery, K. O. and Orr, W. L. (1955). *Deep-Sea Res.* **3**, 23.
Rittenberg, S. C., Emery, K. O., Hülsemann, J., Degens, E. T., Fay, R. C., Reuter, J. H., Grady, J. R., Richardson, S. H. and Bray, E. E. (1963). *J. Sediment. Petrol.* **33**, 140.
Romankevich, E. A. and Urbanovich, I. M. (1971). *Tr. Inst. Okeanol. Akad. Nauk SSSR,* **89**, 106.
Sanders, H. L., Mangelsdorf, P. C., Jr. and Hampson, G. R. (1965). *Limnol. Oceangr. Suppl.* **10**, 216.
Sayles, F. L. and Manheim, F. T. (1975) *Geochim. Cosmochim. Acta,* **39**, 103.
Sayles, F. L., Waterman, L. S. and Manheim, F. T. (1972). "Initial Reports of the Deep Sea Drilling Project," Vol. 13, pp. 801–811. U.S. Gov. Printing Office, Washington, D.C.
Sayles, F. L., Manheim, F. T. and Waterman, L. S. (1973). Initial Reports of the Deep Sea Drilling Project, Vol. 15. U.S. Gov. Printing Office, Washington, D.C. *Spec. Geochem. Sec.* **20**, 783.
Sayles, F. L., Wilson, T. R. S., Hume, D. N. and Mangelsdorf, P. C. (1973), *Science, N.Y.* **181**, 154.
Schmidt, G. W. (1973), *Bull. Amer. Assoc. Petrol. Geol.* **57**, 321.
Shoikhet, P. A. and Sakhnovskaya, N. D. (1956), *Tru. Azer. Nauch.-Issled. Inst Dobyche Nefti,* 323.
Scholl, D. W. (1965), *Nature, Lond.* **207**, 284.
Seibold, E. (1962). *Sedimentology,* **1**, 70.
Sevastyanov, V. F. and Volkov, I. I. (1967). *Tr. Inst. Okeanol, Akad. Nauk SSSR,* **83**, 135.
Sharma, G. D. (1970). *J. Sediment. Petrol.* **40**, 722.
Shchukarev, S. A. (ed.), (1937). "Fiziko-khimiya mineral'nykh vod i lechebnykh gryazei (Physico-chemistry of mineral waters and therapeutic muds)," Biomedgiz.
Shepard, F. P. and Moore, D. G. (1955). *Bull. Amer. Assoc. Petrol. Geol.* **39**, 1463.
Shishkina, O. V. (1955). *Tr. Inst. Okeanol. Akad. Nauk SSSR,* **13**, 94.
Shishkina, O. V. (1957). *Dok. Akad. Nauk SSSR,* **116**, 259.
Shishkina, O. V. (1958). *Tr. Inst. Okeanol. Akad. Nauk SSSR,* **26**, 109.
Shishkina, O. V. (1959). *In* "K poznaniyu diageneza osadkov" (N. M. Strahov, ed.), pp 29–50. Izdat. Akad. Nauk SSSR, Moscow.
Shishkina, O. V. (1959). *Tr. Inst. Okeanol. Akad. Nauk SSSR,* **33**, 146.
Shishkina, O. V. (1959), *Tr. Inst. Okeanol. Akad. Nauk SSSR,* **33**, 178.
Shishkina, O. V. (1961). *Okeanologiya,* **1**, 646.
Shishkina, O. V. (1961). *Dok. Akad. Nauk SSSR,* **139**, 1218.
Shishkina, O. V. (1962). *Tr. Inst. Okeanol. Akad. Nauk SSSR,* **54**, 47.
Shishkina, O. V. (1964). *Geokhimiya,* **6**, 564; (Engl. Ed. 522–527).
Shishkina, O. V. (1966). *In* "Khimiya Tikhogo Okeana" (S. V. Bruevich, ed.), pp. 289–307. Nauka, Moscow.

Shishkina, O. V. (1966). *Geokhimiya,* **2**, 152.
Shishkina, O. V. (1966a). *In* "Khimicheskiye protsessy v moryakh i okeanakh" (S. V. Bruevich, ed.), pp. 26–34. Izdat. Nauka, Moscow.
Shishkina, O. V. (1969). *In* "Gidrofizicheskiye i gidrokhimicheskiye issledovanie" (L. I. Belyaev, ed.), pp. 56–60. Naukova Dumka, Kiev.
Shishkina, O. V. (1972). "Gekhimiya morskikh i okeanocheskikh ilovykh vod." Nauka, Moscow.
Shishkina, O. V. and Bykova, V. S. (1964). *Tr. Morsk. Gidrofiz. Inst., Akad. Nauk Ukr. SSR,* **25**, 187.
Shishkina, O. V. and Zheleznova, A. A. (1964). *Tr. Inst. Okeanol. Akad. Nauk SSSR,* **54**, 144.
Shishkina, O. V. and Bykova, V. S. (1966). *In* "Issledovania v Atlanticheskom okeane" 2. Naukova Dumka, Kiev. Cited in Shishkina (1972).
Shishkina, O. V., Pavlova, G. A. and Bykova, V. S. (1969). "Geokhimiya galogenov v okeanskikh osadkakh i ilovykh vodakh." Nauka, Moscow.
Siever, R., Garrels, R. M., Kanwisher, J. and Berner, R. A. (1963). *Science, N.Y.* **134**, 1071.
Siever, R., Beck, K. C. and Berner, R. A. (1965). *J. Geol.* **73**, 39.
Sillén, L. G. (1961). "Oceanography". (M. Sears, ed.). *Amer. Assoc. Adv. Sci. Publ.* **67**, 549.
Skopintsev, B. A. (1962). *Deep-Sea Res.* **9**, 349.
Smirnov, S. I. (1971). "Proiskhozhdenie solenosti podzemnykh vod sedimentatsionnykh basseinov," Nedra, Moscow.
Smith, R. I. (1955). *J. Mar. Biol. Assoc. U.K.* **34**, 33.
Spencer, D. W. (1973). *Earth Planet. Sci. Lett.* **16**, 111.
Stainer, R. Y., Doudoroff, M. and Adelberg, F. A. (1970). "The Microbial World." Prentice Hall, Englewood Cliffs, New Jersey.
Starikova, N. D. (1956). *Dokl. Akad. Nauk SSSR,* **108**, 492.
Starikova, N. D. (1959). *In* "K poznaniyu diageneza osadkov" (N. M. Strakhov, ed.), 72–91. Izdat. Akad, Nauk SSSR, Moscow.
Starikova, N. D. (1959). *Tr. Inst. Okeanol. Akad. Nauk SSSR,* **33**, 165.
Starikova, N. D. (1961a). *Tr. Inst. Okeanol. Akad. Nauk SSSR,* **50**, 130.
Starikova, N. D. (1961b). *Dokl. Akad. Nauk SSSR,* **140**, 1423.
Strakhov, N. M. (1965)(ed.). "Postsedimentatsionnye izmeneniya chetvertichnykh i pliotsenovykh glinistykh otlozgenii Bakinskogo arkhipelaga," Nauka, Moscow.
Sukhrev, G. M. and Krumbolt, T. S. (1962). *Doklady Akad. Nauk SSSR,* (Engl. Ed.), **145**, 166.
Tageeva, N. V. (1965). *Dokl. Akad. Nauk SSSR,* **163**, 1477.
Tageeva, N. V. and Tikhomirova, M. M. (1961), "Geokhimiya porovykh vod pri diageneze morskikh osadkov." Izdat. Akad. Nauk SSSR, Moscow.
Tageeva, N. V., Tikhomirova, M. M. and Korunova, V. V. (1961). *In* "Sovremennye osadki morei i okeanov" (N. M. Strakhov, ed.), pp. 560–576. Izdat. Akad. Nauk SSSR, Moscow.
Tageeva, N. V. and Tikhomirova, M. M. (1962), "Gidrogeokhimiya donnykh osadkov Chernogo morya." Izdat. Akad. Nauk SSSR, Moscow.
Takahashi, T., Prince, L. A. and Felice, L. J. (1973). Initial Reports of the Deep Sea Drilling Project." Vol. 15. U.S. Gov. Printing Office, Washington, D.C. *Spec. Geochem. Sec.* **20**, 865–876.
Thompson, G. (1973), *Eos,* **54**, 1015.
Thorstenson, D. C. and MacKenzie, F. T. (1971). *Nature, Lond.* **234**, 543.

Trask, P. D. and Rolston, I. (1950). *Science, N.Y.* **111**, 666.
Truesdell, A. H. and Jones, B. F. (1973). *Nat. Tech. Inform. Ser. PB 220464* (computer program).
Turekian, K. K. (1969). *In* "Handbook of Geochemistry" (K. H. Wedepohl, ed.), Vol. 1, pp. 297–320. Springer-Verlag, Berlin.
Valyashko, M. G. (1963). *In* "Chemistry of the Earth's Crust" (A. P. Vinogradov, ed.), pp. 263–290. Transl. Israel Program Sci. Transl., publ. 1966.
Verigo, A. M. (1881). "Issledovaniya Odesskikh tselebnykh limanov i gryazei." Otchety o deyatel'nosti Odesskago balneologicheskogo obshchestva s 1877–1880. Addendum, 107 pp, Odessa.
Vinogradov, A. P. (1939), *Tr. Biogeokhim. Lab., Akad. Nauk SSSR,* **5**, 19 (Engl. ed. pp. 33–49).
Volker, A. (1961). *Econ. Geol.* **56**, 1045.
Waterman, L. S., Sayles, F. L. and Manheim, F. T. (1972). "Initial Reports of the Deep Sea Drilling Project," Vol. 14, pp. 753–762. U.S. Gov. Printing Office, Washington, D.C.
Werner, A. E. and Hyslop, W. F. (1968). *Fish. Res. Bd. Can. Ms. Rep. Ser.* **958**.
White, D. E., Hem, J. D. and Waring, G. A. (1963). *U.S. Geol. Surv. Prof. Pap.* **440-F**.
Wolfe, R. (1971). *Advan. Microb. Physiol.* **6**, 107.
Wollast, R. and De Broeu, F. (1971). *Geochim. Cosmochim. Acta,* **35**, 613.
Zaitseva, E. D. (1954). *Dokl. Acad. Nauk SSSR,* **94**, 289.
Zaitseva, E. D. (1956). *Dokl. Acad. Nauk SSSR,* **111**, pp. 144.
Zaitseva, E. D. (1959). *In* "K poznaniyu diageneza ozadkov" (N. M. Strakhov, ed.), pp. 51–71. Izdat. Akad. Nauk SSSR, Moscow.
Zaitseva, E. D. (1962). *Tr. Inst. Okeanol. Akad. Nauk SSSR*, **54**, pp. 58.
Zaitseva, E. D. (1966). *In* "Khimiya Tikhogo okeana" (S. V. Bruevich, ed.), pp. 271–288. Nauka, Moscow.
Zheleznova, A. A. and Shishkina, O. V. (1964). *Tr. Inst. Okeanol. Akad. Nauk SSSR,* **64**, 236.

Chapter 33

The Mineralogy and Geochemistry of Near-shore Sediments

S. E. CALVERT

Institute of Oceanographic Sciences,
Wormley, Surrey GU8 5UB, England

33.1. Introduction

Marine sediments are highly fractionated crustal materials supplied to the ocean from a number of different sources. This fractionation has been brought about by both chemical and physical processes at the earth's surface, and as Goldschmidt (1937) has pointed out it can be likened to the separation scheme undertaken in classical chemical analysis of silicates. Since the separation is rarely complete, the chemical compositions of many sediments are highly variable and it is difficult to formulate generalizations about average compositions and the processes which control the composition of a particular sediment.

This chapter is concerned essentially with "non-pelagic" sediments, that is, sediments accumulating in the very wide variety of environments to be found close to the continents. Here, physical, chemical and biological conditions are much more variable than they are in the deep sea. Consequently, near-shore sediments are very heterogenous and cover the entire range of sediments accumulating in the ocean at the present time. They include, for example, highly calcareous muds, siliceous oozes, pure sands, phosphorites and ferro-manganese nodules.

Near-shore sediments have been studied for a considerable period by geologists interested in determining the environments of deposition of the ancient sediments preserved in the geological record. Hence, a reasonable amount of information on the textural, structural, mineralogical and biological aspects of recent sediments from near-shore areas is available, and lengthy syntheses of the sedimentary facies in several modern basins of deposition have been published (see e.g. Van Andel and Postma, 1954; Emery 1960; Shepard *et al.*, 1960; Strakhov, 1961; Van Andel and Shor, 1964; Lisitsyn, 1966; Van Andel and Veevers, 1967). However, until relatively recently, the geochemistry of these types of sediments has not attracted the same amount of attention. In contrast, pelagic sediments have been studied in much greater detail by marine geochemists because of their unusual compositions and because of the great time span covered by moderately long cores.

Recently, it has been suggested that near-shore environments are important removal sites for several elements from sea water because the accumulation rates are very much higher and the physico-chemical conditions are different from those found in the open-ocean. This topic will be discussed in detail after the range of chemical compositions of near-shore sediments has been examined.

33.2. Sedimentological Considerations

Sediments accumulating in near-shore areas are found in a wide variety of environments; including estuaries, fiords, bays, lagoons, deltas, tidal flats, the continental terrace and marginal basins. These environments, which will be collectively referred to as the *continental margin*, are geologically and oceanographically diverse. Many attempts have been made to identify specific characteristics of the sediments which can be used to distinguish the various sedimentary environments (see e.g. Shepard *et al.*, 1960). When textural, mineralogical and faunal criteria have been used this has only proved possible in those instances in which the climatic, geological or oceanographic environments have been similar. However, much more success has been achieved in identifying the sources and distribution paths, for the components of modern sediment in near-shore areas, by studying the mineralogy of the sediment. For example, Van Andel and Postma (1954) have determined the sources of the sediments in the Gulf of Paria by means of a detailed analysis of the mineralogical compositions of the gravel, sand and clay fractions of the surface sediments. Van Andel and Poole (1960) restricted their attention to the heavy mineral fraction ($\rho > 2{\cdot}89 \text{ g cm}^{-3}$) of the surface sediments of the northern Gulf of Mexico and were able to delineate several different sources of sand-grade sediment and to identify a series of mixtures of these on the shelf. The techniques used by these authors were derived from the work of Edelman (1933) and the application of the concept of the "petrological province" to the determination of the sources of sediments has been extensively documented (see e.g. Van Andel, 1959, 1964).

The distribution of sediments on the continental margins of the World Ocean has been fundamentally affected by a series of Quaternary fluctuations in sea level. Almost all near-shore environments have therefore undergone alternating periods of exposure and drowning, or a series of transgressions and regressions, over the past million years or so. Since the beginning of the last recession of the glaciers, which began approximately 20 000 B.P., sea level has risen approximately 120 m (see Curray, 1960, 1965; Jelgersma, 1961; Shepard, 1963; Milliman and Emery, 1968).

Recent marine sedimentation, therefore, started at different times over the past 20 000 years on different parts of the continental shelves and nearshore zones of the world depending upon local structure and topography. The most important factors controlling the conditions under which Recent sediments are deposited are the rates of sediment supply, the rate of sea level rise (or fall) and the intensity and persistence of local water movements which reworked the unconsolidated material and supplied new material. Hence, near-shore marine sediments are variable in texture and composition, not merely because they accumulate in environments where physical conditions are themselves quite variable, but also because they occur in areas where very recent transgressions have taken place. Shelf sediments may, therefore, be composed of old, transgressive (reworked) or modern sediment, or mixtures of these three components which may lie exposed side by side, or be superimposed one upon another in vertical sequences. The recognition of modern sediments, that is sediments deposited under present-day conditions, is, therefore, a fundamental consideration in the interpretation of the mineralogical and chemical compositions of unconsolidated sea-floor deposits. Early workers had assumed that all surface sediments represented modern deposits until Shepard (1932) pointed out that Pleistocene and low sea-level sediments are still exposed on many parts of the continental shelf, often in deeper water than the present-day sediment. This old material is referred to as *relict* sediment and is distinguishable from *modern* or *equilibrium* sediment on the basis of textural, compositional and palaeontological evidence (Curray, 1965).

The distribution of relict and modern sediment facies in the northern Gulf of Mexico, an area of relatively low energy conditions, has been mapped by Curray (1960), and Emery (1968) has extended this to cover most of the shelves of the world. In environments of higher energy, bottom sediment consisting of complex mixtures of relict and modern sediment components of all size grades may be in transit (Stride, 1963; Belderson *et al.*, 1971). This material may be thicker than the underlying deposit, which may itself be relict or modern, and may occur as small patches (Kenyon and Stride, 1970). Because of clear evidence that relict sediment is undergoing modification in the present-day environment Swift *et al.*, (1971) have reappraised the concept of modern vs. relict sediments. It is, therefore, essential that in the interpretative geochemistry of modern near-shore sediments, a distinction should be made between modern, relict and transitory deposits.

33.3. Composition of Marine Sediments

Sediments and sedimentary rocks consist of a number of different components which may, or may not, have different sources. There have been several attempts to classify these components, including those by Krynine (1948), Goldberg (1954) and Arrhenius (1963).

The most important of these components are those of *detrital*, *authigenic* and *biogenous* origin. *Detrital* components consist of rock fragments and minerals derived from the continents by weathering, and supplied to the ocean by rivers, ice, or wind. The *authigenic* fraction comprises inorganically derived precipitates of a number of types which are formed in the water in near-shore and off-shore areas, and which can also form after deposition of the bulk sediment. The *biogenous* fraction is derived from the inorganic skeletal remains of marine organisms or from dispersed organic material produced by the degradation of organic tissues. This fraction may be derived from the overlying water or may be supplied from continental areas. In near-shore sediments these three principal components occur in all possible mutual proportions to yield a wide range of sediment types which may be distinguished one from another by their mineralogy, geochemistry or biochemistry.

The concentration of a given element in a sediment is governed by its concentration in, and the relative proportions of, the various components discussed above. Many elements show tendencies to be partitioned between two or more mineral components of a sediment, and the distribution of a given element cannot be described in terms of, for example, mineralogy alone. There is also a very important elemental partitioning effect brought about by a textural control on the mineralogy of sediments. This factor is of critical importance for near-shore sediments in view of their extreme textural variability.

Data given by Grout (1925), who carried out partial chemical analyses of separated grain size fractions of a number of glacial and lacustrine sands and muds, first illustrated how important textural control is. More recent data for some surface sediments from the shelf area of the Barents Sea are shown in Table 33.1. In these sediments: (i) the SiO_2/Al_2O_3 ratios decrease from the sand to the clay fractions indicating a larger proportion of free quartz relative to clay minerals in the coarser fractions; (ii) the sand fraction contains more K_2O, relative to Al_2O_3, denoting an enrichment in K-bearing feldspar; (iii) the silt fraction is enriched in TiO_2 which is probably present in Ti-bearing heavy minerals; (iv) the clay fraction is enriched with Fe_2O_3 and MgO which reside predominantly in clay mineral lattices; and, (v) almost all the minor elements listed in Table 33.1 appear to be selectively partitioned into either the silt fraction or the clay fraction. The greatest differences between the fractions appear to be for zirconium, which is much more abundant in the silt (probably occurring in zircon), and for Rb, Zn, Ni and Cu which are much more abundant in the clay, and probably occur in the lattices of mica and chlorite (Wright, 1972). It may be concluded, therefore, that sediments should not be compared on the basis of chemical composition unless they have similar textural characteristics. The chemical reactions which occur within Recent sediments after deposition, which are collectively known as *diagenesis*, are important in bringing about profound changes in the bulk chemical compositions. Most of these chemical changes take place at, or close to, the sediment/water boundary and for this reason the compositions of the surficial sediments in many near-shore environments are not preserved on burial. This topic is treated fully in Chapters 30 and 32.

The partitioning of elements between the various mineral components of sediments and sedimentary rocks has been studied using several different methods. For example, Hirst and Nicholls (1958), Hirst (1962b), Arrhenius (1963), Chester and Hughes (1967) and Piper (1971) have used various chemical leaching methods to extract quantitatively the different fractions of pelagic sediments, near-shore sediments and sedimentary rocks. Other approaches are to examine data on the overall chemical composition of a sediment, using regression and correlation methods to identify covarying groups of elements, and also to apply multivariate techniques to identify the factors controlling the distribution of individual elements.

TABLE 33.1.

Chemical composition of various grain size fractions from Barents Sea sediments[a] *(from Wright, 1972).*

	Sand[b]	Silt[c]	Clay[d]
SiO_2	83·00	73·87	48·73
Al_2O_3	7·57	11·08	19·12
TiO_2	0·16	0·71	0·98
Fe_2O_3[e]	1·60	2·53	9·48
MgO	0·53	1·37	3·17
CaO	0·82	1·80	0·89
Na_2O	—	2·48	—
K_2O	1·97	2·23	3·49
MnO	0·043	0·058	0·094
P_2O_5	0·07	0·12	0·37
CO_2	—	1·74	0·63
As	—	7	20
Ba	—	769	671
Ce	—	61	121
Cu	—	20	96
La	—	34	66
Ni	—	22	83
Pb	—	17	53
Rb	—	65	173
Sr	—	280	177
Th	—	10	18
Y	—	34	28
Zn	—	49	245
Zr	—	393	131

[a] Major element oxides as wt.%, minor elements in ppm. Analyses by X-ray fluorescence spectroscopy.
[b] Mean of 6 analyses.
[c] Mean of 50 analyses (Na_2O based on 15 analyses)
[d] Mean of 60 analyses
[e] Total Fe as Fe_2O_3.

The factors controlling the chemical composition of marine sediments are complex, and for this reason it is convenient to deal initially with sediments whose compositions are dominantly controlled by one of the important components, i.e., detrital, biogenous or authigenic.

33.4. Detrital Sediments

Studies of the chemical composition of essentially detrital sediments from near-shore environments have been carried out by, for example, Hirst (1962a, b), Moore (1963, 1968), Nota and Loring (1964), Loring and Nota (1968) and White (1970). A compilation of the average sediment compositions from some of the environments studied by these authors is given in Table 33.2.

Table 33.2.
Chemical compositions of some near-shore detrital sediments[a]

	1	2	3	4
SiO_2	78·50	55·02	—	64·8
Al_2O_3	6·12	16·61	6·95	13·0
TiO_2	0·45	0·72	0·50	0·9
Fe_2O_3[b]	4·69	7·26	2·12	4·3
MgO	0·89	2·19	0·63	3·0
CaO	6·70	1·57	1·19	3·3
Na_2O	1·18	2·76	1·73	2·9
K_2O	1·14	2·32	1·73	1·9
MnO	0·09	0·26	0·02	0·08
P_2O_5	0·11	0·18	—	—
CO_2	2·17	2·25	—	—
B	60	81	59	58
Ba	301	394	310	661
Co	8	12	3	12
Cr	31	93	33	191
Cu	7	17	18	28
Ga	7	22	9	20
Ni	16	31	11	100
Pb	13	22	25	10
Rb	47	76	—	—
Sr	147	210	125	296
V	79	146	45	93
Y	—	—	30	32
Zr	413	169	265	176

1. Gulf of Paria. Mean of 12 platform sands (SiO_2 based on 3 analyses). Analyses by wet chemical methods (major elements), and emission spectroscopy (minor elements) (from Hirst, 1962a, b).
2. Gulf of Paria. Mean of 6 basin clays (SiO_2 based on 3 analyses). Analytical methods as in 1 (from Hirst 1962a, b).
3. Buzzards Bay, Massachusetts. Mean of 125 sands and muddy sands. Analyses by emission spectroscopy (from Moore, 1963).
4. Continental shelf, Oregon–Washington. Mean of 45 sands and silts. Analyses by X-ray fluorescence spectroscopy (major elements) and emission spectroscopy (minor elements) (from White, 1970).

[a] Major element oxides as wt.%; minor elements in ppm.

[b] Total Fe as Fe_2O_3.

Hirst (1962a, b) examined the geochemistry of sediments of the Gulf of Paria whose sedimentary characteristics had been previously described in detail in the classic work of Van Andel and Postma (1954). The samples studied ranged from deltaic and platform sands to open Gulf clays. Hirst concluded that, for the most part, the major and minor elements were located within the lattices of the detrital minerals. Thus, the range of chemical compositions observed in the sediments was controlled by variations in the proportions of the major minerals, i.e. quartz, feldspar, illite, montmorillonite, kaolinite and in those of the minor amounts of heavy minerals. Exceptions to this were found for the distributions of iron, calcium and manganese.

The controls exerted by bulk mineralogy can be seen from Table 33.2 and from Figs 33.1, 33.2 and 33.3. (i) The platform sands have higher SiO_2/Al_2O_3

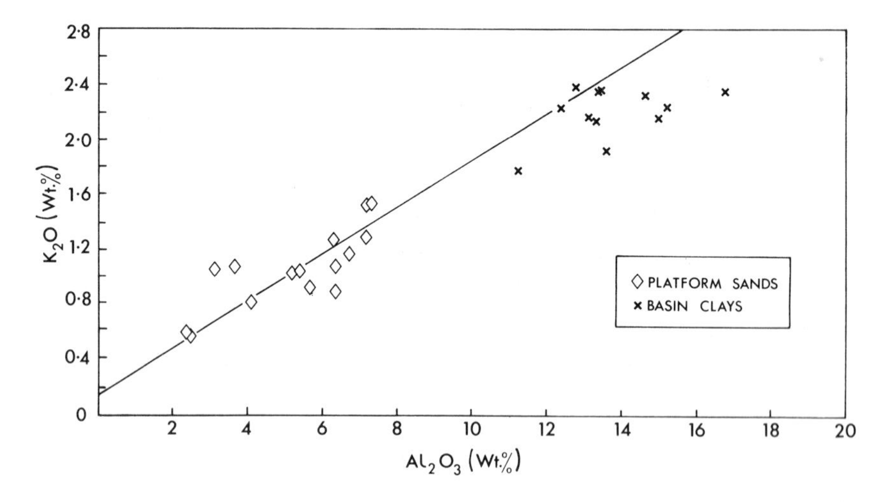

FIG. 33.1. Correlation between K_2O and Al_2O_3 in some Recent sediments from the Gulf of Paria (data from Hirst, 1962a).

ratios than do the clays because of the higher ratio of quartz/clay minerals in the sands. (ii) The platform sands have higher K_2O/Al_2O_3 ratios than do the clays because of the higher feldspar content of the former. (iii) The relatively high proportion of TiO_2 in the sandy sediments is most probably contributed by titanium-bearing detrital minerals, notably ilmenite (Hirst, 1962a), but also anatase and rutile which sometimes make up the second most abundant transparent heavy mineral in the sand fractions (Van Andel

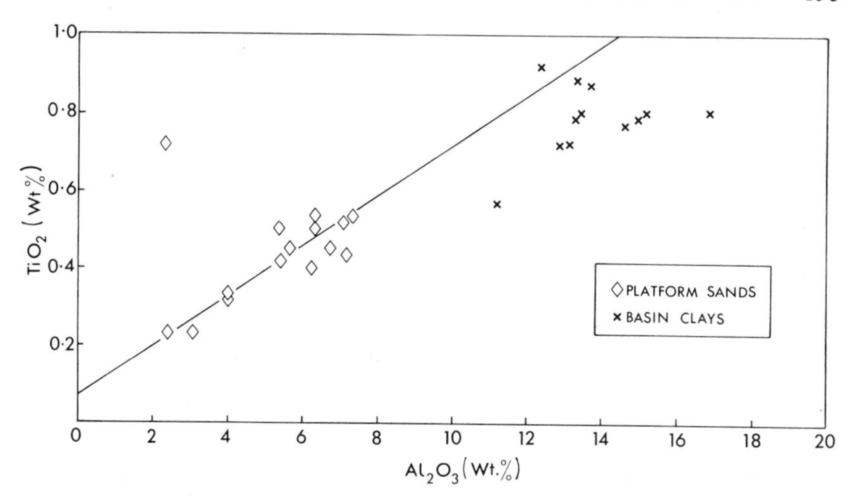

FIG. 33.2. Correlation between TiO_2 and Al_2O_3 in some Recent sediments from the Gulf of Paria (data from Hirst, 1962a).

and Postma, 1954, Table D-2). (iv) the basin clays generally have higher MgO/Al_2O_3 ratios which can be explained by the greater relative abundance of montmorillonite in these clays.

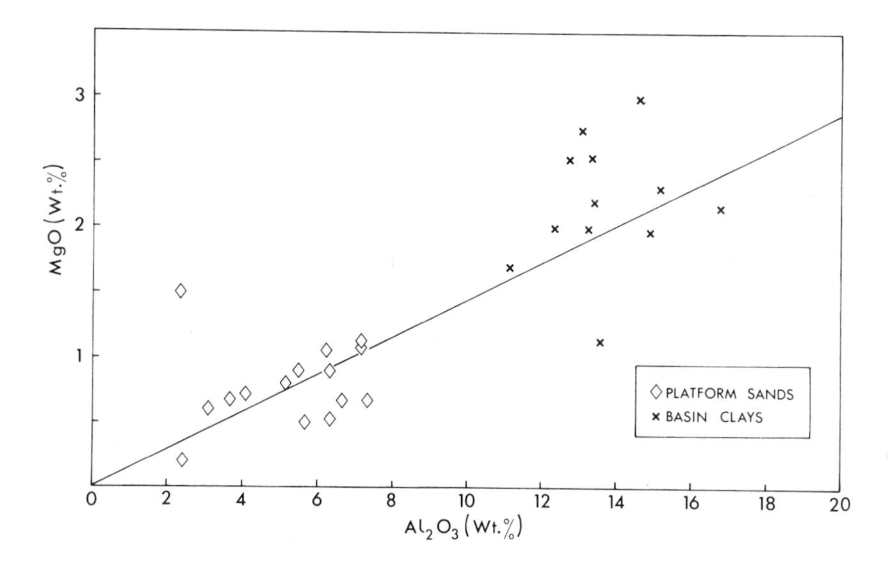

FIG. 33.3. Correlation between MgO and Al_2O_3 in some Recent sediments from the Gulf of Paria (data from Hirst, 1962a).

The influence of non-detrital factors on bulk sediment composition can be seen from the distribution of Fe_2O_3 (Fig. 33.4). The relationship between Fe_2O_3 and Al_2O_3 is complicated by the presence of authigenic Fe-bearing minerals, such as limonite (in the form of residual concretions), glauconite or (since the K:Fe ratio is low) reconstituted illite. In sediments free from these minerals, the Fe_2O_3/Al_2O_3 ratio is reasonably constant ($\sim$0·52), suggesting that the iron is held in structural positions within the aluminosilicates. Any iron exceeding that represented by the average Fe_2O_3/Al_2O_3 ratio can therefore be used, to a first approximation, as a measure of the amount of authigenic minerals present in the sediment.

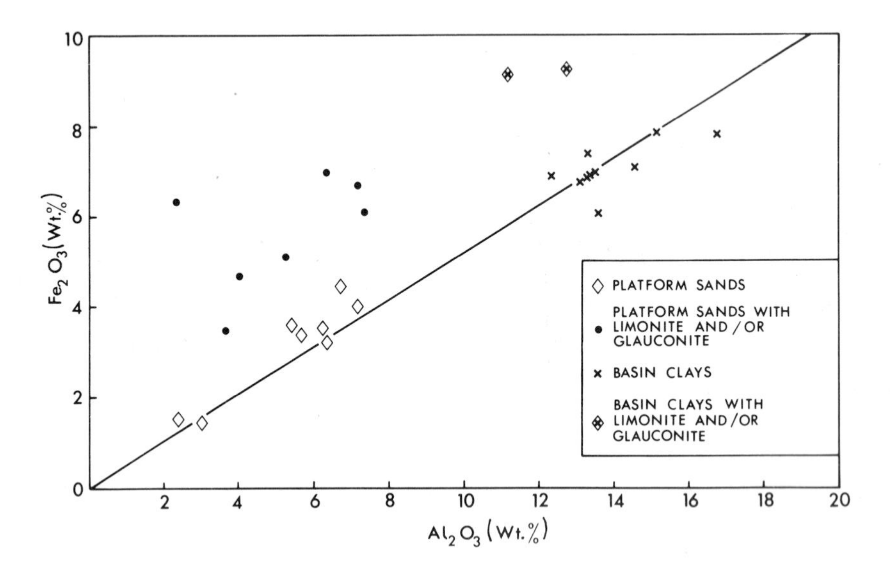

FIG. 33.4. Correlation between Fe_2O_3 and Al_2O_3 in some Recent sediments from the Gulf of Paria (data from Hirst, 1962a).

The considerable variations in the CaO content of the sands and clays in the Gulf is a consequence of the wide variations in the concentration of carbonate shell debris in them. When the carbonate contents are low, the CaO and Al_2O_3 correlate reasonably well suggesting that much of the non-carbonate CaO is held in the aluminosilicates.

The amount of P_2O_5 in the Gulf of Paria sediments is rather low, and clays on average contain more than do the sands. In the clays the P_2O_5 probably occurs in finely-divided apatite (Hirst 1962a), or as a ferri-phosphate in the dispersed iron oxides on mineral grain surfaces, which is suggested by the significantly higher P_2O_5 contents of the iron-rich fractions separated from some of the platform sands. Thus, the P_2O_5 contents (0·46 – 0·89%) of these concentrates (total Fe_2O_3 = 23·1 – 41·0%) were considerably higher than those of the original samples (ave. P_2O_5 = 0·11%). In the Gulf sediments, manganese is concentrated in the basin clays relative to the platform sands. In the sediments the manganese is held in lattice positions in the aluminosilicates and/or as dispersed oxides in close association with the dispersed iron phases discussed above. There is, in fact, a direct correlation between the manganese concentration in the sediments and the "excess iron" calculated from the total Fe_2O_3:Al_2O_3 ratio (see above and Fig. 33.4). In this connection it is interesting to note that iron-rich fractions separated from platform sands are significantly enriched with manganese (0·16 to 0·49% (Hirst, 1962a, Table 1)).

Although there are some subtle differences produced by contributions from biogenous and authigenic sources, the geochemistry of the minor elements in the Gulf of Paria is mainly dominated by transport and deposition of these elements within the lattices of the detrital minerals. Thus, the minor element contents correlate to a greater or lesser extent with one or more of the major elements. The detrital character of most of the minor elements is shown by the high degree of correlation of B, Ba, Cr, Cu, Cs, Ga, Li, Ni, Pb, Rb, and V with Al_2O_3 which is itself largely detrital in origin. All of these elements are considered by Hirst (1962b) to be present in the clay minerals, feldspars (Ba, Sr and Rb), glauconite (B) or heavy minerals (Cr, Ni, Cu and Pb). Zirconium appears to be the only element which is not associated with a major phase probably because of its independent distribution as zircon.

In addition to these mineral-element associations there are other groupings which are not controlled by the detrital fraction. Strontium appears to be correlated with CO_2, thus suggesting that much of the strontium is present in carbonate debris, which in this area is mainly aragonitic (Hirst, 1962a). Strontium levels in carbonate-poor sands and clays ($\sim$140 ppm) probably represent the concentrations of this element in structural positions in clay minerals and feldspars. In carbonate-free sands and clays the Sr/Rb ratios (where Rb is used as a measure of the aluminosilicate contribution) are significantly higher in the sands indicating that the strontium is most probably held in a plagioclase feldspar rather than in clay minerals.

Some minor elements, notably, Co, V and Zr, appear to correlate fairly well with the total Fe_2O_3; this suggests that these elements are associated with dispersed oxide phases. This is supported by their significantly higher concentrations in the iron-rich fractions separated from the sandy sediments.

Although the major and minor element geochemistry of the recent sediments of the Gulf of Paria is largely dominated by the texture and the detrital mineralogy (Hirst, 1962a, b), biogenous and authigenic contributions also have a significant effect. An insight into this partitioning of both major and minor elements can be obtained using statistical techniques. For example, Fig. 33.5 summarizes the associations of the major and minor elements in a cluster diagram constructed from the Kendall rank correlation coefficient between all pairs of elements. This approach suggests that four main interrelated groups of elements are responsible for the geochemical variability in the sediments of the Gulf of Paria. Zirconium is the only element which does not fall conveniently into these groups, probably because it occurs exclusively in zircon.

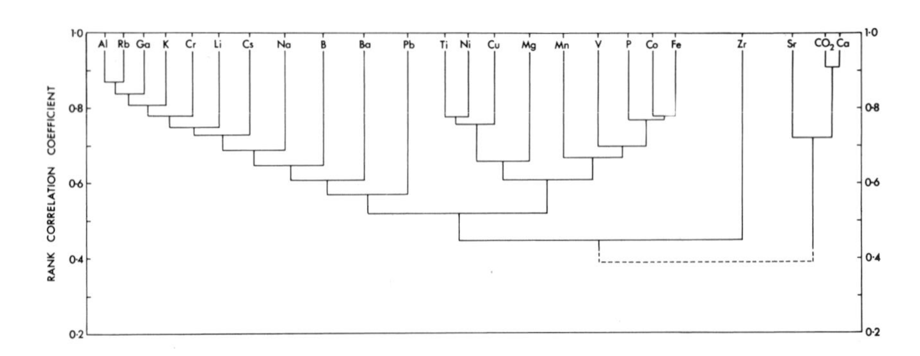

FIG. 33.5. Cluster diagram of the chemical compositions of some Recent sediments from the Gulf of Paria (data from Hirst, 1962a, b).

Three of the four groups shown in Fig. 33.5 can be readily identified with the controlling factors discussed previously. The group containing Al, K, Na, Rb, Ga, Cr, Li, Cs, B, Ba and Pb is controlled by the aluminosilicates; that containing Fe, Co, P, V and Mn is controlled by the dispersed oxide phases; that consisting of Ca, Sr and CO_2 is controlled by the carbonates. The remaining group, containing Ti, Ni, Cu and Mg was not recognized previously. It may be an ilmenite/rutile group consisting of ferroan and magnesian varieties of these two minerals which also contain appreciable concentrations of Ni and Cu as well as Ti.

Spencer *et al.* (1968) have used R-mode factor analysis to re-examine the original data of Hirst (1962a, b) on the geochemistry of the sediments of the Gulf of Paria. Their analysis showed that five factors could account for almost 90% of the total variance in the data. These five factors are as follows.

1. A quartz dilution factor, which affects Ti, Al, K, Cs, Cr, Ni, Ga, Cu, Li and Rb. Plus a negative loading for Zr.
2. A carbonate factor which affects Ca, Sr and CO_2.
3. An oxide factor, which affects Fe, Mn, V and Co.

4. An illite/glauconite factor, which affects Na and B.
5. A montmorillonite factor, which influences Mg, Ba and Be.

Spencer *et al.* also suggested that the significant loadings ($>0{\cdot}2$) which they found for phosphorus and iron on Factor 2 could be accounted for by the contribution of apatite and ferric hydroxide to the carbonate factor. However, it has been suggested above that the correlation of P_2O_5 with excess Fe_2O_3 in the Gulf of Paria sediments may also be explained by the presence of a ferriphosphate or by phosphate adsorption onto the hydrated ferric oxides which are present as thin coatings on sand grains and shell debris in the platform sands (Van Andel and Postma, 1954).

The application of statistics to the chemical analyses of the sediments of the Gulf of Paria has revealed that although the chemical composition is primarily controlled by the detrital minerals, biogenous and authigenic phases can also be recognized as being important.

The chemical compositions of sediments from Buzzards Bay, Massachusetts, have been presented by Moore (1963). The sediments are predominantly sands and muddy sands containing major proportions of quartz, feldspar and clay minerals, together with minor amounts of shell debris. Moore concluded that the distributions of Al, Cr, Co, Cu, Fe, La, Mg, Mn, Ni, Pb, Sc, Ti, V and Y were governed entirely by the amounts of fine-grained aluminosilicates, whereas Ba, Ca, Na and K were related to both the aluminosilicates and the feldspar. Boron appeared to be controlled largely by the amount of tourmaline; zirconium by the amount of zircon; and strontium by the amount of shell debris present. It appeared that authigenic minerals made no contribution to the geochemistry of the sediments, with the exception of a single muddy sample in which molybdenum was relatively high.

A factor analysis (Table 33.3) of the data given by Moore (1963) allows the interpretations given above to be refined. Factor 1 may be interpreted as an aluminosilicate factor, and it dominates the distributions of Al, Fe, Mg, Na, Ti, Cr, Co, Ga, La, Ni, Sc, V and Y. Factor 2 is most probably a carbonate factor and controls the geochemistry of Ca and Sr. Factor 3 is a bipolar factor with significant loadings for K, Na and Ba, and a negative loading for B. This is probably a combined feldspar (K, Na and Ba)—tourmaline (B) factor in which the feldspar is negatively correlated with the tourmaline. Factor 4 is probably a heavy mineral factor which controls the geochemistry of Zn and Ti. This statistical analysis confirms the conclusions of Moore but shows that a discrimination between different terrigenous components can also be made from bulk chemical analyses.

It is also possible to discriminate between the various components of dominantly terrigenous sediments by analysing separate size fractions as illustrated by Price and Wright (1971) and Wright (1972). These workers separated a series of terrigenous sands, silty sands and muds from the south

TABLE 33.3.
Quartimax factor matrix of chemical compositions of the Buzzards Bay sediments analysed by Moore (1963)[a].

Variable	1	2	3	4	Communality
Al	−0·908	—	—	—	0·877
Ti	−0·686	—	—	−0·449	0·897
Fe	−0·948	—	—	—	0·951
Mg	−0·939	—	—	—	0·849
Ca	0·023	−0·941	—	—	0·869
Na	−0·837	—	0·425	—	0·972
K	−0·752	—	0·497	—	0·851
Mn	−0·671	—	—	−0·322	0·763
B	—	—	−0·610	—	0·457
Ba	−0·447	—	0·414		0·600
Co	−0·935	—	—	—	0·914
Cr	−0·966	—	—	—	0·983
Cu	—	—	—	—	0·043
Ga	−0·970	—	—	—	0·981
Ni	−0·953	—	—	—	0·986
Pb	−0·836	—	—	—	0·701
Sr	—	−0·909	—	—	0·793
V	−0·935	—	—	—	0·962
Y	−0·835	—	—	—	0·869
Zr	—	—	—	−0·825	0·836
Percentage of variance	78·9	12·0	5·1	4·0	

[a] Loadings less than 0·300 omitted

western Barents Sea into sand (>63 μm), silt (8–63 μm) and clay (<2 μm) fractions, and carried out mineralogical and chemical analyses on each of them. The sand fraction consisted predominantly of quartz with minor amounts of feldspar. The silt fraction was composed of quartz, feldspar, mica, calcite and traces of hornblende. The minor element data on this fraction were used to outline areas of the shelf having distinctive mineralogies which are described below.

1. The distributions of zirconium and thorium were used to define areas of heavy mineral enrichment; these were found on banks and in near-shore areas.
2. Areas of feldspar- and mica-rich sediments were distinguished on the bases of Ba/Rb and Sr/Rb ratios, and also of the distinction between potash feldspar and plagioclase feldspar, using these same ratios in conjunction with those of Ba/K_2O/Al_2O_3.
3. Areas of apatite enrichment could be delineated using the distributions of Ce, La and Y.

The clay fractions contained illite and chlorite with minor amounts of kaolinite and montmorillonite, and the elements Ba, Cu, Ni, Pb and Zn were preferentially enriched in one or other of these minerals.

33.5. Biogenous Sediments

Sediments containing substantial amounts of biogenous material are found in most types of near-shore environments, such as restricted basins and fiords, estuaries, marginal basins and areas of the continental shelf. This material may consist of calcareous and/or siliceous skeletal debris together with finely dispersed organic material. In general, the calcareous sediments are sand-sized, whereas the organic and siliceous ones belong to the clay and silt size classes. Because of this, textural factors, which often complicate our understanding of detrital sediments, do not greatly hinder the interpretation of the geochemistry of the calcareous sediments.

According to Emery (1968) biogenous sediments are the equilibrium sediments to be expected, at least on continental shelves, in relatively low latitudes. By biogenous sediments Emery (1968) actually meant calcareous sediments, and shelf carbonates are indeed traditionally, regarded as being characteristic of low latitudes. However, Chave (1967) and later Lees and Buller (1972), have shown that calcareous skeletal sands are widespread at latitudes as high as 60°. Carbonates from tropical and temperate latitudes can be distinguished from those of high latitudes on the basis of the skeletal grain types; thus, corals and calcareous algae are mainly restricted to low latitudes, whereas high latitude carbonates are dominantly composed of foraminifera and molluscs. In contrast, high concentrations of organic material, with, or without, siliceous skeletal debris can occur on continental shelves and in marginal basins at virtually all latitudes.

33.5.1. Siliceous-Organic Sediments

Sediments containing relatively large amounts of biogenous silica and dispersed organic matter are found in the Sea of Okhotsk (Bezrukov, 1955), the Sea of Japan (Solov'yev, 1960), the Southern California Borderland basins (Emery, 1960), the Gulf of California (Byrne and Emery, 1960; Van Andel, 1964; Calvert, 1966), the Bering Sea (Lisitsyn, 1966), Saanich Inlet, British Columbia (Gross, 1967), the Cariaco Trench (Dorta and Rona, 1971), and on the South East Arabian (Lees, 1937) and South West African (Calvert and Price; 1970a, 1971a) shelves. Organic matter accumulates in the sediments of many other areas with restricted circulations, but siliceous skeletal material does not. Examples of such areas are: the Black Sea (Murray, 1900; Strakhov, 1962); the Baltic Sea (Manheim, 1961a); and many Norwegian fiords (Strøm, 1936; Doff, 1969; Piper, 1971). Estimates of accumulation rates, together with organic matter contents, for some near-shore biogenous sediments are shown in Table 33.4. In the areas listed above the bulk of the organic material is probably derived from planktonic sources. However, some near-shore areas do accumulate terrestrial organic debris (Scholl, 1963; Carrigy, 1956),

TABLE 33.4.

Accumulation rates and organic carbon contents of some near-shore sediments.

Area	Accumulation rate (mm yr^{-1})[a]	Organic carbon[b] (wt.%)	$CaCO_3$[b] (wt.%)
Santa Barbara Basin[c]	1·14[d]	3·2	10·5
Santa Catalina Basin[c]	0·29	5·0	12·8
Gulf of California, Guaymas Basin[e]	2·73	2·91	2·9
Gulf of California, Western slope	0·60	6·55	11·6
South West African shelf[f]	0·27	22·3	15·4

[a] All data obtained by radio-carbon dating of either organic or carbonate carbon in sediment subsamples
[b] Carbon and carbonate values for the surface sediment only
[c] From Emery (1960, Table 19), and Emery and Bray (1962, Table I)
[d] An accumulation rate of 3·9 mm yr^{-1} has recently been obtained for a sediment core in the central Santa Barbara Basin using the ^{210}Pb method (Koide *et al.*, 1972)
[e] From Van Andel (1964, Table XVII)
[f] From S. E. Calvert and N. B. Price, unpublished.

but these are very localized in extent. Rates of accumulation in these areas are high and the deposited organic material is relatively rapidly buried, and for this reason is preserved. An important additional feature of near-shore biogenous sediments, resulting from their rapid accumulation rates, is the formation of anoxic conditions below the oxidised surface layer. A discussion of oxic and anoxic sediments is given in Section 33.7.

The factors controlling the chemical composition of biogenous sediments can be illustrated by considering those deposited on the South West African continental shelf. Some of the sediments of this area have been described by Marchant (1928) and Copenhagen (1934, 1953) who both drew attention to the area of diatomaceous, sulphide-rich, azoic sediment on the continental shelf off Walvis Bay. The distribution of several different sediment facies on this shelf has been described by Calvert and Price (1970a, 1971a). Organic- and diatom-rich muds, containing sulphide, occur on the inner shelf area off Walvis Bay (Fig. 33.6): terrigenous sands and gravelly sands occur on the inner part of the shelf to the north and south of Walvis Bay; and calcarenites and muddy calcarenites occur on the central and outer shelf and also on the

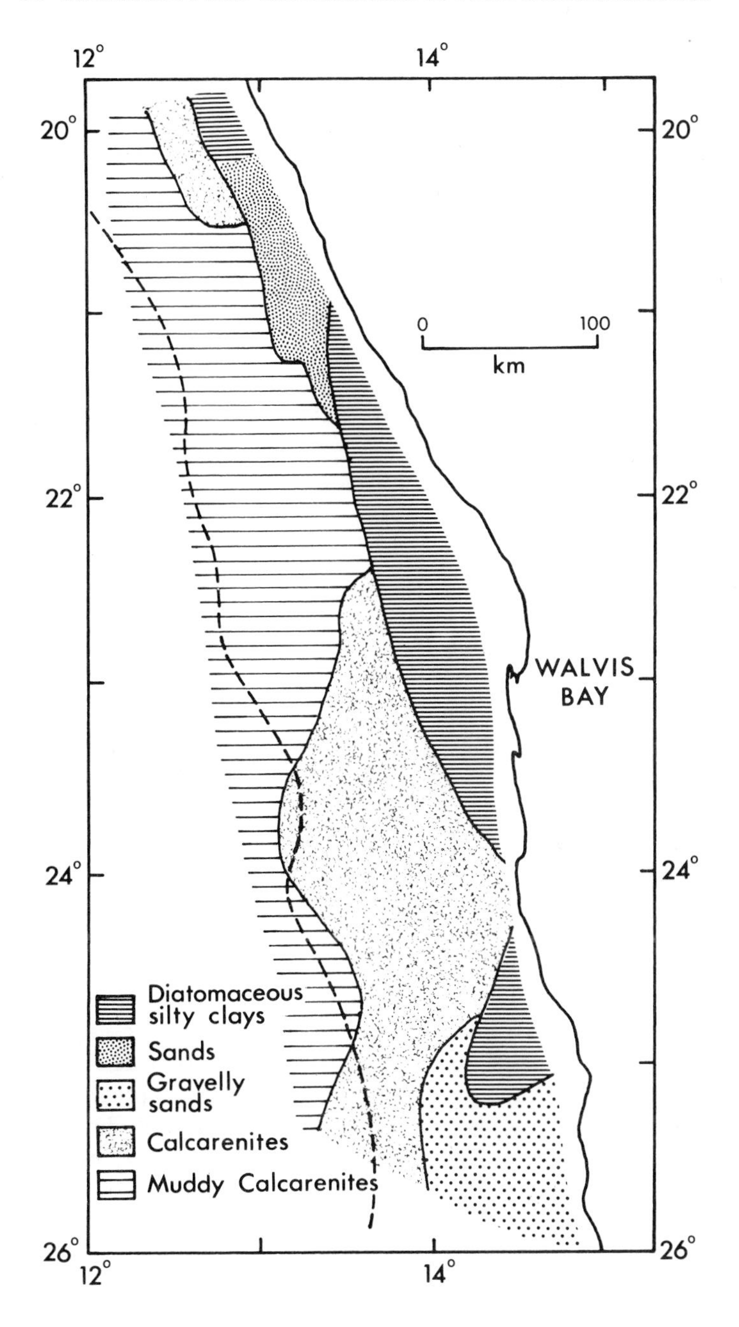

FIG. 33.6. Recent sediment facies of the South West African continental shelf.

upper slope. Some sediments also contain glauconite and phosphorite concretions. The average compositions of these three facies which are shown in Table 33.5 indicate that the proportions of biogenous and terrigenous components vary widely.

TABLE 33.5.

Average compositions of the principal facies of the South West African shelf sediments.[a]

Facies	$CaCO_3$	Organic matter[b]	Opal[c]	Phosphorite[d]	Terrigenous
Diatomaceous silty clays	8·8	15·9	3·62	4·2	34·9
Terrigenous sands	12·2	3·2	0·0	1·4	83·2
Calcarenites	71·5	7·6	0·0	2·8	18·1

[a] Values in wt.% of total dried sediment, corrected to a salt-free basis. (unpublished data by S. E. Calvert and N. B. Price)
[b] Organic carbon ×1·8
[c] Diatomaceous silica determined by infra-red spectrophotometric methods (Chester and Elderfield, 1968)
[d] Estimated from P_2O_5 values.

The near-shore diatomaceous, organic-rich muds are probably the most recent equilibrium sediments on the shelf and contain up to 25% (wt.) of organic carbon (Fig. 33.7). Although the other sediments contain some modern constituents they consist mainly of reworked materials from the Holocene transgression which are principally broken and abraded macro- and microfaunal shell material together with glauconite and phosphorite.

The chemical compositions of the organic-rich sediments are shown in Table 33.6 together with comparative data for the other sediment facies. Relative to the calcarenites and the terrigenous sands, the diatomaceous muds have: (i) high SiO_2/Al_2O_3 ratios resulting from the presence of biogenous silica (Fig. 33.8); (ii) relatively high TiO_2/Al_2O_3 ratios as a consequence of the presence of fine-grained titanium-bearing detrital minerals; (iii) high Fe_2O_3/Al_2O_3 and S/Al_2O_3 ratios because of the presence of pyrite; (iv) high P_2O_5/Al_2O_3 ratios as a result of the presence of phosphorite (see Section 33.6.1.1); and (v) high MgO/Al_2O_3 ratios. Fe_2O_3 and K_2O are occasionally enriched in some outer shelf samples because of the presence of glauconite (Figs. 33.9 and 33.10). Phosphorite occurs in some calcarenites of the central shelf.

The relationship between the concentrations of MgO and Al_2O_3 in the entire suite of sediments from the South West African shelf is difficult to interpret, in part because of the poor precision of the MgO determinations

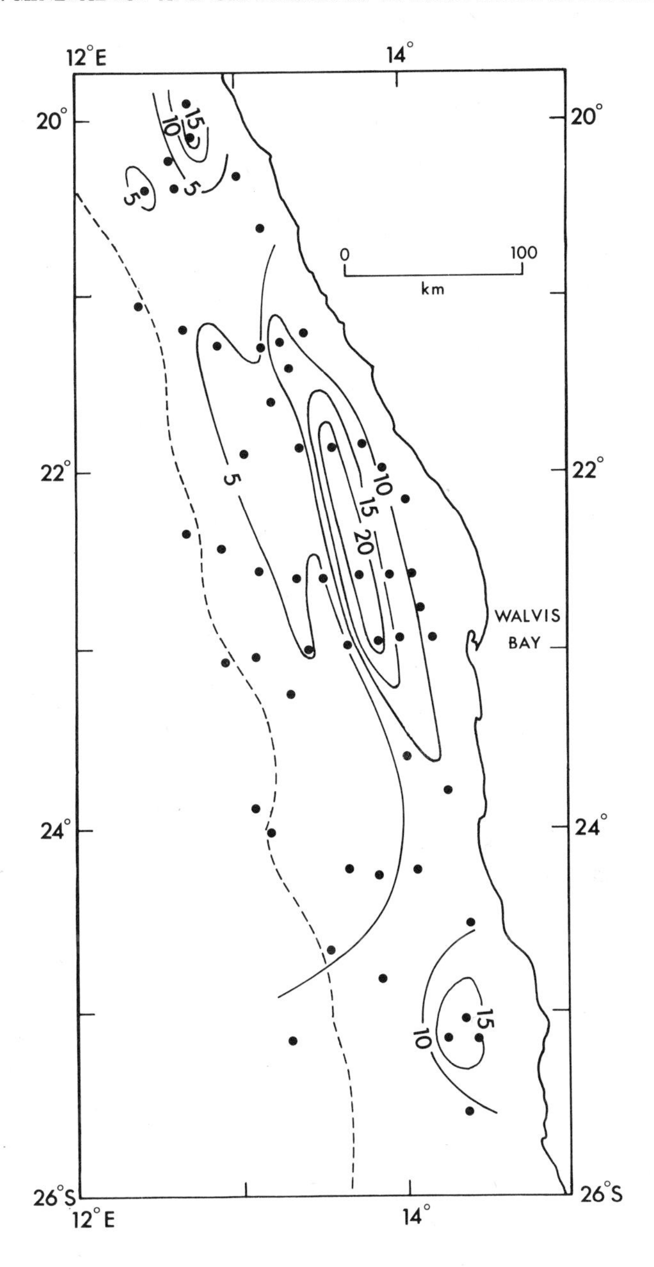

FIG. 33.7. Distribution of organic carbon (%wt) in some Recent sediments from the South West African continental shelf (from Calvert and Price, 1970a).

TABLE 33.6. *Chemical compositions of the sediments of the South West African shelf*[a]

	Diatomaceous clays	Calcarenites	Sands
SiO_2	51·23	9·20	65·60
Al_2O_3	2·92	2·24	4·74
TiO_2	0·23	0·12	0·30
Fe_2O_3[b]	1·62	1·13	1·44
CaO	7·26	43·40	9·60
MgO	1·33	0·68	0·88
K_2O	0·71	0·48	1·36
P_2O_5	1·59	1·08	0·54
S	1·59	0·50	0·37
CO_2	3·77	31·50	5·40
C_{org}	9·35	4·60	1·90
Ba	198	285	279
Cu	68	37	20
Mo	53	5	15
Ni	108	62	37
Pb	12	7	12
Rb	49	30	62
Sr	523	1082	363
U	41	10	7
Y	23	28	27
Zn	68	38	29
Zr	78	52	186

[a] Major element oxides as wt.%, minor elements in ppm (unpublished data by S. E. Calvert and N. B. Price). Analyses by X-ray fluorescence spectroscopy and gasometric methods (C and CO_2).
[b] Total Fe as Fe_2O_3.

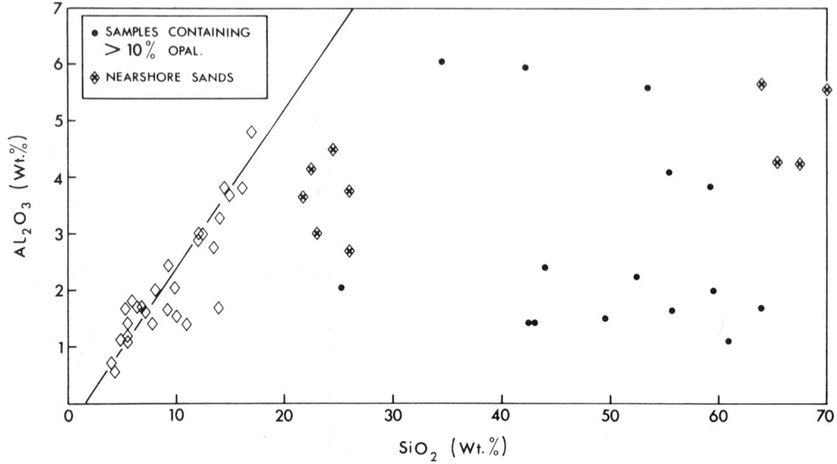

FIG. 33.8. Correlation between Al_2O_3 and SiO_2 in some Recent sediments from the South West African continental shelf (Calvert and Price, unpublished).

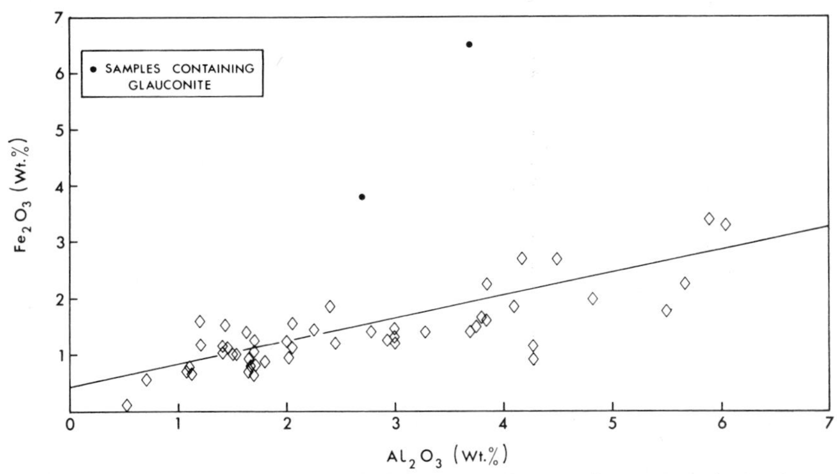

FIG. 33.9. Correlation between Fe_2O_3 and Al_2O_3 in some Recent sediments from the South West African continental shelf (Calvert and Price, unpublished).

made by X-ray fluorescence methods. Nevertheless, some of the scatter (Fig. 33.11) is real, and results from the inclusion of samples of organic-rich anoxic sediments having unusually high MgO/Al_2O_3 ratios. In this context, it is

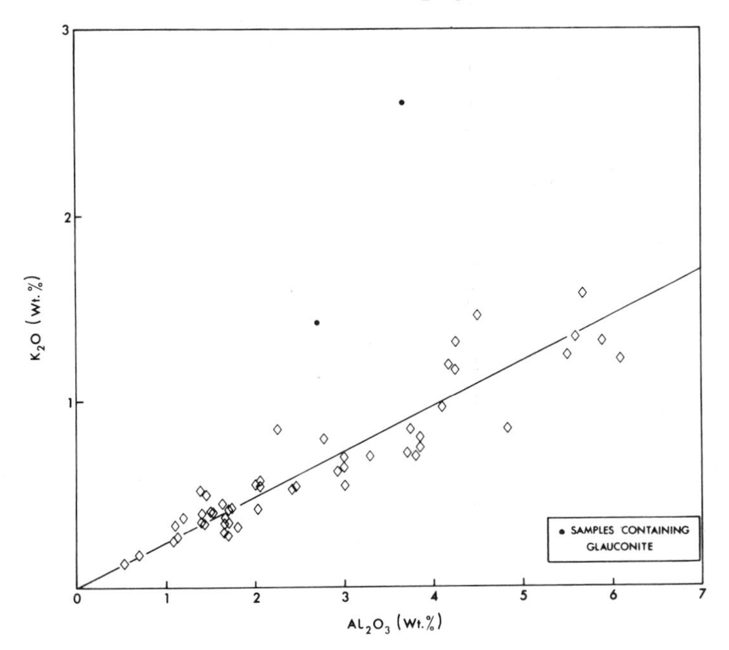

FIG. 33.10. Correlation between K_2O and Al_2O_3 in some Recent sediments from the South West African continental shelf (Calvert and Price, unpublished).

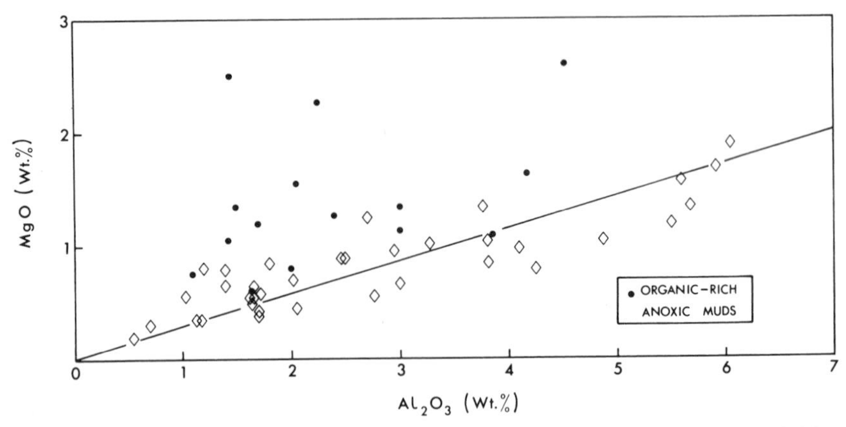

FIG. 33.11. Correlation between MgO and Al_2O_3 in some Recent sediments from the South West African continental shelf (Calvert and Price, unpublished).

interesting to note that for anoxic sediments Drever (1971) has described Mg enrichment resulting from the exchange of Mg for Fe in the clay minerals, a process which may also have operated on the South West African shelf. Alternatively, dolomite might be the source of the Mg in these sediments; however, it could not be detected by X-ray diffraction.

Among the minor elements listed in Table 33.5 Cu, Mo, Ni, U, and Zn appear to be enriched in the diatomaceous muds relative to the other sedimentary facies. The Cu, Ni, Pb and Zn contents correlate well with the amounts of organic carbon in the sediments (Calvert and Price, 1970a), and the same appears to be true for molybdenum. However, the distribution of uranium is complicated because it is present in phosphorite as well as in organic matter (see Section 33.7.5).

An R-mode factor analysis of the chemical data for the South West African shelf sediments shows that five factors account for more than 88% of the total variance of the data (Table 33.7). Factor 1, with high positive loadings of C, Mo, Cu, Ni, Zn, S, Mg, Rb and Pb, is represented by the near-shore organic-rich muds containing above average S and Mg. Rubidium values are also relatively high because the fine-grained nature of the sediments means that fine aluminosilicate material is reasonably abundant, and because rubidium substitutes for potassium in such material. However, unlike rubidium, potassium does not appear in Factor 1 because it is also present in feldspars. This is illustrated by the loadings on Factor 2, where potassium is included with Al, Ti, Zr and Rb. This factor is best interpreted as a general terrigenous one because those elements which are partitioned between both fine-grained and coarse-grained aluminosilicates and both silt- and sand-sized detritals, all appear in it. Factor 3 is clearly a pure carbonate factor which

TABLE 33.7.

Quartimax factor matrix of chemical data for 53 surface sediment samples from the South West African shelf.[a]

Variable	Factor 1	2	3	4	5	Communality
SiO_2	—	0·345	−0·901	—	—	0·985
Al_2O_3	—	0·914	—	—	—	0·952
TiO_2	—	0·803	−0·334	—	—	0·901
Fe_2O_3	—	0·434	—	—	0·792	0·893
CaO	−0·510	—	0·790	—	—	0·993
MgO	0·698	—	—	—	—	0·689
K_2O	—	0·743	—	—	0·555	0·942
P_2O_5	—	—	—	0·982	—	0·989
S	0·772	—	—	—	—	0·766
CO_2	−0·547	−0·305	0·738	—	—	0·978
C_{org}	0·958	—	—	—	—	0·961
Ba	—	—	—	—	—	0·842
I	—	—	0·504	—	—	0·981
Zr	—	0·725	—	—	—	0·731
Y	—	—	—	0·726	—	0·821
Sr	—	—	0·656	0·585	—	0·963
Rb	0·433	0·723	—	—	—	0·994
Zn	0·601	—	—	—	—	0·845
Mo	0·868	—	−0·355	—	—	0·935
U	—	—	—	0·869	—	0·948
Cu	0·846	—	—	—	—	0·849
Ni	0·801	—	—	—	—	0·826
Pb	0·426	—	—	—	—	0·701

[a] Loadings less than 0·300 are omitted.

also has high loadings of strontium and iodine, an element which is relatively abundant in the middle shelf calcarenites (Price and Calvert, 1973). Factor 4 is a phosphorite one which contains a high strontium loading and accounts for most of the variance of yttrium and uranium. Factor 5 which has high loadings of Fe and K can be interpreted as a glauconite one, and indeed samples containing abundant glauconite have high factor scores for it.

This analysis clearly demonstrates that several elements are partitioned between different sediment components; notably strontium between carbonate and phosphorites, iron and potassium between different aluminosilicates, and Si, Fe and Rb between diatomaceous and terrigenous sediments.

33.5.2. METAL ENRICHMENT IN ORGANIC-RICH SEDIMENTS

The concentrations of certain minor metals in the organic-rich diatomaceous sediments described below are considerably higher than those in near-shore detrital (Section 33.4) and pelagic sediments (Chapter 34). Evidence that the metals which are enhanced in the organic-rich sediments off South West Africa do not have a terrigenous origin is based on the facts that: (a) there are very low proportions of terrigenous components in many of the diatomaceous muds (Table 33.4); and, (b) the minor metal contents of the total sediments can be higher than those of many near-shore terrigenous sediments (see Table 33.2).

The minor element contents of a selection of near-shore organic-rich sediments (including those accumulating on the South West Africa shelf) are shown in Table 33.8. As is to be expected, the concentrations of those ele-

TABLE 33.8.

Minor metal contents of selected near-shore fine-grained organic-rich sediments (in ppm).

	1	2	3	4	5	6
Ba	198	—	270	—	—	—
Co	—	7	—	4	5	12
Cr	—	54	42	30	50	84
Cu	68	45	64	33	—	30
Mo	53	26	—	1	23	33
Ni	108	26	146	30	46	67
Pb	12	—	51	—	—	24
Rb	49	—	—	—	—	—
Sr	523	—	233	—	—	—
U	41	—	—	—	—	15
V	—	66	95	71	152	98
Y	23	21	—	—	—	—
Zn	68	71	—	—	—	147
Zr	78	76	—	—	—	82

1. S.W. Africa (S. E. Calvert and N. B. Price, unpublished). Analyses by X-ray fluorescence spectroscopy. Data presented on a salt-free basis.
2. Saanich Inlet, British Columbia (Gross, 1967). Analyses by emission spectroscopy.
3. Gulf of California (S. E. Calvert, unpublished). Analyses by emission spectroscopy.
4. Sea of Okhotsk (Strakhov and Nesterova, 1968). Analytical methods not described.
5. Mljet, Adriatic (Seibold *et al.*, 1958). Analyses by emission spectroscopy.
6. Black Sea (Glagoleva, 1961; Kochenov *et al.*, 1965; Pilipchuk and Volkov, 1966; Glagoleva, 1970; Belova, 1970; Lubchenko, 1970). Analytical methods not described.

ments associated principally with the detrital fractions of the sediments (Ba, Co, Cr, Rb and Zr) are low compared with those of fine-grained detrital sediments (Table 33.2). In contrast, certain elements appear to be relatively enriched in some organic-rich sediments, even when considered on a total-sediment basis. This is particularly true of Mo, Ni and Cu.

It was stated above that the concentrations of Cu, Mo, Ni, Pb and Zn in the South West African sediments can be significantly correlated with the amount of organic carbon present. Such correlations have frequently been taken as evidence that these minor metals are directly associated with the organic carbon in rocks (see Curtis, 1966). However, this may not in fact be the case because organic carbon is normally negatively correlated with grain-size in sediments (Trask, 1953; Van Andel, 1964). The increased minor metal contents of organic-rich sediments may, therefore, be the result of the presence of fine-grained metal-bearing components (for example sulphides) which may not be of direct biogenous origin. Until analytical methods are developed which are capable of distinguishing and evaluating the contributions of metals from these sources it will not be possible to establish their importance (see Calvert and Price, 1970a). However, estimates of the amount of "excess metals" in the organic-rich sediments of the South West African shelf may be derived by deducting the probable metal contribution from terrigenous, opaline and calcareous components from the total concentration in the sediment (Table 33.9). In many samples the amounts of "excess metals" calculated in this way are higher than the combined contributions from terrigenous and calcareous components. The most extreme enrichment is found for molybdenum which normally occurs in detrital sediments only at very low concentrations.

The data in Table 33.9 provide a measure of the relative enrichments of the various metals in these organic-rich sediments, but do not give any indication of their host minerals. One way in which such indications can sometimes be

TABLE 33.9.

Estimates of the contributions of metals (ppm) from terrigenous and calcareous components in the organic-rich sediments of the South West African shelf.

	1	2	3	4
Cu	68	15	4	81
Mo	53	2	2	96
Ni	108	20	5	75
Pb	12	25	8	67
Zn	68	18	4	78

1. Mean content (ppm) of metals in total, dried, salt-free sediment containing 11·2% organic carbon, 12·7% $CaCO_3$ and 25·2% opal. Mean of 17 analyses.
2. Mean contributions, in percent, of metals from the terrigenous fraction of the same sediment samples.
3. Mean contribution, in percent, of metals from the carbonate fraction of the same sediment samples.
4. Mean percentages of excess metals in the sediments not accounted for by the terrigenous and carbonate fractions.

obtained is by estimating the tentative maximum permissible metal contents of the organic and the sulphide phases on the basis of the contents of organic carbon, sulphur and the "excess metals" listed in Table 33.9. Such estimates are shown in Table 33.10 and are based on the simplifying assumption that

TABLE 33.10
Estimated minor metal contents (ppm) of organic matter and sulphide fraction of the organic-rich sediments of the South West African shelf.

Element	Organic matter[a]	Sulphide phase[b]
Cu	269	2105
Mo	1644	9389
Ni	446	3219
Pb	34	319
Zn	351	2770

[a] Estimated from excess metal contents of sediments shown in column 4 of Table 33.9 on the assumption that all the excess metal is associated with organic matter (organic carbon × 1·8)
[b] Estimated from excess metal content of sediments shown in column 4 of Table 33.9 assuming that all the excess metal is associated with the sulphide phase (calculated as FeS_2 from the sulphur content).

the "excess metals" are entirely associated with either the organic or the sulphide phase. The latter is assumed to be pyrite (FeS_2), a recognizable component in the organic-rich sediments. With the exception of molybdenum, the metal contents of the sulphide phase are high, but are not entirely unreasonable in light of the data tabulated by Fleischer (1955). However, the molybdenum value is far in excess of its concentrations in sedimentary pyrites (5–20 ppm) (Noddack and Noddack, 1931; Hegemann, 1949). On this limited evidence, therefore, molybdenum, at least, is unlikely to be present in association with the FeS_2.

The projected metal contents of the organic components of the sediments are, apart from molybdenum and nickel, within the ranges of metal contents of plankton reported by a number of workers (see Table 33.11). These comparisons are not strictly valid because the organic material in the sediments is, at least partially, degraded planktonic organic matter, whereas the plankton analyses are those of freshly collected material. A considerable amount of nutrient regeneration from settling planktonic materials in the shelf waters off South West Africa can be inferred from the high concentrations of dissolved phosphate and silicate in these waters (Calvert and Price, 1971b). It is likely that during partial degradation a rapid exchange of nutrient elements and metals takes place between the organic debris and the waters

TABLE 33.11.

Selected minor metal contents (ppm) of marine organic matter and South West African sediment organic fraction.

Metal	Plankton[a]	Zoo-plankton[b]	Brown algae[c]	S.W. African shelf sediment[d]
Cu	200	238	11	269
Mo	10	3	0·45	1644
Ni	36	65	3	446
Pb	5	83	8	34
Zn	2600	—	150	351

[a] Black Sea plankton. From Vinogradova and Koval'skiy (1962)
[b] North Atlantic plankton. From Nicholls *et al.* (1959)
[c] From Bowen (1966, Table 5.4)
[d] Derived from regressions of excess metals (Table 33.9) on organic carbon.

which subsequently upwell. It is also probable that organic material will adsorb available metals (see e.g. Kee and Bloomfield, 1961; Szilagyi, 1967) either in the water column, or in the bottom sediments, so that the minor metal content of this organic material would be increased. The problem of metal enrichment in anoxic sediments is discussed more fully in Section 33.7.

33.5.3. CALCAREOUS SEDIMENTS

Although the grain size distributions of the sandy calcareous sediments on many parts of the continental shelf are well known (Lees and Buller, 1972) little is known of their chemical compositions, and the only data available are for South West African and Somalian shelf sediments.

In the calcarenites off South West Africa CaO, CO_2 and Sr concentrations are high, whereas those of elements associated with the detrital or the biogenous fractions are low, compared with the other facies on the shelf. The concentrations of these other elements (see Table 33.6) are considered to reflect admixtures of the calcareous sediments with varying proportions of other components since the foraminiferal and macrofossil shells contribute only very small amounts of these metals to the sediments (Krinsley, 1960; Arrhenius, 1963).

The relationship between CaO and CO_2 in the calcarenites is shown in Fig. 33.12. Points representing samples without a significant content of phosphorite fall along a line representing pure $CaCO_3$ with an intercept at 2·5% CaO which gives an indication of the calcium content of the detrital silicates. Samples containing phosphorite depart significantly from this line because of the presence of some CaO in the phosphorites which contain little CO_2 (see Section 33.6.1.2).

The distribution of strontium in the South West African sediments is

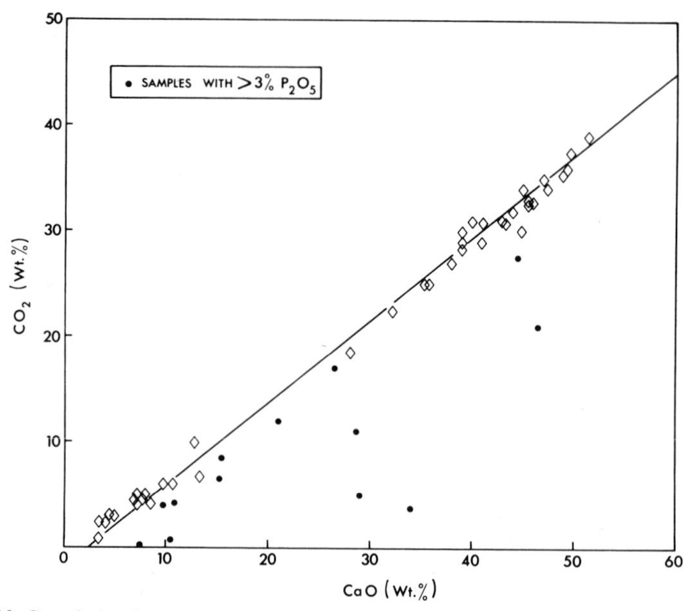

Fig. 33.12. Correlation between CO_2 and CaO in some Recent sediments from the South West African continental shelf (Calvert and Price, unpublished).

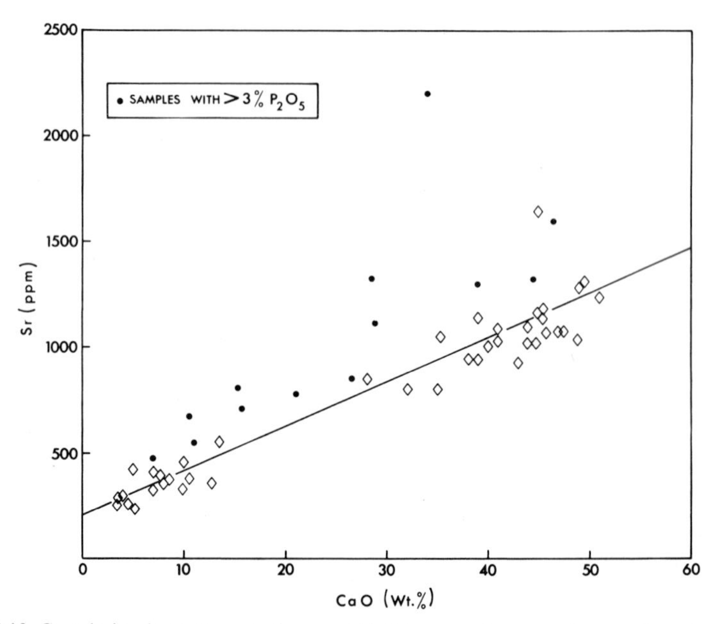

Fig. 33.13. Correlation between strontium and CaO in some Recent sediments from the South West African continental shelf (Calvert and Price, unpublished).

dominantly controlled by carbonate minerals of which calcite is by far the most important (Fig. 33.13). Points representing samples containing little phosphorite fall along a line approximating to that for foraminiferal carbonate with ~1400 ppm of strontium (cf. Turekian, 1964). The intercept of ~200 ppm of strontium corresponds to that associated with the detrital fraction of the sediments. Samples containing significant contents of phosphorite are enriched in strontium, consistent with the enrichment of many rock phosphorites with strontium (Gulbrandsen, 1966; Swaine 1962). This is discussed more fully in Section 33.6.1.1.

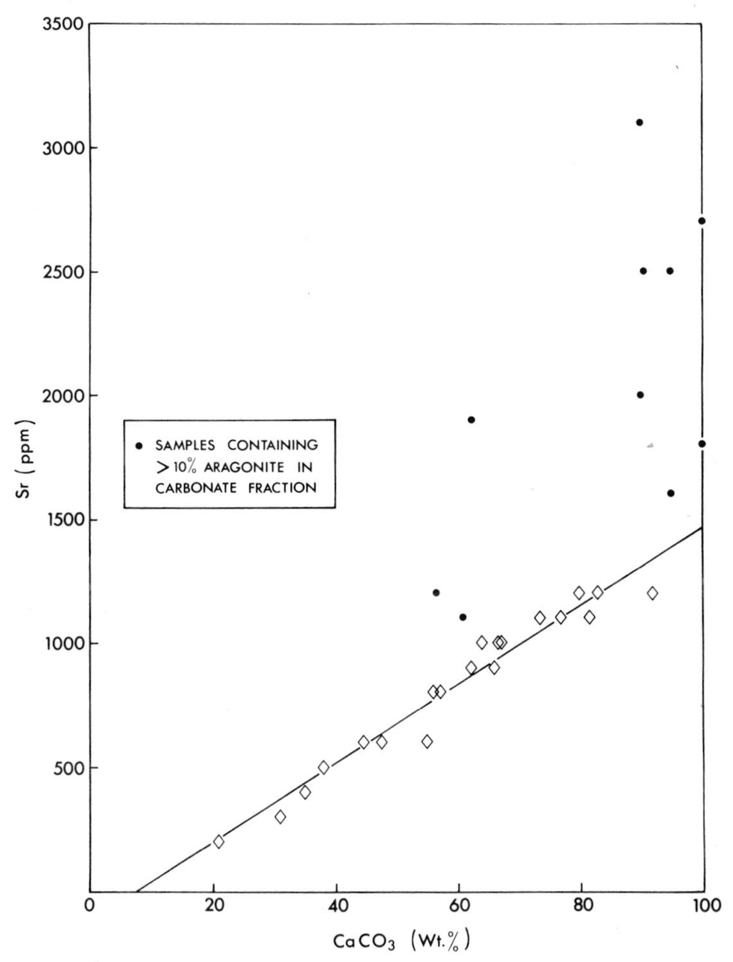

FIG. 33.14. Correlation between strontium and $CaCO_3$ in some Recent sediments from the Somalian continental shelf (data from Müller, 1967).

Strontium levels are much higher in calcareous sediments containing appreciable amounts of aragonite than in the calcite-rich sediments of the South West African shelf. This accords with the observations of Thompson and Chow (1955) and Odum (1957) for calcareous shells. This effect is clearly demonstrated by data on the strontium content of the sediments from the continental shelf off Somalia (Fig. 33.14). Sediments having abnormally high strontium contents contain coralline reef debris (Müller, 1967) which may contain up to 8000 ppm of Sr (Kinsman, 1969). It should be noted that the regression line in Fig. 33.14 has an intercept on the abscissa rather than the ordinate; this implies that strontium is absent from the terrigenous fractions of these sediments. However, at 100% $CaCO_3$, the strontium content is approximately 1450 ppm, in agreement with other previously cited data for aragonite-free skeletal carbonates. Similar enrichments of strontium (up to 1% Sr) have been found for aragonite-bearing sediments from the continental shelf off Western India (Marchig, 1972).

33.6. Authigenic Components in Near-shore Sediments

The occurrence and distribution of authigenic components in near-shore sediments have been extensively described by a number of workers. These components include phosphorites, iron silicates (glauconite and chamosite), iron sulphides and iron and manganese oxides and carbonates.

33.6.1. Phosphorite

Phosphorite occurs in a variety of continental shelf areas including: the eastern U.S. (Gorsline, 1963; Lutenauer and Pilkey, 1967; Pevear and Pilkey, 1966; Gorsline and Milligan, 1963); California and Baja California (Dietz *et al.*, 1942; Emery, 1960; d'Anglejan, 1967, 1968); N. W. Africa (Tooms and Summerhayes, 1968; Summerhayes *et al.*, 1972); S.W. Africa (Baturin, 1969a, 1971a; Parker, 1971; Summerhayes *et al.*, 1971); western Australia (Von der Borch, 1970); eastern New Zealand (Norris, 1964; Summerhayes, 1967); and western India (Murty *et al.*, 1968). Maps showing the distribution of phosphatic sediments, and areas of "potential" phosphorite formation, have been produced by McKelvey (1967) and Tooms *et al.* (1969).

The phosphorite in these shelf sediments occurs either as nodules, ranging from small pebbles to boulder grade material, or as sand grade pellets. The nodules have a wide variety of textures (including conglomeratic and oolitic ones), or they may be structureless. They range in colour from pale cream through dark brown to black. They are frequently admixed with glauconite (Summerhayes, 1971; Parker, 1971), manganese and iron oxides (Pratt and McFarlin, 1966; Parker, 1971), and minor amounts of opal, pyrite, organic matter, skeletal carbonate and terrigenous debris. The phosphatic pellets are usually smooth black ovoid grains, containing a substantial amount of organic material (d'Anglejan, 1967); some of them have a crude internal oolitic structure.

Marine phosphorites are considered by most modern authors to be carbonate fluorapatite (Altschuler *et al.*, 1952; McConnell, 1952). This mineral is extremely complex with mono- and di-valent cations (Na^+, Mg^{2+}, Ba^{2+}, Sr^{2+}, Pb^{2+}), lanthanides and uranium substituting for calcium, and a variety of oxyanions ($VO_4{}^{3-}$, $AsO_4{}^{3-}$, $CO_3{}^{2-}$) substituting for $PO_4{}^{3-}$. In addition, F^- enters vacant oxygen sites created by the substitution of $CO_3{}^{2-}$ for $PO_4{}^{3-}$. Altschuler *et al.* (1958) suggested that the approximate composition of sedimentary carbonate fluorapatite is $Ca_{10}(PO_4, CO_3)_6F_{2-3}$. The carbonate ion is present as an integral part of the structure, and not as a contaminating carbonate mineral (Silverman *et al.*, 1952; Rooney and Kerr, 1967; Ames, 1959, 1960; McConnell, 1970; Kolodny and Kaplan, 1970a). In the geological literature the mineral names *collophane* and *francolite* are used to describe optically isotropic and anisotropic carbonate fluorapatite respectively.

Baturin (1969a, 1971a) has described an occurrence on the South West African shelf of unconsolidated and friable phosphorite which has almost certainly been precipitated under essentially recent conditions, and Veeh *et al.* (1973) have established that concretionary phosphorite on the Peru continental margin is also modern. It is, therefore, convenient to distinguish between consolidated phosphorite in nodular or pelletal form and modern unconsolidated varieties.

In addition to the inorganically formed phosphate minerals described above; Lowenstam (1972) has shown that francolite and *dahlite* (carbonate hydroxyapatite) are precipitated by some species of brachiopoda. It is also of interest to note that some species of polychaeta, polyplacophora, gastropoda, bivalvia, malacostraca and holothuroidea precipitate amorphous ferric phosphate hydrogels and that crystalline fluorite (CaF_2) appears to be an important constituent of the gizzard plates of the gastropod *Scaphanda lignarius.*

33.6.1.1. *Composition of nodular phosphorites*

Chemical compositions of some representative nodular phosphorites are given in Table 33.12. The major element data show that the phosphorites are relatively impure rocks, thus the P_2O_5 content is less than that characteristic of commercially-mined phosphate deposits (Lehr *et al.*, 1967). It can be readily seen that the marine phosphorites are impure because of: (i) their relatively high concentrations of Fe_2O_3, K_2O and Al_2O_3, which may indicate the presence of glauconite; (ii) their high Fe_2O_3 content which may also signify the presence of ferric oxides; (iii) their high SiO_2/Al_2O_3 ratios, indica-

TABLE 33.12.

Chemical compositions of rock phosphorites from the north west and south west African continental margins.[a]

	1	2	3	4
SiO_2	6·20	15·30	2·78	14·33
Al_2O_3	1·13	2·14	—	—
TiO_2	0·06	0·10	0·03	0·21
Fe_2O_3[b]	1·40	5·58	0·59	5·48
CaO	47·02	37·04	50·40	50·40
MgO	0·97	1·34	0·80	3·15
Na_2O	0·62	0·78	—	—
K_2O	0·43	1·57	0·13	1·21
MnO	0·01	0·01	0·01	0·34
P_2O_5	14·82	17·89	25·30	13·40
S	0·31	0·46	—	—
F	2·12	2·21	—	—
CO_2	19·00	9·50	—	—
As	—	—	10	3·7
Ba	—	—	10	103
Co	—	—	6	38
Cr	—	—	107	38
Cu	—	—	21	18
Ni	—	—	40	127
Pb	—	—	37	73
Sr	—	—	1067	816
V	—	—	67	173
Zn	—	—	70	85

1. Tertiary micro- and macro-fossiliferous phosphorite, Agulhas Bank, South Africa (Parker, 1971). Analyses by X-ray fluorescence spectroscopy for all elements except CO_2 which was estimated by the method of Hülsemann (1966) and F which was determined using a fluoride specific electrode.
2. Tertiary glauconitic nodular microfossiliferous phosphorite, Agulhas Bank, South Africa. (Parker, 1971). Analytical methods as for 1.
3. Tertiary pelletal conglomeratic phosphorite, North West African continental shelf (Summerhayes, 1971). Analyses by direct reading emission spectroscopy, except for As and P_2O_5 which were determined by wet chemical methods.
4. Miocene/Eocene glauconitic ferruginous-dolomitic phosphorite, North West African continental shelf (Summerhayes, 1971). Analytical methods as in 3.

[a] Major element oxides as wt.%, minor elements in ppm.
[b] Total Fe as Fe_2O_3.

tive of the presence of free quartz; and (iv) their high MgO content characterizing the presence of dolomite. In addition, many of the phosphorites contain relatively high concentrations of biogenous carbonate which enhances the CaO and CO_2 contents. Information about the major element composition of the phosphorite minerals themselves cannot, therefore, be gained from an examination of the bulk chemical analyses of the nodules (Parker, 1971), and in fact only the P_2O_5 and the fluorine are associated solely with the carbonate fluorapatite structure.

Values of the F/P_2O_5 ratio of marine phosphorite deposits (Table 33.13) are generally not significantly different from those of commercial deposits, but are quite different from those of carbonate-free fluorapatite (Rooney and Kerr, 1967). This is consistent with the role of fluorine in compensating for the charge imbalance resulting from the substitution of CO_3^{2-} for PO_4^{3-} in carbonate fluorapatite (Gulbrandsen *et al.*, 1966). Phosphorites having high F/P_2O_5 ratios should have relatively high CO_2 contents and this is supported by the data given by Gulbrandsen (1969) and Parker (1971).

TABLE 33.13.

Ratios of F/P_2O_5 in marine phosphorites.

Area	Ratio	Source
Agulhas Bank	0·109–0·151	Parker (1971)
California borderland	0·106–0·113	Dietz *et al.* (1942)
Miocene, North Carolina	0·106–0·116	Rooney and Kerr (1967)
Permian Phosphoria formation	0·091–0·124	Gulbrandsen (1966)
Various commercial phosphorites	0·087–0·132	Lehr *et al.* (1967)
Ideal fluorapatite	0·089	Rooney and Kerr (1967)

Substitutions of Na^+ and Mg^{2+} for calcium have been detected in commercial phosphorites by McClellan and Lehr (1969). Gulbrandsen (1966) has suggested that Na^+ and SO_4^{2-} are involved in coupled substitution in Phosphoria Formation phosphorites (see below), but the data of McClellan and Lehr (1969) do not support this. The results presented by Parker (1971) strongly suggest that $SO_4^{2-}-S$ and P_2O_5 values are positively correlated. It would, of course, be expected that $SO_4^{2-}-S$ and P_2O_5 would be negatively correlated if SO_4^{2-} is substituting for PO_4^{3-} in these rocks; however, P_2O_5 values are much more indicative of the extent of contamination of the phosphorites with non-phosphate bearing material. The substitution of Mg^{2+} ions in the marine phosphorites cannot be demonstrated with the analyses available because of the much larger amounts of MgO present in the contaminating aluminosilicates and dolomite.

The minor element composition of marine rock phosphorites has been reviewed by Tooms *et al.* (1969) who considered patterns of enrichment of elements relative to their crustal and sea water abundances. According to these authors Ag, As, Cd, Cr, I, La, Mo, Pb, Sb, Se, Sn, Sr, U and Y are all strongly enriched in phosphorites. However, this conclusion was based on average data for a very wide range of phosphorites, the composition and amounts of contamination of which varied very widely (Krauskopf, 1955). It is therefore perhaps better to examine minor element data of individual sets of phosphorite analyses in order to detect enrichments or depletions.

From an analysis of the compositional data for the Phosphoria Formation phosphorite of the Western United States Gulbrandsen (1966) concluded, following Krauskopf (1955), that the concentrations of most of the minor elements could be accounted for by presence of organic matter in the rocks. Elements accounted for in this fashion included As, Cd, Cr, Cu, Mo, Ni, Sb, Sc, V, Zn and possibly Ag, and this group of elements indeed appeared to be enriched in those phosphorites which had higher than average organic contents. One corollary to Gulbrandsen's argument would be that for a phosphorite rock containing 1000 ppm of Cr and 3·6% organic matter, the latter would need to contain 2·8% of Cr if the element was confined to this fraction alone. If this reasoning is also applied to the other elements listed above the same difficulty may arise and the composition of the included organic material therefore warrants further detailed examination.

Gulbrandsen has also suggested that in phosphorites Sr, U, Y and lanthanides (e.g., La., Nd and Yb) are associated with the carbonate fluorapatite component, substituting for calcium; this is consistent with crystal chemistry (Altschuler *et al.*, 1958; Altschuler *et al.*, 1967). This hypothesis is supported by analytical data on carbonate fluorapatite separated from North West African phosphorites (Summerhayes, 1971). As an example of this type of enrichment, two samples were found to contain Sr (1298 and 1754 ppm), U (36 and 147 ppm), Y (9 and 142 ppm), Ce (3 and 26 ppm) and La (4 and 48 ppm). In addition, the relatively high contents of As (51 and 69 ppm), Cr (105 and 290 ppm), Cu (356 and 559 ppm), Br (1311 and 1314 ppm), I (312 and 565 ppm) and V (75 and 151 ppm) suggest that several other elements are also located in the carbonate fluorapatite lattice.

The pattern of lanthanide distribution in marine phosphorites has been discussed by Altschuler *et al.* (1967) and Summerhayes (1971). Analyses of separated carbonate fluorapatites show that the atomic ratio of Ce/La is less than 1 (0·53 to 0·91) which compares with sea water values ranging from 0·16 to 0·91 (Høgdahl *et al.*, 1968). In contrast, the Ce/La atomic ratio in bulk phosphorites ranges from 1·43 to 2·05 (Altschuler *et al.*, 1967) which is similar to that for shales (Wildeman and Haskin, 1965). The cerium and lanthanum pattern in the carbonate fluorapatite lattices thus follows the pattern in sea water, whereas, because of admixture of contaminating detrital materials, the pattern in bulk phosphorite follows the normal terrestrial abundance pattern. Alternatively, it has been suggested that there is significant fractionation of rare earth elements in phosphatic materials during their residence on the sea floor (Arrhenius and Bonatti, 1965), but the data are not sufficiently detailed to establish this definitely.

33.6.1.2. *Composition of Recent phosphorites*

Descriptions and compositional data for phosphorites accumulating in the Recent sediments off South West Africa and Chile have been presented by Baturin (1969a; 1971a, b) and Baturin *et al.* (1970). Off South West Africa

TABLE 33.14.

Chemical compositions of Recent phosphorites from the South West African shelf[a]

	1	2	3	4	5
SiO_2	45·69	42·90	4·23	1·40	4·80
Al_2O_3	1·85	2·07	0·36	0·37	1·21
TiO_2	0·08	0·20	0·005	0·06	0·10
Fe_2O_3[b]	1·06	1·14	0·30	0·88	2·92
CaO	8·22	20·70	42·30	51·25	44·20
MgO	2·49	1·10	1·25	0·60	0·90
K_2O	—	0·54	—	0·13	0·39
P_2O_5	8·32	11·25	29·37	32·10	28·30
S	1·46	—	0·90	—	—
F	0·34	1·30	3·00	2·38	2·42
CO_2	1·40	—	5·44	—	—
C_{org}	3·59	—	0·73	—	—
Ba	900	—	300	—	—
Ce	—	43	—	111	252
Cr	37	—	—	—	—
Cu	40	38	20	18	27
La	—	—	—	15	150
Mo	5	29	1	33	18
Ni	10	112	1	17	74
Rb	—	26	—	5	11
Sr	1300	1004	4500	1965	1920
U	—	68	—	128	115
Y	—	10	—	57	266
Zn	—	46	—	96	35
Zr	20	53	10	36	59
V	40	—	30	—	—

1. Phosphatised diatomaceous ooze (Baturin *et al.*, 1970). Analytical methods not given.
2. Unconsolidated phosphatic lenses in diatomaceous ooze (S. M. Brown, N. B. Price and S. E. Calvert, unpublished). Analysed by X-ray fluorescence spectroscopy, except for F which was determined by the method of Peck and Smith (1964).
3. Friable phosphorite concretions (Baturin *et al.*, 1970).
4. Consolidated phosphorite concretions (S. M. Brown, N. B. Price and S. E. Calvert, unpublished). Analytical methods as for 2.
5. Black sand-grade phosphorite pellets from middle-shelf calcarenites (S. M. Brown, N. B. Price and S. E. Calvert, unpublished). Analytical methods as for 2.

[a] Major elements as wt.%, minor elements in ppm.
[b] Total iron expressed as Fe_2O_3.

the phosphorites are found in the near-shore diatomaceous muds both as unconsolidated phosphatized sediment masses and as friable concretions (Baturin, 1969a). Phosphorite also occurs, however, as more consolidated concretions and as black pellets in calcarenites on the central part of the shelf (Calvert and Price, 1971). Uranium disequilibrium measurements (Thurber, 1962) on these phosphorites made by Baturin *et al.* (1972) have confirmed that the phosphatized sediment and the relatively soft and friable concretions are Recent. Baturin (1971b) has, moreover, suggested that the different forms of phosphorite on this shelf represent an age sequence in which the phosphatised sediment became progressively more consolidated with age with the result that it appears in the older sediments as hard reworked grains. The young age of the unconsolidated phosphorites has recently been confirmed, and an age greater than 700 000 years has been established by Veeh *et al.* (1974) for the consolidated concretions and pellets.

The chemical compositions of the various forms of phosphorite off South West African are shown in Table 33.14. The X-ray diffraction patterns (S. M. Brown, N. B. Price and S. E. Calvert, unpublished) of all of these phosphorites are similar to those of the sedimentary carbonate fluorapatites described by Rooney and Kerr (1967). The phosphatised muds and the unconsolidated phosphatic lenses are relatively impure and their contents of SiO_2, Al_2O_3, TiO_2, Fe_2O_3, MgO, K_2O, Rb and Zr suggest that they contain large amounts of diatomaceous silica and variable amounts of terrigenous materials. In addition, Fe_2O_3/Al_2O_3, MgO/Al_2O_3 and K_2O/Al_2O_3 ratios are all lower in the unconsolidated phosphatized muds (and in those organic-rich muds lacking phosphorite) relative to the friable or more consolidated concretions. Unless the nature of the included terrigenous material is different in the different forms of phosphorite this implies that Mg, Fe and K are substituting in the ageing fluorapatite lattice.

The F/P_2O_5 ratios of both the unconsolidated and the consolidated phosphorites off South West Africa (ranging from 0·041 to 0·116) are much lower than those of most land-based commercial phosphorites (see Lehr *et al.*, 1967). This fluorine deficiency is confirmed by the fluorine contents of the carbonate fluorapatites calculated from the analyses in Table 33.14. The unit cell, on average, contains only 1·90 atoms of fluorine rather than 2·0, and probably the charge deficiency is made up by OH^- ions. Similar calculations suggest that the CO_2 content is high, and this has been confirmed by infra-red spectrophotometric determinations of the CO_2 content of the carbonate fluorapatite, using the method of Tuddenham and Lyon (1960), and also by comparing the unit cell parameters given by Baturin *et al.* (1970) with those of commercial phosphorites given by Lehr *et al.* (1967).

If allowance is made for the presence of silicate impurities, the Zn, Cu and Mo contents of the phosphorites are similar to those of the diatomaceous clays. However, Ce, La, Ni, Sr, Y and U are all significantly enriched. These enrichments, apart from that of Ni, have been discussed previously. It is noteworthy that the Ce/La ratio is considerably greater than unity in these phosphorites, and is much greater than that of shales. It is unlikely, therefore, that the high ratio results from the presence of aluminosilicate impurities, and this is confirmed by the very high ratio in the consolidated concretion which contains less impurities than do the other phosphorites (Table 33.14, analysis 4).

The high strontium contents of the phosphorites confirm the conclusions drawn previously on the basis of analyses of the bulk sediments (Fig. 33.14) that the element is enriched in these minerals. Relative to P_2O_5, the uranium contents of the unconsolidated phosphorites are much higher than those of the consolidated varieties. In contrast, Ce, La and Y appear to behave conversely. Some of the enhancement of uranium in the Recent phosphorites may be due to its association with included organic material (Section 33.7.5), although when allowance is made for this effect these phosphorites still contain relatively high concentrations of uranium. This suggests that uranium is fixed in the carbonate fluorapatite at an early stage of its formation whereas the concentrations of Ce, La and Y are initially low and increase in older phosphorites.

33.6.1.3. *Age of sea-floor phosphorites*

Very little attention has been devoted to the problem of the age of the marine phosphorites, or even to whether they represent modern authigenic materials in equilibrium with present-day conditions. However, it has been suggested that many phosphorites are fossil deposits (see *e.g.* Dietz *et al.*, 1942; Emery, 1960). In spite of this, many workers, notably McKelvey (1967) and Sheldon (1964), have persisted in tacitly assuming that rock phosphorites on the continental shelves are entirely Recent, and that they occur in these localities because of a direct causal link between phosphorite formation and the prevailing oceanographic conditions.

Evidence for the ages of many sea-floor phosphorites has been reviewed by Tooms *et al.* (1969) and by Parker (1971). Palaeontological, sedimentological and structural criteria suggest that extensive deposits off California, the eastern United States, north-west and south-west Africa, New Zealand and on many Pacific Ocean seamounts are pre-Pleistocene in age, most probably late Mesozoic to Cenozoic. Kolodny (1969) and Kolodny and Kaplan (1970b) have used the uranium disequilibrium dating technique (Thurber, 1962) to confirm these conclusions by showing that these widely scattered deposits have ages greater than 8×10^5 yr. Indeed, these authors found evidence from the $^{234}U/^{238}U$ ratios that rather than recent uranium deposition, extensive leaching of uranium from the phosphorites had probably occurred. This evidence shows that many of the extensive deposits of phosphorite which are generally considered to be modern (Tooms, 1967; McKelvey, 1967) are, in fact, not Recent. They are probably related to former periods of phosphorite accumulation, presumably under conditions somewhat different from those prevailing in these areas at the present time (Kolodny, 1969).

The discovery that the unconsolidated phosphorites on the South West African, Chilean and Peruvian continental shelves have essentially Recent radiometric ages (see Section 33.6.1 and Baturin *et al.*, 1972, Veeh *et al.*, 1973, 1974) allows criteria to be established for the recognition of truly Recent phosphorites. However, care is needed in establishing these criteria because even in these areas Recent and old phosphorites are found within the same sediment.

33.6.1.4. *Origin of marine phosphorites*

Discussions of the origin of marine phosphorites have been dominated for the past 35 years by the hypothesis of Kazakov (1937) which relates the precipitation of phosphorite to upwelling of nutrient-rich waters. Thus, McKelvey *et al.* (1953), McKelvey (1967), Sheldon (1964) and Gulbrandsen (1969) have assumed that phosphorites are formed predominantly at low latitudes on the western sides of the continents where upwelling is frequent, and where apatite precipitates inorganically from phosphate-rich water (McKelvey *et al.*, 1953) or in the microenvironment associated with degrading organic material (Gulbrandsen, 1969). The mechanism which has been postulated is essentially simple and involves the supersaturation of sea water with respect to idealized "apatite" as a result of the increase in pH and temperature which occurs when the relatively deep phosphate-rich water rises to the sea surface. Many elaborate schemes have been devised to account for the preferential precipitation of apatite rather than $CaCO_3$ (which is also assumed to precipitate under these conditions) (see for example, Gulbrandsen, 1969). Calculations of equilibrium solubilities with respect to idealized fluorapatite ($Ca_5(PO_4)_3F$) are not valid as the phosphatic mineral in phosphorites has a very complex, and far from adequately documented, composition (McConnell, 1970).

The possibility that phosphorites are derived by the replacement of calcareous material was first suggested by Murray and Renard (1891), and following the work of Ames (1959), has received a great deal of attention (see for example, Simpson, 1964, 1967; Pevear, 1966, 1967; d'Anglejan, 1968; Parker 1971; Summerhayes, 1971). Many Recent sediment pore waters contain relatively higher concentrations of dissolved phosphate than do normal sea waters (Rittenberg *et al.*, 1955; Brooks *et al.*, 1968; Baturin *et al.*, 1970), and for this reason it might be expected that interactions would occur between the solid calcareous phases and the phosphate-rich pore waters. In addition, as suggested by the models by McKelvey and Sheldon, the increased phosphate activity might lead to purely inorganic precipitation of phosphorite. This phenomenon is therefore much more likely to occur within the pore water than in the water column itself.

The discovery that phosphorite precipitation has occurred in the modern sediments off South West Africa and Chile has resolved many of the problems mentioned above. Baturin (1971a, b) and Baturin *et al.* (1970) have suggested that the phosphorite is formed in the organic-rich diatomaceous muds by the precipitation of phosphate derived from the breakdown of large quantities of planktonic organic material. This is possible because in the relatively shallow water off, for instance, Walvis Bay (water depth ~ 100 m), a considerable amount of decaying plankton reaches the sediment before the phosphorus is entirely regenerated and returned to the euphotic zone. That some regeneration of phosphorus and other nutrients takes place within the water column is shown by the very high dissolved phosphate concentrations in the shelf waters in this region (Calvert and Price, 1971b). Moreover, C/P ratios in the sediments range between 30 and 70 whereas in diatomaceous plankton they average 24 (Lisitsyn, 1966), indicating some loss of phosphorus either during settling, or shortly after deposition.

In spite of these losses sufficient organic phosphorus is evidently deposited off Walvis Bay to provide an adequate source for phosphorite formation. The phosphorus is regenerated from decomposing organic material within the sediment yielding dissolved inorganic phosphate-phosphorus concentrations of 6·5–9·7 μg-at l^{-1} in the pore waters (Baturin *et al.*, 1970) which are at least three times higher than those in the bottom waters on the shelf. This phosphorus is then fixed in an unknown way as phosphorite in small lenses and laminae within the organic-rich sediments and these presumably become progressively more phosphatized with time. It is uncertain at present whether the formation of the phosphorite involves purely inorganic precipitation or replacement of carbonate debris.

The phosphorites of the South West African shelf also occur as concretions and pellets which may be either friable or consolidated (see above), and Baturin (1971c) has suggested that these different forms represent an evolutionary sequence of nodular and pelletal phosphorite formation. Thus, after the initial precipitation of carbonate fluorapatite in fine-grained organic-rich sediments this mineral gradually becomes cemented into more coherent masses and is finally concentrated into sandy deposits during periods of sediment reworking. This may have taken place on the South West African shelf during the Pleistocene sea level fluctuations (Section 33.2) since black, pelletal phosphorites occur in reworked, relict, calcarenites on the central part of the shelf (Calvert and Price, 1971a).

The mechanism of phosphorite formation presented above, appears to offer the most self-consistent explanation of a number of features of those phosphorites which occur as concretionary, nodular and pelletal varieties. It also appears to vindicate the faith of many workers in the hypothesis of Kazakov because modern phosphorites are indeed forming in areas of intensive upwelling and high primary production. However, as Baturin (1971a, b) has emphasized, the present day phosphorite is forming *diagenetically*, and not by direct precipitation from upwelled water. A very rapid deposition of planktonic material, which has undergone minimal degradation and therefore minimal phosphate regeneration, is necessary in order to provide the large source of phosphate which is necessary for recognizable phosphorite accretion to occur. For this reason, at the present time this process is occurring only in shallow water shelf areas in which extremely intensive primary production is taking place, and in which rapid modern (i.e. post-Pleistocene) sedimentation has been established.

33.6.2. Glauconite

Glauconite, the name of both a mineral and a rock, is applied to the pale to dark green sand-sized pellets which are found in many Recent sediments generally in near-shore environments. It also occurs in rocks ranging from Precambrian to Pleistocene in age (Cloud, 1955), and is perhaps the most frequently reported authigenic mineral in the geological literature.

The mineralogical identity of glauconite was uncertain until Burst (1958) pointed out that the term embraces an extremely wide range of different types of material, some of which may not, indeed, be glauconitic. Thus, the pale green pellets found in many rocks and Recent sediments have been casually referred to as glauconite, and although considerable attention has been paid to their morphologies (Pratt, 1963; Triplehorn, 1966; Birch, 1971) and to their probable modes of formation, their actual mineralogical nature is unknown.

Following the suggestions of Burst (1958) and of Bentor and Kastner (1965) glauconite is defined as a hydrous, potassium-rich, illitic mineral of the 1M or 1Md polymorph type (see Yoder and Eugster, 1955). It is generally dioctahedral, contains more iron than aluminium in its octahedral layer, and its Fe^{3+}/Fe^{2+} ratio lies between 3 and 9. This definition should be restricted to the material which Burst (1958) refers to as *mineral glauconite*.

In many instances, green pellets from both Recent and ancient sediments which have been described as glauconites, have been shown by X-ray diffraction techniques to possess a structure quite different from that of mineral glauconite. These pellets are, in fact, composed of either interlayered clay minerals (in which measurable swelling on treatment with glycerol or ethylene glycol occurs), or clay mixtures containing, for example, kaolinite, chlorite and/or mica. It is clear, therefore, as emphasised by Burst (1958), that the name glauconite has often been very loosely applied, and that its true identification should rest on it fulfilling the criteria given above. Further, Burst (1958) has suggested a classification scheme for glauconite pellets which is based on both X-ray diffraction and chemical parameters. There are four classes in this scheme. Class 1 consists of well ordered potassium-rich minerals having sharp symmetrical X-ray diffraction peaks with a 10 Å periodicity. Class 2 contains disordered potassium-poor minerals having broad, asymmetrical X-ray diffraction peaks with a 10 Å periodicity. Class 3 comprises very disordered, potassium-poor minerals with expandable, montmorillonitic layers. Class 4 consists of mixed mineral glauconites.

Further work on the mineralogy of glauconites by Hower (1961) has shown that all glauconites are interlayered to some degree. The interlayering involves expandable montmorillonitic layers and non-expandable layers having a 10 Å spacing. The glauconites examined by Hower contained between 5 and 40% of expandable layers. Glauconites with a low percentage of expandable layers can be classified as the 1M polymorph, whereas, those with more expandable layers are considered to be the 1Md polymorph.

33.6.2.1. *Chemical composition*

The chemical compositions of some glauconites are shown in Table 33.15.

TABLE 33.15

Chemical composition of some glauconites.[a]

	1	2	3	4
SiO_2	48	50·85	51·5	48·66
TiO_2	0·3	—	—	—
Al_2O_3	13·5	8·92	23·6	8·46
Fe_2O_3	19·0[b]	24·40	10·0	18·80
FeO	—	1·66	3·7	3·98
CaO	1·1	1·26	0·06	0·62
MgO	3·2	3·13	2·0	3·56
K_2O	2·9	4·21	4·8	8·31
Na_2O	0·3	0·25	0·62	0·0
P_2O_5	—	—	—	—
H_2O^+	9	5·55	4·7	6·56
H_2O^-	—		—	—

1. "Recent" glauconite, Niger Delta (Porrenga, 1966). The sample analysed contained some quartz impurity. Analysis by emission spectroscopy, except for K_2O (X-ray fluorescence spectroscopy).
2. Glauconite from Challenger Expedition Station 164, eastern Australian shelf (Murray and Renard, 1891, p. 387). Analyses by wet chemical methods.
3. Poorly-ordered 1 Md glauconite, California (Burst, 1958). Analyses by microchemical wet methods and spectrographic techniques (Na and K).
4. A well-ordered 1M type Cambrian glauconite, Missouri (Burst, 1958). Analytical methods as for 3.

[a] Data given as wt.%

[b] Total Fe as Fe_2O_3.

According to these analyses the glauconites have variable compositions which can only be partly explained by the presence of impurities within the pellets. Some of the variability may be caused by the substitution of Fe^{3+}, Fe^{2+} and Mg for Al in the octahedral layer, and of Fe^{3+} and Al for Si in the tetrahedral layer.

The chemical composition and the mineralogical nature of glauconite are related (Burst, 1958). Well-ordered 1M glauconites contain more K_2O than do poorly-ordered or mixed-layer mineral types (see Table 33.15, analyses 3 and 4). When the K-atom equivalent falls to ~1·4 per unit cell stacking within the structure becomes more disordered. Hower (1961) and Manghnani and Hower (1964) have provided additional data on a wide range of glauconites which show that the variation in the K_2O content extends continuously from ~8% K_2O in minerals with virtually no expandable layers to ~4% K_2O in those with ~50% of expandable layers. In addition, the cation exchange capacity of glauconites is directly proportional to the amount of expandable layers in the structure (Manghnani and Hower, 1964).

Minor element data for glauconites are sparse. The analyses by Porrenga (1966) show that the concentrations of Cr, Mn, Sr and U are higher in "modern" glauconites from the Niger Delta than they are in the enclosing clay matrix, whereas the converse is true for Ba and Cu. Bentor and Kastner (1965) have used statistical methods to examine the correlations between K_2O and several minor elements in Cretaceous–Eocene glauconites and found that of the elements Ba, Cs, Sr, U and Ti, only the last was present in the glauconite itself. Chromium and rubidium showed no correlation with K_2O. The fact that the concentration of rubidium, which ranged from 170 to 300 ppm was quite high compared with that in fine-grained, glauconite-free sediments (Wedepohl, 1971) suggested that some of it may well be present in the glauconite. This is supported by the data of Hower (1961) which showed that rubidium is negatively correlated with the proportion of expandable layers in glauconites, which suggests that it proxies for potassium as it does in many aluminosilicates (Taylor, 1965). Hower (1961) has also shown that the strontium concentrations in glauconites are significantly positively correlated with the proportion of expandable layers, thus indicating that this element probably occupies an exchangeable cationic site. This is not necessarily at variance with the observations by Bentor and Kastner (1965) regarding strontium (see above) because their glauconites were entirely well-ordered 1M types with very few expandable layers.

Data on the distribution of the lanthanides in glauconites have been given by Balashov and Kazakov (1968). Compared with shales (Coryell *et al.*, 1963), the glauconites off the California coast are considerably enriched in cerium and the lighter lanthanides. This is in marked contrast to sea water (Goldberg *et al.*, 1963; Høgdahl *et al.*, 1968) in which these elements are depleted.

The distribution of the lanthanides in glauconite is therefore somewhat similar to that in carbonate fluorapatites (see Section 33.6.1).

33.6.2.2. *Composition, structure and age of glauconite*

The chemical composition and mineralogical structure of glauconite appear to be related to its geological age, a phenomenon first noted by Smulikowski (1954), and subsequently confirmed by Burst (1958) and Hower (1961). Thus, Recent and Tertiary glauconites have much lower potassium contents than do older ones, and Lower Palaeozoic glauconites contain fewer expandable layers than do younger varieties. Porrenga (1966) has suggested that Recent glauconites are markedly potassium-deficient compared with older well-crystallized ones. The interpretation of the evidence that younger glauconites contain a higher proportion of mixed mineral assemblages is made difficult by the lithology of the enclosing sediment, the apparent correlation between order/disorder in the glauconite structure, and the presence of other clay minerals. However, Hower (1961) has concluded that non-glauconite minerals (usually kaolinite and chlorite) are "eliminated" from glauconite pellets as they age.

33.6.2.3. *The age of glauconites in recent sediments*

The problem of the origin of glauconite in Recent sediments bears many similarities to that of the origin of phosphorite (see above) because it is uncertain whether it is contemporaneous with the sediment. Glauconite is widely distributed in sediments of both the continental shelves and the upper part of the continental slope at water depths ranging from 50 to 2000 m. A considerable proportion of the glauconite is present as polished dark-green to black pellets which are enclosed in, or lie upon, extensive sheets of well-sorted medium to fine-grained sands. These sediments consist almost entirely of reworked materials (Swift *et al.*, 1971), many of which have characteristics of relict sands deposited during the last sea-level rise (Curray, 1960; Emery, 1968). It is probable therefore that these glauconites are geologically old, or at least have formed under conditions quite different from those now prevailing. Failure to make a distinction between derived "old" glauconites and those that are more probably Recent has led to many misconceptions about the origin of this mineral.

In contrast to the types of glauconite just described, many examples of pale-green to yellow glauconites occurring as aggregates (Birch, 1971) and as replacements of foraminferal, gastropod and bivalve shells (Murray and Renard, 1891; Cayeux, 1897, 1935; Ehlmann *et al.*, 1963; Heath (in Van Andel and Veevers, 1967)) may represent modern glauconite in the process of transformation from a precursor to true mineral glauconite. Unfortunately, these various forms of glauconite are rarely distinguished in terms of their relative ages. However, the criteria proposed by Burst (1958) and Hower (1961) may provide a way of drawing an unequivocal distinction between modern and old glauconites on the basis of their mineralogical and chemical compositions.

Bell and Goodell (1967) have made an extensive mineralogical study of glauconites, and of their associated Recent sediments, from several localities in an attempt to distinguish between modern and relict forms of this mineral. They assumed that glauconite forms initially from a sediment substrate. Consequently, if a given glauconite sample contained clay minerals similar to those of the substrate they concluded that it was probably indigenous to the area and had probably not been derived from another area having a different type of sediment. Bell and Goodell (loc. cit.) claimed that in many areas the glauconites were forming within the Recent sediments. However, most of the data on the glauconites studied show that they were predominantly of only expandable varieties (Burst's Class 3), and were associated with a remarkably wide range of sediment types containing one or more of the following minerals: illite, chlorite, montmorillonite, kaolinite, mixed-layer illite-montmorillonite, amphibole, quartz and calcite. Conversely, heterogeneous glauconites (Burst's Class 4), which consist of either montmorillonite, illite and chlorite, or only mixed-layer illite-montmorillonite, occurred in sediments containing montmorillonite, illite, chlorite and kaolinite. These data are therefore rather confusing and cannot be adequately interpreted until the mode of formation of glauconite is known.

33.6.2.4. *Origin of Glauconite*

Much of the discussion in the literature on the formation of glauconite has been concerned with the identity of the starting materials. This began with the observation by Takahashi and Yagi (1929) of the glauconitization of faecal pellets in Recent sediments, and also with that by Galliher (1935; 1939) who was able to detect several stages in the alteration of detrital biotite grains and the growth of glauconite in Recent sediments off California. Subsequent work has shown that glauconite can be formed by the alteration of a wide range of detrital minerals, including muscovite, quartz, augite, hornblende, pyroxene and volcanic glass as well as a wide range of biogenous materials, including calcitic and siliceous tests.

A wide range of environments for glauconite formation has been inferred from the present-day distribution of this mineral. This range embraces both mildly oxidizing and mildly reducing conditions, and thus includes environments with both restricted and open circulations (Cloud, 1955). These conclusions have generally been reached on the basis of pellet morphology, the nature of the enclosing sediment and the general physical conditions of the given area, such as water depth, temperature, turbulence etc. However, it should be mentioned that (as suggested above) many glauconites are not modern and so are not found under their original depositional conditions.

On the basis of a detailed series of observations of a wide range of glauconites Burst (1958) has suggested that the formation of the mineral merely requires a layered silicate lattice in a suitable environment in which a supply of iron and potassium and a suitable redox potential exist. This latitude covers the wide range of specific origins that have been proposed for the formation of this mineral which can now be viewed as an end product of a slow reaction between a layer lattice aluminosilicate and sea or pore water. The nature of the initial product of the glauconitization process is therefore of interest. Information on the initial phase of this process can be obtained from a consideration of the data on the mineralogy of glauconites from the Sahul Shelf published by Heath (in Van Andel and Veevers, 1967). These glauconites occur as pellets and as replacements of echinoderm, bryozoan and foraminiferid fragments. The compositions of the replacement glauconites vary from those of interlayered forms to those of iron-rich montmorillonite associated with small amounts of iron-rich chlorite. The pellets consist of iron-rich montmorillonite, iron-rich chlorite or a mixture of chlorite and interlayered glauconite. The pellets appear to have formed within the shell chambers, and are mineralogically more mature than the pellet ones. Heath observed that a clear yellow, pure montmorillonite, which is subsequently transformed into green glauconite, is invariably associated with the early stages of the alteration of shell debris.

Somewhat similar observations on several possible stages in glauconite formation have been reported by Ehlmann *et al.* (1963). The foraminiferal sands in Recent sediments of the south-eastern United States continental shelf contain variable amounts of glauconite in the form of pale yellow to green moulds and fillings of foraminiferids, and of medium to dark green pellets. The compositions of these forms of glauconite were quite distinct one from another; the pale yellow varieties being interlayered (11·6 Å spacing) and containing, on average, 4·31% K_2O; the dark-green varieties containing no interlayers (10 Å spacing) and an average of 7·60% K_2O. In contrast, the iron contents of both types of glauconite had a much more restricted range, viz. from 19·6 to 24·3% as Fe_2O_3. Similarly, on the California banks and shelf areas, the glauconites range from montmorillonitic "proto-glauconites" (with a basal spacing of 14–15 Å which swell to 17 Å on glycolation), to well-ordered glauconites having a non-swelling basal spacing of 10 Å (Pratt, 1963). In these areas the expandable material occurs as replacements for foraminiferid shells and rock fragments, whereas the non-expandable glauconite occurs as pellets.

On the basis of the observations presented above, the earliest stage in the formation of glauconite is probably the precipitation of, or the replacement of a clastic particle by, an iron-rich expandable aluminosilicate which may be montmorillonite. This mineral which contains as much as 20% Fe_2O_3 and <5% K_2O subsequently takes up and fixes potassium in interlayer sites and also loses water with the result that its expandability is decreased. Finally, the internal filling, or the replacement, is released by dissolution, and/or breakage, of the shell and the process of alteration of the initial precipitate to true glauconite then takes place. This process is marked by a distinct colour change, from light to dark green, and by the oxidation of iron(II) (Ehlmann *et al.*, 1963). It is at this stage, if at all, that the newly formed pelletal glauconite is most likely to be mixed with and masked by older reworked glauconite.

Pratt (1963) and Ehlmann *et al.* (1963) have discussed some of the difficulties in accounting for the presence of clay infillings in foraminiferid and gastropod tests in pure shelly sands. Since shelf sands are very often reworked relict deposits (see below), the clay material may simply represent material derived from the muddy environment in which the initial impregnation, filling or replacement took place. The final stages in the formation of glauconite, therefore, may frequently occur in an environment quite different from that in which the initial reactions took place. This means that any similarity between the mineralogy of glauconite and that of the *presently*-associated sediments, an association which Bell and Goodell (1967) considered to be a criterion for the recognition of Recent glauconite, would be purely fortuitous.

33.6.3. Chamosite

In contrast to glauconite, chamosite (the berthierine of French authors), the major iron-bearing silicate in many post-Precambrian sedimentary ironstones, occurs only rarely in Recent sediments. Modern work has shown, however, that chamosite is forming in several near-shore shallow-water areas in which the mineralization of faecal pellets is known to be occurring (Porrenga, 1966, 1967; Rohrlich *et al.*, 1969; Caillière and Martin, 1972).

Chamosite is considered to be an iron-rich septo-chlorite (Nelson and Roy, 1954) indicating that it has a chemical composition similar to that of chlorite, but a 7 Å periodicity. The approximate formula (Brindley, 1951) is:

$$(Fe^{2+}, Mg)_{4\cdot6}(Fe^{3+}, Al)_{1\cdot4}(Si_{2\cdot8}Al_{1\cdot2})O_{10}(OH)_8.$$

The structure is similar to that of kaolinite, but with considerably more iron (together with magnesium and aluminium) in the octahedral layer, and with aluminium substituting for silicon in the tetrahedral layer. Identification of the mineral by X-ray diffraction methods is difficult if iron-rich chlorite is also present (see for example; Schoen, 1964; Rohrlich *et al.*, 1969).

33.6.3.1. *Chemical composition*

The chemical composition of some Recent chamosites, and comparative data for some ancient chamosites, are given in Table 33.16. The Recent minerals have a very high iron content, much of it being in the 2-oxidation state in

TABLE 33.16

Chemical composition of some chamosites[a]

	1	2	3	4
SiO_2	50	32·00	20·62	23·81
TiO_2	0·4	0·23	0·72	—
Al_2O_3	8	10·05	11·96	23·12
Fe_2O_3	20[b]	38·10[b]	5·11	0·23
FeO	—	—	25·98	39·45
CaO	0·5	0·72	11·50	—
MgO	8·3	4·70	3·07	2·72
K_2O	0·5	1·30	0·00	—
Na_2O	0·3	—	0·09	—
P_2O_5	—	0·81	5·11	—
MnO	0·22	1·68	0·30	—
H_2O^+	11·0	—	8·36	10·67
H_2O^-	0·4	—	2·56	—
Ba	120	210	—	—
Ce	—	420	—	—
Co	90	—	—	—
Cr	110	10	—	—
Cu	30	160	—	—
Nb	—	20	—	—
Nd	—	15	—	—
Ni	150	76	—	—
Pb	—	130	—	—
Rb	—	11	—	—
Sr	120	48	—	—
V	280	—	—	—
Y	—	31	—	—
Zn	—	380	—	—
Zr	—	0	—	—

1. Recent chamosite, Niger Delta (Porrenga, 1966). Goethite and quartz present as impurities. Analyses by emission spectroscopy.
2. Recent chamosite, Loch Etive, Scotland (Rohrlich *et al.*, 1969). Composition from analyses of untreated and acid-leached material by X-ray fluorescence spectroscopy.
3. Jurassic chamosite, Raasay, Scotland (MacGregor *et al.*, 1920). Analyses by wet chemical methods.
4. Jurassic chamosite, Northamptonshire, England (Brindley and Youell, 1953). Analyses by wet chemical methods. Recalculated to a carbonate- and sulphate-free basis.

[a] Major element oxides as wt. %, minor elements in ppm.

[b] Total Fe as Fe_2O_3.

contrast to that in glauconite. The Fe/Al ratio of the Recent chamosite from Loch Etive, Scotland is significantly higher than that in ancient chamosites. The silica contents of both Recent samples are higher (partly because of the presence of free quartz), and the alumina contents are lower than those of ancient chamosites (Table 33.16, analyses 3 and 4). The K_2O content of the Loch Etive chamosite is relatively high, but is still much lower than that of modern glauconites (Porrenga, 1966).

The minor element data for chamosites, given in Table 33.16, are variable. The Cu, Ni and Zn contents of the chamosite from Loch Etive are much higher than they are in the associated sediments (Rohrlich *et al.*, 1969), and cerium and copper are much richer than in the Niger Delta chamosite, whereas, Cr, Ni and Sr are more concentrated in the latter. The high level of cerium in the Loch Etive chamosite is particularly interesting in view of the reported cerium enrichments in glauconites (Balashov and Kazakov, 1968) and in phosphorites (Altschuler *et al.*, 1967).

Empirical formulae of some of the chamosites listed in Table 33.16 are given in Table 33.17. The Niger Delta chamosite cannot be represented in this way

TABLE 33.17

Empirical formulae of some chamosites[a]

1. $(Al_{1\cdot2}Fe^{3+}_{0\cdot0x}Fe^{2+}_{3\cdot4}Mg_{0\cdot85})(Si_{3\cdot8}Al_{0\cdot2})O_{10}(OH)_8$
2. $(Al_{1\cdot15}Fe^{3}_{0\cdot58}Fe^{2+}_{3\cdot26}Mg_{0\cdot67}((Si_{3\cdot07}Al_{0\cdot93}(O_{9\cdot75}(OH)_{8\cdot25}$
3. $(Al_{1\cdot54}Fe^{3+}_{0\cdot02}Fe^{2+}_{4\cdot05}Mg_{0\cdot49})(Si_{2\cdot19}Al_{1\cdot81})O_{9\cdot29}(OH)_{8\cdot71}$

1. Chamosite, Loch Etive (Rohrlich *et al.*, 1969). In the absence of data on water content, the formula was calculated assuming a small amount of ferric iron in the octahedral layer, and an ideal number of oxygen atoms and (OH^-) ions.
2. Chamosite, Raasay (MacGregor *et al.*, 1920).
3. Chamosite, Northamptonshire (Brindley and Youell, 1953).

[a] Calculated assuming that the tetrahedral layer contains exactly 4 atoms, and assigning the excess Al to the tetrahedral layer.

because the very high SiO_2 value implies that there are more than 4 silicon atoms in the tetrahedral layer. In comparison with ancient chamosites the Loch Etive example has a higher Si/Al ratio in its tetrahedral layer, but has similar proportions of these atoms in the octahedral layer.

33.6.3.2. *Origin of chamosite*

The consensus among geologists who have studied ancient sedimentary ironstones is that chamosite forms by primary precipitation in marine environments in which there is a higher than normal supply of iron from continental denudation (Pettijohn, 1957; James, 1966). Borchert (1965) has outlined a scheme to account for the origin of several different iron minerals (including oxide, silicate, carbonate and sulphide phases) in marine basins having waters quite different in composition from that of present day sea water.

Strakhov (1959) and Curtis and Spears (1968) have argued that chamosite forms diagenetically in marine sediments through reactions between dissolved or colloidal iron, aluminium and silicate in pore waters. Evidence for the diagenetic origin of oriented chamositic coatings on faecal pellets and detrital silicates in Loch Etive has been presented by Rohrlich *et al.* (1969). The chamositic pellets occur in very fine-grained organic-rich sulphide-bearing muds, the pore waters of which contain relatively high concentrations of dissolved iron. Thus, the primary oolitic structure, so common in ancient chamositic ironstones, can also be produced within Recent sediments. In contrast, the chamositic pellets which occur in sandy sediments off the Niger Delta are characterized by iron oxide rinds (Porrenga, 1966). The pellets are therefore most probably not Recent, but are, in fact, old ones which have been concentrated by reworking.

The occurrences of both chamosite and glauconite in several near-shore areas have been discussed by Porrenga (1967). On the continental terrace off equatorial West Africa and north of Paria, Venezuela, chamosite occurs at water depths generally shallower than 80 m, whereas glauconite occurs in deeper water (see also Ciresse, 1965; Caillière and Martin, 1972). Porrenga (1967) has suggested that both minerals are being formed contemporaneously in the environments in which they are now found, and that their distribution with depth reflects the different temperatures at which they are being generated, i.e. 25–27°C for chamosite and 10–15°C for glauconite. However, it seems very unlikely that such a small temperature difference would lead to the formation of two very distinct mineral species. In addition, the formation of chamosite in Loch Etive, at water temperatures varying little from 8°C, shows that the physicochemical environment, and not simply the temperature, is important. The different depth distributions of chamosite and glauconite noted by Porrenga may also be partly the result of reworking during the last sea level rise which concentrated glauconite at water depths slightly greater than the minimum sea level. In contrast, chamosite has probably formed at shallower depths in environments which have been established since sea level reached its present position.

33.6.4. IRON SULPHIDES

The iron sulphides which may be present in marine sediments are pyrite (FeS_2), marcasite (FeS_2) and the various monosulphides which have been collectively referred to as hydrotroilite (Doetler, 1926). Pyrite is probably the most abundant form and occurs as micro-crystalline framboidal concretions dispersed in unconsolidated sediments or within microfossil cavities. Hydrotroilite is extremely fine-grained and often poorly crystalline. It is widely distributed in the subsurface sediments of tidal flats (van Straaten, 1954), restricted basins such as fiords, the Black Sea (Strøm, 1936; Ostroumov, 1953) and many eutrophic lakes (Doyle, 1968). This poorly-crystalline material, which has not been accorded a valid mineral name, is probably the precursor of pyrite (Berner, 1970a).

The identities of the various iron sulphides which have been found in marine sediments are listed in Table 33.18. Pyrrhotite (hexagonal FeS), which is the thermodynamically stable sulphide species relative to greigite and mackinawite, is not shown in the table because it has not been identified in Recent sediments.

TABLE 33.18
Iron sulphides in Recent sediments

Mineral	Formula	Crystal structure
Pyrite	FeS_2	Cubic
Marcasite	FeS_2	Orthorhombic
Greigite[a]	Fe_3S_4	Cubic
Mackinawite[b]	$Fe_{1+x}S$[c]	Tetragonal

[a] Greigite is the cubic dimorph of smythite (rhombohedral Fe_3S_4) and is referred to as *melnikovite* in Russian literature.
[b] Included within this term are kansite (Meyer *et al*., 1958), tetragonal FeS (Berner, 1962), "amorphous" FeS and "precipitated" FeS.
[c] $Fe_{1.04}S$ to $Fe_{1.07}S$ (Clark, 1966).

The relatively low total S content of many organic-rich reducing sediments (Table 33.6) means that the concentrations of iron sulphides in Recent sediments are generally small. Table 33.19 lists some estimates of the amounts of S present as FeS and FeS_2 in some Recent sediments studied by Berner (1964a, 1970a, b). A relatively small amount of monosulphide imparts a distinctive black colouration to a sediment. However, sediments in which pyrite is the predominant sulphide are grey to dark green in colour according to the grain size.

The distribution of iron sulphides in some modern marine sediments was systematically studied by van Straaten (1954) who found a regular distribution of different sulphides in the tidal flat sediments of the Netherlands. A surface zone of iron hydroxides was underlain by a black monosulphide zone, 20–40 cm thick, which was itself, in turn, underlain by a grey disulphide zone. Van Straaten interpreted this distribution in terms of the sequential formation of iron sulphides by the reduction of buried surface oxides, the formation of

TABLE 33.19

Concentrations of iron sulphides in some Recent near-shore sediments[a]

Locality	FeS[b]	Pyrite	Colour of sediment
Long Island Sound[c]	0·29	0·79	Black
Black Sea[d]	0·58	0·30	Black
Black Sea[d]	0·01	1·99	Grey
Gulf of California[e]	—	1·45	Olive green

[a] Data expressed as weight %S in the form of either monosulphide or pyrite.
[b] Includes amorphous FeS, mackinawite and greigite.
[c] Berner (1970a).
[d] Berner (1970b).
[e] Berner (1964a).

monosulphide in an anaerobic zone and the transformation of the monosulphides into pyrite. This was easily confirmed because samples of the hydroxide layer kept out of contact with atmospheric oxygen turned black owing to the formation of the monosulphide. Conversely, when samples of the black sediments were exposed to air they rapidly lost their colour as a result of the oxidation of the monosulphide.

Studies of sulphides in sediments which do not contain black monosulphides have been concerned mainly with the amounts of pyrite (or marcasite) present, the form of the sulphide grains, the presence or absence of free H_2S in the sediments and the possible ways in which the sulphide minerals could have formed (Galliher, 1933; Emery and Rittenberg, 1952; Volkov, 1961). Because the monosulphides are extremely unstable and very poorly crystalline, and also because it is difficult to concentrate any of the iron sulphides from Recent sediments, attempts have been made to produce such sulphide minerals experimentally under conditions simulating those in the natural environment. This has led to a reasonably coherent explanation of the mechanisms of sulphide formation which can be tested by application to modern sediments.

33.6.4.1. *Experimental sulphide formation*

Berner (1964b) has investigated the products formed in reactions of H_2S with metallic iron, goethite and dissolved iron(II). He showed that under a range of experimental conditions the initial reaction products were black, tetragonal, or cubic iron monosulphides, which will be subsequently referred to as mackinawite and greigite, respectively (Berner, 1967). Rickard (1969a) performed a series of similar experiments and claimed that mackinawite is the immediate precursor of greigite. Comparison of the X-ray diffraction data for the synthetic monosulphides produced by Berner and Rickard with published X-ray data for natural black iron sulphides (Berner, 1964b; Volkov, 1961) indicates that the natural material consists of one or more of the species mackinawite, greigite or non-crystalline FeS.

The formation of pyrite in the experiments referred to above takes place after the formation of the monosulphides by their reaction with sulphur; a type of reaction proposed by Ostroumov (1953), and confirmed experimentally by Berner (1964b, 1970a) and Rickard (1969a). The reactions are:

$$FeS_{mackinawite} + S^{\circ} \rightarrow FeS_2$$

$$Fe_3S_{4\,greigite} + 2S^{\circ} \rightarrow 3FeS_2$$

Berner has maintained that pyrite cannot form in the absence of free sulphur, or of reactions which produce free sulphur. However, Rickard (1969a) has argued that pyrite is also formed by the reaction between iron(II) or mackinawite and polysulphide ions if atmospheric oxygen is rigorously excluded from the reaction. Berner (1970a) has been unable to confirm Rickard's results and finds that pyrite is only produced when excess elemental sulphur is present. Pyrite formed in this way occurs as small crystals embedded in the surface of the sulphur grains.

Roberts *et al.* (1969) have proposed a slightly modified reaction path on the basis of a series of experiments similar to those performed by both Berner and Rickard. They suggested that sulphur, produced by the oxidation of H_2S by iron(III), reacts with sulphide ions to produce disulphide ions and these, in turn, react with iron(II) ions to produce pyrite directly:

$$S^{2-} + S^{\circ} \rightarrow S_2^{2-}$$

$$Fe^{2+} + S_2^{2-} \rightarrow FeS_2$$

33.6.4.2. *Pyrite formation in Recent sediments*

For simplicity, the formation of pyrite in marine sediments may be discussed with respect to the sources of sulphide and iron.

The ultimate source of sulphide in Recent marine sediments is the sulphate in sea water. The reduction is brought about by sulphate-reducing bacteria, as postulated by Murray and Irvine (1893) and later confirmed by Miller (1950) and Thode *et al.* (1951) who both used sulphur isotope techniques. Most of the reduction must take place close to the sediment surface because bacterial populations are highest there (Zobell, 1946; Emery and Rittenberg, 1952; Sorokin, 1964). Non-biological reduction of sulphate does not apparently occur (Rickard, 1969b). By using the bacterium *Desulfovibrio desulfuricans* Rickard (1969b) was able to show that the sulphide minerals produced in the reaction between bio-chemically produced sulphide and iron salts or goethite are indistinguishable from those produced inorganically.

The source of the sulphate which is reduced bacterially is the sea water overlying the sediments. Total sulphur in the sediments exceeds that initially present in the pore waters, and therefore sulphate must diffuse into the sediment (Kaplan *et al.*, 1963). Moreover, the isotopic composition of the sulphur present in the sediments shows that it must have been supplied from a very large reservoir and this is present only at the sediment surface.

The vertical distribution of sulphate in the pore waters of marine sediments often shows a steady decrease with depth (Ostroumov, 1953; Emery and Rittenberg, 1952; Kaplan *et al.*, 1963; Berner, 1964c). In some instances the decrease appears to be exponential, and can only have arisen as a result of a steady state balance between the rates of sulphate reduction and that of the downward diffusion of sulphate. From these distributions Berner (1964c) has been able to derive reasonable estimates of the diffusion coefficient of sulphate, and of the rate of sulphate reduction at the sediment surface. In many cases the concentration of sulphate does not decrease to zero, but instead approaches a steady value which is a function of the downward decrease in the amount of metabolizable organic material as well as of the rate of diffusion of sulphate from above.

Sulphide produced in the laboratory by the bacterial reduction of sulphate has the same $^{32}S:^{34}S$ ratio as that in naturally occurring sedimentary pyrite (Thode *et al.*, 1951; Jones and Starkey, 1957; Kaplan *et al.*, 1960; Kaplan and Rittenberg, 1964; Kemp and Thode, 1968). However, ^{32}S is only enriched to the extent of 5–40‰ in sulphides produced synthetically through the agency of bacteria, whereas in natural pyrite enrichments of up to 63‰ are common. This difference in isotopic fractionation is apparently produced within the bacterial cells during the rate-controlling sulphate reduction process. The higher enrichments in ^{32}S found in sediments are most probably the result of at least the uppermost layers of marine sediments being an open system (Kemp and Thode, 1968); this allows sulphur species to undergo successive oxidation and biochemical reduction and leads to considerable fractionation of the sulphur isotopes (Kaplan *et al.*, 1960).

An important factor determining the total amount of sulphide that can be produced in a sediment is the quantity of organic material present in a form that can be metabolized by sulphate-reducing bacteria. This, in turn, will control the total amount of pyrite that can be produced in a given sediment (Berner, 1970a). It would be expected, therefore, that relative to organic-poor sediments those ones which contain high concentrations of organic material will also have larger contents of dissolved sulphide and higher concentrations of pyrite. Berner has shown that, in a suite of sediments from coastal Connecticutt, organic carbon (a measure of organic material) is indeed a limiting factor in the formation of pyrite. In the uppermost 2 cm of the sediments, the amount of pyrite is linearly related to the amount of organic carbon

(Fig. 33.15), the regression line passing through the origin as would be expected. Most of the pyrite present in the sediments is produced in the near-surface region, a conclusion previously reached by Kaplan *et al.* (1960). The degree of pyritization (see Berner, 1970a) reaches a constant value at depths below about 5 cm because no more iron can react with H_2S as there is insufficient H_2S in the sediment pore water.

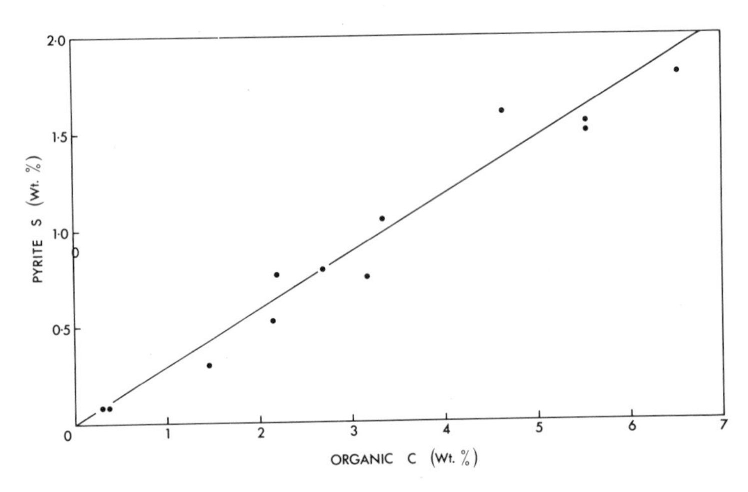

FIG. 33.15. Correlation between pyrite sulphur and organic carbon in the uppermost 2 cm of some near-shore sediments (from Berner, 1970a).

The availability of iron could also theoretically limit the amount of pyrite produced in a sediment if sulphide concentrations were high (Berner, 1970a). This situation is apparently very rare because of the abundance of particulate forms of iron in most sediments; this iron occurs both as detrital silicates and as oxide coatings on clay minerals (Carroll, 1958). Ferric oxides and hydroxides may also be made available by bacterial reduction to ferrous iron (Oborn and Hem, 1961). It is therefore only in sediments, such as highly calcareous ones, to which detrital iron contributions are small, that pyrite formation is likely to be limited by a supply of iron.

According to the experimental work of Berner, (1964b; 1970a), the formation of pyrite from iron monosulphides can only take place in the presence of free sulphur. This sulphur is readily formed in sediments by the reaction between H_2S and dissolved oxygen brought into the sediments by physical or biological disturbance from the overlying waters if these are oxygenated. This reaction may also be brought about by sulphur-oxidizing bacteria (e.g. *Thiobacillus*) which, in addition, are capable of oxidizing the sulphur produced to sulphate. In sediments deposited in truly anoxic waters (Chapter 16) the source of elemental sulphur may be more restricted because sulphur-oxidizing bacteria are probably absent from the sediment (Zobell, 1946) and the waters are devoid of O_2. One possible mechanism for supplying elemental sulphur to the sediments under these conditions is the sedimentation of particulate sulphur from the anoxic/oxic boundary higher up in the water column (Ostroumov *et al.*, 1961).

33.6.4.3. *Formation and occurrence of marcasite in marine sediments*

Marcasite, the orthorhombic dimorph of pyrite, has rarely been detected in marine sediments. This may partly be due to the difficulty of distinguishing between the two forms without the use of X-ray diffraction techniques.

Van Andel and Postma (1954) have outlined the distribution of marcasite and pyrite in the Recent sediments of the Gulf of Paria. They found that marcasite was restricted to sediments which have accumulated relatively rapidly and which have relatively low interstitial pH values. In contrast pyrite was dominant where the sedimentation rate was low and the pH of the interstitial water relatively high.

Rickard (1969a) synthesized marcasite by the reaction between iron(II) ions and polysulphide ions at low pH values. At high pH values the ratio of pyrite to marcasite increased until, at pH values greater than 9·5, marcasite did not form at all. In addition, the reaction between mackinawite and elemental sulphur led dominantly to the production of marcasite at 70°C and the formation of increasing proportions of pyrite as the temperature was increased to 150°C. Rickard (1969a) concluded from these observations (see also Allen *et al.*, 1912), that marcasite represents a non-equilibrium sulphide phase which forms preferentially at low pH values, and which, with time, transforms to pyrite, a more stable phase. Rickard (1969a) also suggested that under the different pH conditions at which the two minerals were formed the predominance of either elemental sulphur or polysulphide ions may have played a part in determining which phase was produced. The presence of elemental sulphur would promote the formation of marcasite from mackinawite by means of a solid state reaction whereas polysulphide ions would take part in an exchange reaction leading to the production of pyrite.

33.6.5. FERRO-MANGANESE OXIDES

The frequent occurrence of ferro-manganese concretions in near-shore marine areas has been reviewed by Manheim (1965). They are well known to occur in the Kara, White and Barents Seas (Gorshkova, 1957; Klenova, 1936; 1960; the Baltic Sea (Gripenberg, 1934; Winterhalter, 1966); the Black

Sea (Murray, 1900; Sevast'yanov and Volkov, 1966); the Gulf of Maine (Pratt and Thompson, 1962); the Blake Plateau (Stetson, 1961; Pratt and McFarlin, 1966); and some Scottish (Buchanan, 1891; Calvert and Price, 1970b) and Canadian (Grill *et al.*, 1968) inlets and fiords. The concretions occur in a wide variety of forms, including round nodules ("pea ores"), flat disc-shaped concretions ("penny ones"), flattened "girdles" around shells and pebbles, and flat slabs. These different forms are found on, and in, distinctive types of sediments ranging from muds for round nodules, to sands for flat concretions and to pebbles and boulders for girdles and slabs. The concretions are generally present on sediment surfaces in contact with the overlying water; if they do occur within the sediments, they are restricted to the uppermost sections in which oxidizing conditions prevail (see Section 33.7.6).

33.6.5.1. *Mineralogy*

Manganese nodules contain several different manganese oxide phases, poorly ordered iron oxides and various detrital impurities. Buser and Grütter (1956) and Grütter and Buser (1957) have shown that the manganese oxide phases comprise two forms of "manganese (II) manganite", having basal spacings of 7 Å or 10 Å, and δ-MnO_2. The two manganites have subsequently been referred to respectively by a number of authors as birnessite (Jones and Milne, 1956), and todorokite (Straczek *et al.*, 1960).

Manganese nodules from near-shore environments appear to consist predominantly of 10 Å manganite (Manheim, 1965; Cronan, 1967; Grill *et al.*, 1968; Calvert and Price, 1970b; Glasby, 1972). However, Winterhalter (1966) has presented X-ray diffraction data showing that nodules and crusts from the Baltic consist of 7 Å manganite, in contradiction to the observations by Manheim (1965). As pointed out by Buser and Grütter (1956) the various forms of manganese oxides in nodules have different oxidation grades, increasing in the order 10 Å manganite $<$ 7 Å manganite $<$ δ-MnO_2. On this evidence, the manganese in the manganese oxides of near-shore environments is therefore in a lower oxidation state than is that in open oceanic forms (Buchanan, 1891). Price and Calvert (1970) have used the mineralogy of the nodules to deduce their oxidation grades.

The nature of iron oxide phases in ferro-manganese nodules remains obscure. Thus, Cronan (1967) and Glasby (1972) could not detect any crystalline iron phase in a wide suite of manganese nodule samples. Winterhalter (1966), on the other hand, has suggested that goethite can be detected in some nodules and crusts from the Baltic.

33.6.5.2. *Chemical composition*

The compositions of ferro-manganese nodules from several near-shore

TABLE 33.20

Chemical compositions of some ferro-manganese nodules and crusts from various near-shore environments[a]

	1	2	3	4	5
SiO_2	16·40	11·90	13·72	13·53	8·90
TiO_2	0·22	0·17	0·19	0·35	0·12
Al_2O_3	2·90	3·12	3·22	4·30	3·60
Fe_2O_3	32·10	37·92	8·80	5·60	1·23
CaO	1·70	6·23	1·85	7·79	1·62
MgO	0·96	1·73	3·02	3·10	—
K_2O	0·91	—	1·19	1·24	1·15
Na_2O	0·47	—	1·28	—	—
P_2O_5	1·60	2·60	0·79	0·81	—
MnO	18·10	8·76	42·27	38·96	50·19
S	0·08	—	0·07	—	—
C_{org}	2·50	0·67	—	—	—
CO_2	0·76	5·50	0·56	11·88[b]	—
As	—	687	—	245	—
Ba	2500	—	2857	3090	2700
Co	160	84	157	120	100
Cr	10	16	7	—	—
Cu	48	37	67	17	100
Mo	130	18	231	55	220
Ni	750	281	314	77	450
Pb	38	—	N.D.[c]	34	250
Rb	—	—	—	40	—
Sr	—	—	—	770	1000
U	10	—	—	—	—
V	150	286	157	—	—
Y	—	—	—	28	—
Zn	80	—	23	60	—
Zr	—	42	—	55	230

1. Baltic Sea, composite analysis of nodules and crusts (Manheim, 1965). Analyses by precision spectrochemical method for major elements (Landergren, 1964), gravimetry for CO_2 and S and optical spectroscopy for minor elements.
2. Black Sea manganese nodules, mean of: 15 analysis for Fe_2O_3, TiO_2, MnO, P_2O_5, Ni, Co, Cu, Mo, V and Cr; 8 analyses for Zr (Sevast'yanov and Volkov, 1967; Sevast'yanov, 1967). Analytical methods not described.
3. Jervis Inlet, British Columbia (Grill *et al.*, 1968). Mean of 2 analyses calculated on a composite nodules basis, except for those of S, Co, Cu, Pb, Ni, V and Zn which are for the HCl-soluble fractions only. Analytical methods not described.
4. Loch Fyne, Scotland (Calvert and Price, 1970b). Mean of 2 analyses of bulk nodule samples. Analyses by X-ray fluorescence spectroscopy.
5. Gulf of California manganese nodules (Mero, 1965, Table 30). Analyses by X-ray fluorescence spectroscopy.

[a] Major element oxides as wt.%, minor elements in ppm.

[b] CO_2 and CaO values are high because of the presence of admixed manganoan calcite (see Calvert and Price, 1970b).

[c] Not detected.

environments are shown in Table 33.20. As Manheim (1965) has pointed out, the chemical compositions of near-shore and shallow-water nodules are highly variable, and the analyses presented in the table are simply intended to illustrate these variations. Apart from the variable proportions of detrital silicates present as impurities within them, the main variable appears to be the ratio of manganese to iron, which ranges from 45 in nodules from the Gulf of California to between 5 and 8 for the Loch Fyne and the Jervis Inlet nodules to <1 for those from the Baltic and Black Seas.

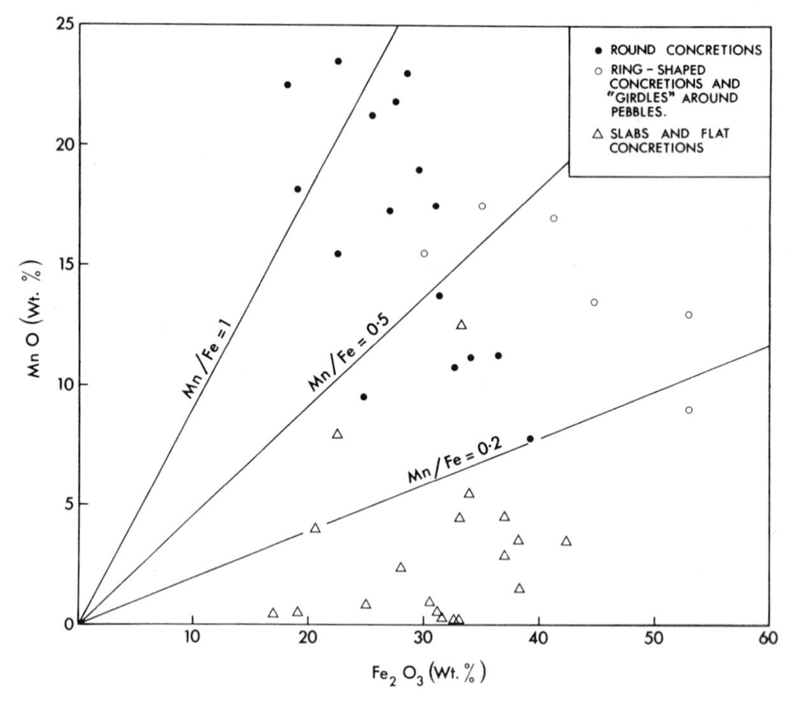

FIG. 33.16. Correlation between MnO and Fe_2O_3 in manganese nodules and concretions in the Baltic Sea (from Winterhalter, 1966).

There are variations in chemical composition not only between nodules from different areas, but also between those in any particular area. This is illustrated by the analyses of Black Sea nodules presented by Sevast'yanov and Volkov (1966) in which the Mn/Fe ratios range from 0·039 to 0·77. Moreover, in the Baltic Sea the composition appears to vary markedly with the morphology of the nodules and crusts (Winterhalter, 1966). Thus, for round pea ores the Mn/Fe ratios generally lie between 0·5 and 1·0, whereas for slabs and flat concretions they are usually less than 0·2 (Fig. 33.16). Ring-shaped concretions, or penny ones, have intermediate values.

The variable amounts of included detrital aluminosilicates present in the nodules and crusts are reflected in the contents of SiO_2, Al_2O_3, MgO, K_2O and Na_2O. It is only possible to correct the compositions of the nodules to a detrital-free basis if the composition of the associated nodule-free sediment is known. Thus, Calvert and Price (1970b) have shown that the nodules from Loch Fyne have, on average, an impurity level of 25%. The SiO_2 and Al_2O_3 values for the other analyses in Table 33.20, suggest that the nodules and crusts from the Baltic, the Black Sea, the Jervis Inlet and the Gulf of California are all contaminated to a similar degree. Manganese nodules from abyssal areas also generally contain proportions of aluminosilicate impurities similar to those of near-shore varieties. Because of this, it is probably valid to make direct compositional comparisons between near-shore and abyssal nodules.

In addition to MnO and Fe_2O_3, near-shore nodules appear to contain more K_2O, TiO_2 and P_2O_5 than can be accounted for by aluminosilicate contaminants. This can also be seen by comparing the compositions of the nodules with those of their associated sediments (Calvert and Price, 1970b). Thus, the manganese nodules from Loch Fyne have significantly higher K/Al, Ti/Al and P/Al ratios than do the surrounding sediments. Furthermore, the K_2O enrichment is greatest in the central Mn-rich part of the nodules, whereas TiO_2 and P_2O_5 have their highest enrichment in the iron-rich rims. The presence of potassium in the Mn-rich, 10 Å manganite phase appears reasonable in view of the compositions of this phase reported by Frondel *et al.* (1960) and Straczek *et al.* (1960). The association of phosphorus with the iron-rich phase is probably the result of either sorption of phosphorus onto iron oxides (Winterhalter and Siivola, 1967) or the presence of an iron phosphate phase (Sevast'yanov and Volkov, 1967). Similar associations between phosphorus and iron are found for the Baltic and Black Sea nodules, and also for many abyssal Pacific nodules (Calvert and Price, unpublished data). The association of titanium with iron in the Loch Fyne nodules suggests that titanium takes part in authigenic oxide precipitation (see Goldberg and Arrhenius, 1958) and that it is not simply a detrital element.

The proportions of CaO and MgO in the nodules (Table 33.20) are probably dominantly controlled by the amount of carbonate present. The Black Sea nodules have relatively high CaO and CO_2 contents, probably due to the presence of shell fragments; the high CaO and CO_2 values of the Loch Fyne nodules are the result of admixture with manganoan calcite (see Section 31.6.6). Near-shore manganese nodules often have higher organic carbon contents than do abyssal nodules (Manheim, 1965) probably because their associated sediments are richer in organic matter than are deep-sea ones.

It can be seen from Table 33.20 that many minor elements are significantly enriched in near-shore nodules compared with their associated detrital sediments. In addition, different groups of elements are either enriched or depleted in the near-shore nodules relative to abyssal ones. Thus, Ba, Co, Mo, Ni, and Sr have their highest contents in the manganese-rich phases of the Loch Fyne nodules, whereas As, Pb, Y and Zn are associated with the iron-rich phase (Calvert and Price, 1970b). Near-shore nodules contain more arsenic and occasionally more strontium, less Co, Cu, Mo, Ni, Pb, Y, Zn and Zr, and roughly similar amounts of barium compared to abyssal Pacific ones (Calvert and Price, unpublished data). The contrasting distributions of Co, Cu, Ni and Pb in near-shore and abyssal nodules have been discussed by Manheim (1965), Price (1967) and Price and Calvert (1970).

Fomina and Volkov (1969) have shown that yttrium and the lanthanides are depleted in ferro-manganese nodules from the Black Sea relative to their associated sediments (Table 33.21). However, there are differences in the

TABLE 33.21

Yttrium and lanthanide contents (ppm) of ferro-manganese nodules and sediments from the Black Sea (from Fomina and Volkov, 1969)

	1	2	3
La	6·2	17·0	18·2
Ce	12·0	28·2	29·7
Pr	2·5	6·7	7·0
Nd	8·2	12·5	12·0
Sm	2·5	2·7	3·7
{Gd, Eu}	2·5	2·7	3·7
{Tb, Y}	14·7	12·7	11·0
Dy	0·5	0·4	0·4
Er	0·7	0·8	0·7
Yb	0·6	0·8	0·77
Ce/La	1·93	1·66	1·63

1. Ferro-manganese nodules, mean of 4 samples.
2. Surface oxidised sediment, mean of 4 samples.
3. Sub-surface reduced sediment, mean of 4 samples.

distribution patterns of the individual lanthanides between the nodules and the sediments. Thus, the cerium and lanthanum contents are lower and the yttrium content is higher in the nodules than in the associated sediment. This lanthanide pattern contrasts markedly with that of oceanic nodules in which the total lanthanide content averages 1350 ppm (Piper, 1974); this is much higher than that in the associated pelagic sediments ($\sim$200 ppm in red clays). In such nodules the Ce/La ratios range between 1·28 and 7·42 (average for sediments, 2·22). Goldberg *et al.* (1963) have drawn attention to the very high cerium content of a Pacific manganese nodule and have contrasted it with the characteristic depletion of this element in ocean water (see also Høgdahl *et al.*, 1968). The Ce/La ratio in the Black Sea ferromanganese nodules is slightly greater than that in the associated sediments.

33.6.5.3. *Rates of accretion*

Manheim (1965) has estimated that the accretion rate of near-shore ferromanganese concretions lies in the range 0·01–1·0 mm yr^{-1}. These values were obtained from considerations of minimum permissible growth rates in areas where post-Pleistocene sedimentation had been established for a known period of time, and by taking into account the fact that the nodules do not survive burial below the uppermost oxidised sediment layer.

Allen (1960) has estimated the deposition rate of manganese on living mollusc shells from the Clyde estuary to be 0·017–0·04 mm yr^{-1} by measuring the thicknesses of manganese crusts on shells of known age. Calvert and Price (1970b) have suggested that, if these rates are applicable to the manganese nodules in Loch Fyne, these nodules would be 500 to 1200 years old.

Ku and Glasby (1972) have used ^{230}Th and ^{231}Pa dating methods to determine the growth rates of ferro-manganese nodules from Loch Fyne and Jervis Inlet, and have found values of about $0{\cdot}3 \times 10^{-3}$ mm yr^{-1} for both areas. Although these rates are considerably greater than those found for deep-sea nodules ($1 - 40 \times 10^{-6}$ mm yr^{-1}; Barnes and Dymond, 1967; Bender *et al.*, 1966; Ku and Broecker, 1969), from geological considerations they are also much lower than those which would be expected for shallow water nodules (Manheim, 1965).

33.6.5.4. *Origin of manganese nodules*

The origin of the ferro-manganese nodules and concretions in near-shore environments has been discussed at length by Manheim (1965) and Price (1967). According to Manheim the data on the types of sediments and the concentrations of dissolved metals in their pore waters and on nodule morphology suggest that the metals in the nodules are derived mainly from the underlying sediments. This is in contrast to the origin of open-ocean nodules for which the metals are largely precipitated from the overlying water onto oxidized surfaces. In this context, Price and Calvert (1970) have proposed that the continuous range of manganese nodule compositions has been produced by variations in the roles of either diagenetic, or primary accretion mechanisms in their formation.

According to Price and Calvert (1970) Mn/Fe ratios can be used to identify manganese nodules which have predominantly diagenetic origins. Nodules from open-ocean areas which have high ratios (i.e. generally > 4) are predominantly diagenetic because these elements are partially separated during post-depositional reactions within the sediments. Evidence to support this is provided by anoxic sediments (see Section 33.7) in which manganese and iron are mobilized in their lower oxidation states. When this occurs iron is partially fixed within the sediment as sulphide whereas manganese migrates to the upper sediment layers (Lynn and Bonatti, 1965) where it is precipitated in the surface oxidized layer. Thus, the resulting oxide concretions are enriched in manganese relative to iron (Cheney and Vredenburg, 1968). In most open-oceanic environments, in which post-depositional redistribution of metals does not occur, Mn/Fe ratios in the nodules and in the dispersed ferro-manganese oxides in the sediments are approximately unity (Price, 1967; Price and Calvert, 1970).

Calvert and Price (1972) and Duchart *et al.* (1973) have shown that manganese is present at relatively high concentration in the pore waters of those sediments in which manganese oxide precipitation is taking place. Using the dissolved manganese profiles Calvert and Price (1972) have presented a model in which rapid recycling of the element between solid and dissolved phases leads to manganese depletion in subsurface reduced sediments, and to manganese enrichment in the surface oxidized layer (see also Lynn and Bonatti, 1965).

The influence of the source of metals on the composition of the final oxide precipitate is well illustrated for the manganese nodules of the Baltic studied by Winterhalter (1966). As stated above, nodules having different morphologies also have different compositions. Round nodules, which occur in muddy sediments, have Mn/Fe ratios of 0·5 to 1·0, whereas flat concretions and slabs, which occur on boulders, have ratios of less than 0·2. The composition of the precipitate which forms directly from Baltic Sea water is probably similar to that of the slabs, and has approximately 1% MnO and 30% Fe_2O_3. Wherever post-depositional processes enhance the supply of dissolved manganese to the sediment surfaces the resulting precipitates have approximately 20% MnO and 25% Fe_2O_3 (see also Strakhov, 1966). This process results in the formation of the round nodules which are found predominantly in the fine-grained sediments that contain relatively high concentrations of organic material and therefore have low redox potentials. Thus, variations in the compositions of near-shore manganese nodules, (see Table 33.20 and Manheim, 1965) are most probably the result of both the different diagenetic effects produced by the various sediment types and the abnormal composition of the sea water in any one of these restricted environments.

The mode of formation of near-shore manganese nodules also affects their minor element compositions (Manheim, 1965). As stated above, near-shore manganese nodules have significantly lower concentrations of Co, Cu, Mo, Ni, Pb, Y, Zn and Zr than do open-ocean nodules. These differences may be the result of a number of processes including: (a) the stabilization of minor elements as insoluble phases (mainly sulphides) in sub-surface reducing (predominantly near-shore) sediments; (b) the stabilization of minor elements as metal–organic complexes in near-shore sediment pore waters; and (c) the dilution of the coprecipitated minor elements in near-shore nodules because of their much higher accretion rate.

Manheim (1965) has suggested that Baltic nodules contain higher concentrations of Co and Ni than of Cu, Pb and Zn because the latter elements are locked up as insoluble sulphides, precipitated within the reducing sediments, and are thus immobile. However, available data on the concentrations of dissolved trace metals in sub-surface reducing sediment pore waters (Duchart *et al.*, 1973) show that Cu, Zn and Pb concentrations are generally higher than are those of Co and Ni. Moreover, Price (1967) has pointed out that sedimentary iron sulphides contain much more Co and Ni than Cu, Pb and Zn (see Minguzzi and Talluri, 1951); this is easily understood on the basis of crystal chemistry (Mohr, 1959). It is, therefore, unlikely that sulphide precipitation in sediments is responsible for the distribution of minor elements in near-shore manganese nodules, and consequently also for the depletion of minor elements in these nodules relative to open-ocean forms.

The influence of organic materials on the behaviour of minor metals in near-shore sediments has been discussed by Price (1967) who has drawn on the experimental work of Krauskopf (1956), Swanson *et al.* (1966) and Kee and Bloomfield (1961). He has pointed out that the chelates of lead and the transition metals produced by the reaction between dissolved metals and the products of degradation of plant material are water-soluble rather than colloidal and that they are stable in air. Chelates, in general, have stability constants which increase in the order Mn < Fe < Co < Ni < Cu < Zn < Pb (Irving and Williams, 1953), whereas their ease of coprecipitation with hydrous ferric oxides increases in the order Fe, Mn, Zn > Co, Ni, Cu and Pb (Kee and Bloomfield, 1961). Therefore, in near-shore sediments having higher concentrations of labile organic material than open-ocean sediments, metal-organic complexes may be coprecipitated on the surface of the ferromanganese oxides to a greater or lesser extent. This results in a general depletion of minor elements in near-shore precipitates compared with those from the deep-sea; a depletion which will differ from element to element. Thus, molybdenum and zinc do not appear to be depleted to the same degree as do cobalt and nickel in near-shore relative to open-ocean nodules from the Atlantic and the Pacific (see Chapter 28). If organic association is involved, the behaviour of zinc is entirely in keeping with that predicted from the work of Kee and Bloomfield (1961). Similarly, the association of molybdenum with organic materials has been pointed out by Krauskopf (1956); however, it is also known that the element coprecipitates readily with manganese oxides.

The significantly higher accretion rates of near-shore ferro-manganese nodules relative to those from the deep-sea (Ku and Glasby, 1972) may also cause minor metal depletion if the metals are taken up at a constant rate by all nodules in the World Ocean. However, in view of the varying degrees of minor metal depletion discussed above it is unlikely that variations in the overall growth rates are the major factor controlling the abundance of minor metals in nodules.

33.6.6. CARBONATES

The authigenic calcium carbonate components of modern calcareous sediments, particularly those of reef structures, will not be discussed here and attention will be confined to those carbonates which form in essentially non-calcareous near-shore sediments; important among these are siderite and manganoan calcite. It has been suggested that siderite occurs, and has been formed, in the sediments of the Baltic Sea (Debyser, 1957; Manheim, 1965), the Adriatic Sea (Hinze and Meischner, 1968) and Oslo Fiord (Doff, 1969). Manheim (1965) observed an X-ray diffraction line which corresponded to the principal line of siderite in some reduced sediments of the central Baltic. Hinze and Meischner (1968) and Doff (1969) have inferred that this mineral is present in subsurface, reduced sediments from the other areas; this deduction was based on the fact that there is excess non-detrital iron which cannot be entirely accounted for as sulphide. Unfortunately, there are no adequate chemical analyses of this excess iron phase, nor has siderite been definitely identified in the sediments.

Manganoan carbonates appear to be more abundant than siderite in near-shore sediments. The compositions of some of these manganese minerals are shown in Table 33.22. The rhodochrosite from the Peru Trench was identified optically, whereas the manganoan carbonates from the Baltic Sea were characterized on the basis of the compositions of sediment leachates and the relationship between the calcium and the CO_2 content. The latter mode of identification, in conjunction with X-ray diffraction measurements, has also revealed the probable occurrence of manganoan carbonate in Oslo Fiord, but its specific composition is not known (Doff, 1969).

Shterenberg *et al.* (1968) were able to obtain sufficient manganoan carbonate from ferro-manganese nodules from the Gulf of Riga to carry out an X-ray diffraction analysis, and deduced the composition of the mineral from the *d* spacing of its strongest line using the compositional diagram of Mn, Mg and Ca carbonates given by Goldsmith and Graf (1960). It should be noted that this method can only give the compositional range, and the composition given by Shterenberg *et al.* (1968) does not correspond with that deducible from the X-ray data given.

The data for the manganoan carbonate (Table 33.22) from Loch Fyne is based on a complete chemical analysis of a large pure carbonate concretion (Calvert and Price, 1970b). This chemical composition was confirmed by X-ray diffraction methods used in conjunction with the data of Goldsmith and Graf (1960). A carbonate with a similar composition also cements ferro-manganese nodules into irregular aggregates.

The formation of manganoan carbonate in reducing sediments from the Eastern Pacific was interpreted by Lynn and Bonatti (1965) as indicating the

TABLE 33.22

Empirical formulae of some manganoan carbonates from recent sediments

Locality	Empirical formula	Source
Peru Trench	$MnCO_3$	Zen (1959)
Baltic Sea	$(Mn_{70}Ca_{30})CO_3$ to $(Mn_{60}Ca_{32}Mg_8)CO_3$	Manheim (1961a)
Baltic Sea	$(Mn_{56\cdot8}Ca_{25\cdot5}Mg_{9\cdot7}Fe_{8\cdot0})CO_3$	Hartmann (1964)
Eastern Marginal Pacific	$(Mn_{50}Ca_{50})CO_3$ to $(Mn_{80}Ca_{20})CO_3$	Lynn and Bonatti (1965)
Gulf of Riga	$(Mn_{40}Ca_{25}Mg_{35})CO_3$	Shterenberg *et al.* (1968)
Loch Fyne	$(Mn_{47\cdot7}Ca_{45\cdot1}Mg_{7\cdot2})CO_3$	Calvert and Price (1970b)

presence of manganese(II) in these sediments. The presence of similar carbonate phases in the sediments of reducing basins of the Baltic and Oslo Fiord may also be an indication of the formation of these phases under reducing conditions when interstitial manganese and carbonate activities are relatively high. Li *et al.* (1969) found that the dissolved manganese and the ΣCO_2 concentrations in the interstitial waters of an Arctic Basin sediment core reached values which corresponded to those for equilibrium with respect to idealized $MnCO_3$. Calvert and Price (1972) have suggested that the precipitation of manganoan carbonate within reducing sediments is a process which can partially account for the observed manganese concentration profiles in some sediment pore waters.

The manganoan carbonates of the Gulf of Riga and Loch Fyne are also intimately associated with ferro-manganese oxide phases. Therefore, reducing conditions under which manganese oxides would be rapidly dissolved, are not an absolute requirement for the formation of the carbonate. Presumably, conditions within sediments close to oxidizing/reducing boundaries are occasionally suitable for the precipitation of such concretionary carbonates although the mechanism of formation is unknown.

33.7 Oxic and Anoxic Sediments

In contrast to pelagic sediments, which characteristically have high redox potentials (Bramlette, 1961; Arrhenius, 1963) many near-shore sediments have a low redox potential either at formation, or following diagenesis. These reducing conditions may arise in two principal ways:

(i) Sediments of this type are found in basins in which the circulation is restricted and in which oxygen is therefore absent from the water column (Richards, 1965; see also Chapter 16). Under these conditions the water frequently contains free H_2S and is azoic. Such conditions are referred to as *anoxic* by Richards (1965), and prevail, for example, in the Black Sea, the Cariaco Trench and many Norwegian fiords. In some other restricted basins, and indeed in some lakes, anoxic conditions may occur intermittently, or seasonally, depending on the rate at which water exchange takes place.

(ii) In many sediments, reducing conditions may be produced beneath a surface oxidized layer. This often occurs in fine-grained sediments which accumulate rapidly and which contain a relatively high concentration of organic material. Because of its rapid burial this organic material is utilized by bacteria and other micro-organisms within the sediments, with the result that oxygen becomes depleted in the pore waters. After all oxygen has been removed production of H_2S by sulphate-reducing bacteria leads to reducing conditions (see Section 33.6.4.2). The boundary in the sediment between the layers which contain free O_2 and H_2S represents a reaction zone in which there is a steady state balance between O_2 which has diffused downwards and H_2S which has diffused upwards. This redox gradient produces its imprint on the redox state of the sediment. The upper oxidized sediment layer is generally pale brown to reddish brown in colour and is easily distinguished from the underlying grey to pale green reduced layer (Klenova, 1938; Gorshkova, 1957; Bezrukov, 1960; Lynn and Bonatti, 1965).

The thickness of the upper oxidized layer is evidently a function of the total rate of accumulation of the sediment. It is very thin ($\sim$1 mm) in many near-shore sediments when oxygen is present in the overlying water, and it increases in thickness seawards (Bezrukov, 1960), presumably because of the decreasing sedimentation rates in the deeper waters. The lower reduced layer eventually disappears in pelagic sediments.

From the discussion above it is apparent that it is necessary to distinguish between two types of reducing sediments – those which are deposited under anoxic, or euxinic, conditions and those which are produced by diagenetic changes.

In some older literature the terms *sapropel* and *gyttja* (Wasmund, 1930; Brongersma–Sanders, 1951) have been used to describe two types of sediment which accumulate in reducing environments. Sapropels are typically found in basins in which the O_2/H_2S boundary lies within the water column. The basin floor is therefore devoid of benthic organisms and the sediment is commonly organic-rich and black in colour because of the presence of iron monosulphides. Gyttja is a rather loose term used to describe those sediments which, although organic-rich, are olive-green in colour, and which contain benthic organisms. In this instance the boundary between oxygenated and sulphide-bearing water ideally coincides with the sediment/water interface. Therefore, although the water contains oxygen the sediment is anoxic.

Sapropels appear to occur mainly in truly anoxic basins, such as the Black Sea and many Norwegian fiords, whereas gyttjas are common in basins in which anoxic conditions are only established intermittently (e.g. Saanich Inlet), or in which the dissolved O_2 levels are extremely low (e.g. Gulf of California, Santa Barbara Basin, South West African Shelf).

In this discussion those sediments which are azoic and black in colour are referred to as *anoxic* sediments, whereas those which contain a fauna, which are oxygenated, and which are red, brown or yellow in colour are termed *oxic* (Doff, 1969). On this basis a useful distinction can be made between sediments accumulating in basins in which the water contains either H_2S or oxygen.

33.7.1. COMPOSITIONS OF SEDIMENTS ACCUMULATING IN ANOXIC BASINS

Although there have been several comprehensive studies of the chemical composition of the water in anoxic basins, and of the changes in composition brought about by the onset of anoxic conditions (Richards, 1960; 1965; Richards and Vaccaro, 1956; Richards and Benson, 1961; Richards *et al.*, 1965; Deuser, 1970, 1971), very little attention has been paid to the composition of the sediments in such basins, and to the ways in which they differ from normal oxic sediments. Some of the published total sediment analyses of very fine grained anoxic sediments from these basins are listed in Table 33.23. The principal differences in the major element composition of these sediments, compared with those in oxic fine-grained sediments (Table 33.2), appears to lie in the concentrations of MnO, P_2O_5, S and C. The anoxic sediments are usually black in colour because of the presence of monosulphides which accounts for both the high sulphur contents and for the Fe_2O_3/Al_2O_3 ratios which are high compared with those of average shale (Krauskopf, 1967) and detrital clayey sediments (Table 33.2). Manheim (1961a) and Doff (1969) have suggested that the high MnO concentrations can be explained by the presence of manganoan calcite, and this is supported by the fact that the sediments contain CO_2 in excess of that required to balance the non-silicate CaO contents.

The high P_2O_5 values of these anoxic sediments have been discussed by Doff (1969) who suggested that they can be partly accounted for by organic

TABLE 33.23
Chemical composition of anoxic sediments[a]

	1[b]	2[c]	3
SiO_2	39·0	34·85	32·5
Al_2O_3	14·2	8·77	8·97
TiO_2	0·72	0·43	0·29
Fe_2O_3[d]	7·7	10·43	4·82
CaO	2·2	7·30	20·92
MgO	2·7	1·83	2·21
Na_2O	0·3	0·93	1·61
K_2O	4·3	2·27	1·72
MnO	5·2	4·78	0·066
P_2O_5	0·23	0·61	0·27
CO_2	4·7	6·45	16·23
S	2·8	4·26	2·90
$C_{org.}$	4·6	5·44	2·25
As	—	71	—
Ba	750	552	265
Co	22	60	—
Cr	90	107	107
Cu	78	135	45
Mo	35	26	26
Ni	43	45	108
Pb	25	167	10
Rb	150	87	72
Sr	130	368	842
V	130	276	—
Y	—	23	7
Zn	110	944	76
Zr	—	112	78

1. Gotland Basin, Baltic Sea (Manheim, 1961a). Analyses by X-ray fluorescence and emission spectroscopy together with standard chemical techniques for C, CO_2 and S.
2. Bunnefjord, Norway (Doff, 1969). Analyses by X-ray fluorescence spectroscopy and gasometric methods (C and CO_2).
3. Black Sea (Hirst, 1973). Analyses by X-ray fluorescence spectroscopy, except for C and CO_2 which were determined by gasometric methods (Groves, 1951).

[a] Major element oxides as wt.%, minor elements in ppm.
[b] Corrected for the composition of the interstitial water.
[c] Corrected for the composition of the interstitial water and the diluting effect of the salt in the dried sample.
[d] Total Fe as Fe_2O_3.

contributions. The ratio of C/P is about 41 for average plankton (Fleming, 1940) and 24 for diatomaceous plankton (Lisitsyn, 1966). If these ratios are typical of those of the organic material in the sediments then organic phosphorus can account for 50–80% of the total phosphorus content of the Oslo

Fiord and Black Sea sediments. The assumption that the C/P ratio in the sediment organic matter is the same as that of plankton is not necessarily valid because phosphorus regeneration from decomposing planktonic material is more rapid than is that of carbon. This can be seen for the Baltic sediment for which the calculated phosphorus contribution exceeds the total phosphorus present. Therefore, the calculated contributions of phosphorus made by the plankton must be regarded as maxima. Consequently, a substantial fraction of the phosphorus, at least in the Oslo Fiord and Black Sea sediments, is present in another form, perhaps as sorbed phosphorus on clay minerals (probably of minor importance), or as a phosphate which has been directly precipitated (Doff, 1969). Evidence for the diagenetic precipitation of phosphorite in the organic-rich sulphide-bearing sediments off South West Africa has already been discussed in Section 33.6.1.4.

The relatively large amounts of organic carbon in anoxic sediments have been noted by Richards (1970). The sediments accumulating in anoxic basins, and including those shown in Table 33.23, are very often fine-grained, and Trask (1953) has pointed out that such sediments invariably contain more organic matter than do coarser-grained sediments from the same area. Therefore, the enhanced organic carbon contents of anoxic sediments may be, at least to some extent, caused by textural controls. However, there is a likelihood that deposited organic matter is more effectively preserved in anoxic environments in which relatively low redox potentials are operative, and from which there is an absence of higher organisms capable of breaking down organic matter by mechanical or enzymatic processes (Richards, 1970). From a survey of the organic carbon and phosphorus contents of anoxic and oxic waters Richards (loc. cit) has concluded that there is, in fact, no clear-cut evidence for enhanced preservation of organic matter in the waters of such environments. He has also suggested that the higher concentrations of organic carbon in the sediments accumulating in anoxic basins may, in fact, be a function of the more rapid sedimentation rates which results in the preservation of organic material which is rapidly buried, and is thus removed from the zone at the sediment surface where anaerobic bacterial populations are highest (Kriss, 1963).

Among the minor elements listed in Table 33.23 Co, Cu, Mo, Pb, V and Zn appear to have higher concentrations in the fine-grained anoxic sediments than they do in fine-grained oxic ones (see Table 33.2). The possible mechanisms controlling such enrichments are discussed below.

33.7.2. COMPARATIVE GEOCHEMISTRY OF OXIC AND ANOXIC SEDIMENTS

Valid comparisons between the geochemistry of oxic and anoxic sediments can only be made if the two types of sediments accumulate in the same basin of deposition, so that differences in the composition and supply of detrital and/or biogenous materials are minimised. Data for such a comparative study have been provided by Doff (1969) who investigated the major and minor element geochemistry of the sediments of Oslo Fiord. Table 33.24 shows the mean concentrations of a range of elements in oxic and anoxic

TABLE 33.24

Minor element concentrations (ppm) in some anoxic and oxic sediments from Oslo Fiord (from Doff, 1969)[a]

	Anoxic[b]	Oxic[c]
As	32	26
Ba	735	768
Br	281	202
Co	27	24
Cr	125	113
Cu	133	54
I	148	413
Mn	5743	3990
Mo	33	5
Ni	55	54
P	1322	1305
Pb	148	94
Rb	149	162
Sr	241	230
V	181	186
Y	37	38
Zn	571	342
Zr	197	235
$C_{org.}$[d]	3·58	2·27
S[d]	1·15	0·22

[a] Analyses by X-ray fluorescence spectroscopy, except for C which was determined by a gasometric procedure. Analyses corrected for salt content.
[b] Mean of 52 samples.
[c] Mean of 62 samples.
[d] Wt.%.

sediments of this basin. A discriminant function analysis of the chemical compositions of the group of 114 samples allowed an objective appraisal of the difference between the two groups to be made (Doff, 1969). Thus, As, Br, C, Co, Cu, Mo, Pb, S and Zn contributed to the discrimination of anoxic sediments at the 99·0% level of significance, whereas I, Rb and Zr were indicators of oxic sediments at the same significance level. The elevated levels of rubidium and zirconium in the oxic sediments can be readily explained by detrital contributions because many of these sediments were collected from relatively shallow water areas. However, the enhanced iodine concentrations appear to be a characteristic feature of near-shore fine-grained oxic sediments (see Section 33.7.6).

Doff (1969) has suggested that the coprecipitation with sulphides (Goldschmidt, 1954; Krauskopf, 1955, 1956; Wedepohl, 1964) and/or adsorption onto degraded organic material may account for the enrichment of several metals in anoxic sediments (see e.g. Krauskopf, 1956; Kee and Bloomfield, 1961; Szilagyi, 1967). However, data on the partitioning of minor metals between organic and sulphide phases in sediments is meagre.

33.7.3. MOLYBDENUM ENRICHMENT IN ANOXIC SEDIMENTS

Molybdenum is not only enriched in the sediments of the Baltic, the Black Sea, Oslo Fiord (Table 33.24) and off South West Africa (Table 33.6), but is also accumulating in a variety of other environments including a small basin in the Adriatic (Seibold *et al.*, 1958), in Saanich Inlet (Gross, 1967), in Lake Nitinat, (Crecelius, 1969) and in the Cariaco Trench (Dorta and Rona, 1971). The overlying waters in these areas are all euxinic, and according to Krauskopf (1955), Vine and Tourtelot (1970) and Bertine (1970), the underlying sediments accumulating under these conditions are characteristically enriched in this element.

The mechanisms for the removal of molybdenum from sea water under anoxic conditions and for its deposition in the sediments has been examined experimentally by Korolev (1968) who showed that it is quantitatively removed from solutions containing thio-molybdate during the precipitation of black FeS at pH 7·2. He also pointed out that natural FeS can contain several hundred ppm of molybdenum whereas natural FeS_2 contained only a few ppm, and concluded that the element will be released during the transformation of FeS into FeS_2. Since this reaction takes place within the sediments molybdenum is therefore available for uptake onto other sediment components.

Bertine (1972) has reported a series of experiments using radioactively-labelled molybdenum which largely corroborate the findings of Korolev (1968). On a short time scale, 70% of the molybdenum was removed from anoxic sea water by co-precipitation with iron sulphide. Reactions of the type

$$\mathrm{FeS(amorph)} + \mathrm{MoO_2S_2^{2-}} + \mathrm{H_2S} + 2\mathrm{H^+} \rightarrow \mathrm{FeS.MoS_3(amorph)} + 2\mathrm{H_2O}$$

$$\mathrm{FeS.MoS_3(amorph)} \rightarrow \mathrm{FeS(Tetr)} + \mathrm{MoS_3}$$

were suggested as possible removal mechanisms. However, Bertine has also shown that, at the pH of sea water, molybdenum was only removed to a slight extent by adsorption onto organic material (peat).

These results contrast markedly with the observations of Crecelius (1969) and Nissenbaum (in Bertine, 1972) that a considerable amount of molybdenum is present in the extractable humic fractions of anoxic sediments (up to 2700 ppm Mo according to Nissenbaum). Pilipchuck and Volkov (1968) have suggested that in the sediments of the Black Sea a large proportion of the molybdenum is bound on organic materials, although in the surface sediments some of it is also present as sulphide. This conclusion is not supported by their data since there was no significant correlation between molybdenum and sulphur in the surface sediments. However, there was a considerably better correlation of molybdenum with carbon. Pilipchuk and Volkov (1968) also showed that up to 40% of the molybdenum in some Black Sea organic-rich sediments was present in the humic fraction, and that no molybdenum was released when FeS was removed from the sediment with 5% HCl.

On balance, the evidence appears to suggest that most of the molybdenum in H_2S-bearing sea water is removed by co-precipitation with FeS, whereas most of that in anoxic sediments is associated with organic materials. Thus, some mechanism of transfer must be operative for molybdenum during diagenetic transformations within the sediments. Pilipchuk and Volkov (1968) showed that the concentration of dissolved molybdenum in the Black Sea waters decreases from an average value of 3·1 $\mu g\, l^{-1}$ in the oxygenated water to 1·0 to 1·1 $\mu g\, l^{-1}$ in the sulphide zone. They did not attribute this decrease to the formation of its sulphide but to co-precipitation with FeS which is known to occur in particulate form in the deep water of the Black Sea (Spencer *et al.*, 1972).

Berrang and Grill (1974) have shown that in the Saanich Inlet molybdenum is removed from sea water by particulate hydrous MnO_2 in the oxygenated zone overlying the sulphide-containing water and is therefore transferred to the anoxic zone by the settling oxide. The manganese is recycled through the oxic/anoxic boundary (cf. Spencer and Brewer, 1971), and this provides a constant mechanism for the removal of dissolved molybdenum from oxygenated water, and consequently for the transport of the element to the anoxic zone and for its concentration there when the manganese oxides dissolve. It is subsequently stripped from the waters of this zone by one of the mechanisms suggested above, and deposited in the sediments. Hence, the molybdenum content of the bottom sediments is probably a good indicator of the existence of anoxic conditions in a basin of deposition, as suggested by Doff (1969).

33.7.4. ENRICHMENT OF OTHER MINOR METALS IN ANOXIC SEDIMENTS

The enrichments of Cu, Ni, Pb and Zn as well as Mo, in the sediments off South West Africa have been discussed above. Since these enrichments are not as great as that of molybdenum (Table 33.9) it may be concluded that the enrichment mechanisms for these elements are different from that of Mo (see above). Nevertheless, the possibility of the sorption of these other elements onto degraded organic material is a possibility. In this connection it is interesting to note that Nissenbaum *et al.* (1972) separated a high molecular weight polymer, consisting mainly of humic materials, from pore waters of sediments from the Saanich Inlet and showed that it contained 2000 ppm Cu, 800 ppm Ni, 500 ppm Pb and 6000 ppm Zn. The total concentration of the polymer in the pore water was low ($\sim$300 mg organic matter l^{-1}), and the contribution of metals from this source was therefore small. However the data do serve to show that metal may be enriched in the organic fractions of anoxic sediment pore waters.

33.7.5. DEPOSITION OF URANIUM IN ANOXIC SEDIMENTS

The enrichment of uranium in many near-shore sediments (see Table 33.27 later) has been emphasized by Strøm (1948), Holland and Kulp (1954), Koczy *et al.* (1957), Baturin (1969b) and Veeh (1967). This enrichment occurs in anoxic sediments which are being deposited in environments thought to be similar to those in which many geologically old uraniferous black shales also accumulated (Swanson, 1961).

The mechanism involved in the enrichment of uranium in these sediments is not certain although several different processes may operate. Under the conditions of low redox potential found in the waters of restricted basins, such as some Norwegian fiords, the Baltic Sea and the Black Sea, dissolved uranium(VI) may be reduced to its 4+ oxidation state and precipitated as insoluble $U(OH)_4$. According to Starik and Kolyadin (1951) this reduction requires a redox potential lower than $-0{\cdot}1$ V; potentials which are in fact found in these basins (Manheim, 1961b; Ignatius *et al.*, 1971; Skopinstev *et al.*, 1966), and this mechanism may therefore account for the enrichment of uranium in the sediments. However, attempts to detect uranium(IV) in the anoxic water of the Black Sea have been unsuccessful (Kolyadin *et al.*, 1960).

The observed correlation between uranium and organic carbon in the sediments of anoxic basins (Kochenov *et al.*, 1965, Baturin *et al.*, 1967) suggests an alternative mechanism, namely that the uranium in these sediments is organically bound; Fig. 33.17 shows such a correlation for the sediments of the Black and Mediterranean seas. In the Black Sea sediments uranium is concentrated in chitinous material to levels of 200 ppm. It is possible that uranium may also be associated with other types of organic materials since it has been established that it is concentrated in the alkali-soluble (humic) fraction of black shale, peat and coal (Szalay, 1958; Breger and Deul, 1956; Vine, 1962; Manskaya and Drozdova, 1968).

There is also a third possibility which is that the uranium in some organic-rich anoxic sediments is associated with a phosphatic phase, either that of fish debris or phosphorite. For example, in the sediments of the South West African shelf there are two phases which contain enhanced uranium contents. These are fish scales, bones and teeth, (up to 700 ppm U), and both modern and reworked phosphorites (10–140 ppm U) (Baturin *et al.*, 1971; S. M. Brown, N. B. Price and S. E. Calvert, unpublished).

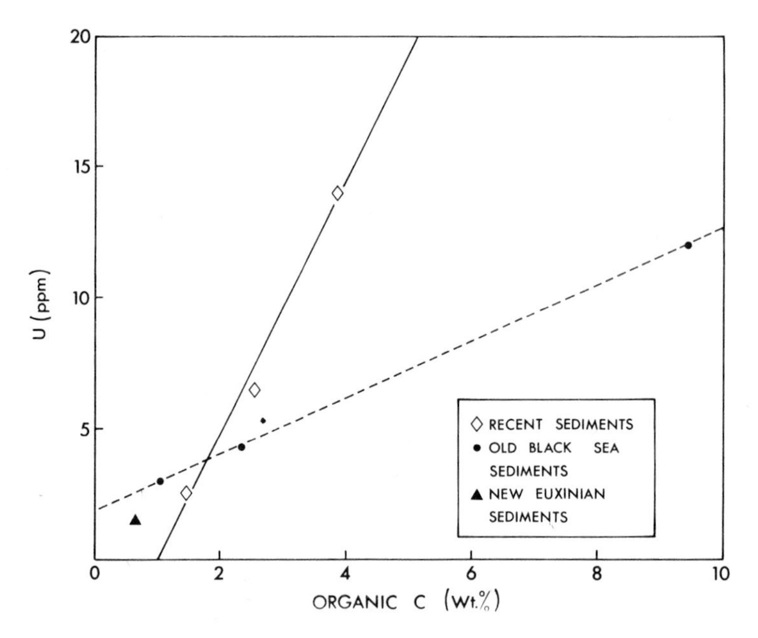

FIG. 33.17. Correlation between uranium and organic carbon in some sediments from the Black Sea (from Baturin *et al.*, 1967).

The correlation between uranium and P_2O_5 in the South West African shelf sediments (Fig. 33.18) shows that the organic-rich inner shelf muds contain excess uranium over that indicated by the estimated regression line. This may be interpreted as an indication of the degree of partition of uranium between the phosphatic and the organic phases; the phosphorites containing ~150 ppm uranium (but occasionally higher) and the organic material up to 70 ppm uranium. Similar conclusions can be reached using the relationship between uranium and yttrium in these sediments since yttrium is present only in the phosphorites (Table 33.14).

The distribution of uranium in the sediments of Saanich Inlet, an intermittently anoxic fiord, has been determined by Kolodny and Kaplan (1973) using successive leaching with NH_2OH and H_2O_2. They were able to show that 45–90% of the total uranium in the sediments is authigenic and

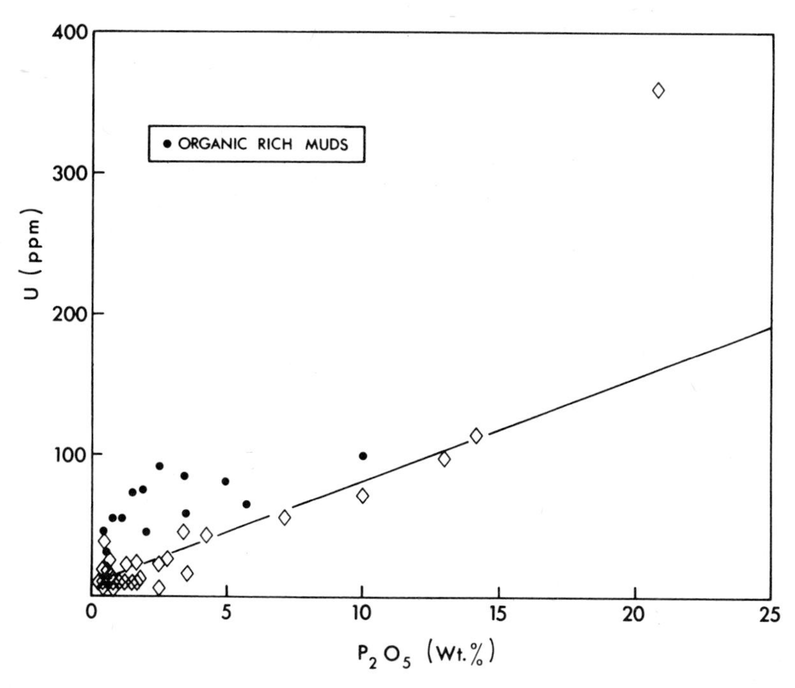

FIG. 33.18. Correlation between uranium and P_2O_5 in some Recent sediments from the South West African continental shelf. In calculating the regression line organic-rich muds have been excluded (Calvert and Price, unpublished).

has a $^{234}U/^{238}U$ ratio of approximately 1·15. More than 50% of this fraction is probably incorporated into organic material. The remaining, non-extractable, uranium is probably either adsorbed onto organic material or present in the 4+ oxidation state in an unspecified form. Kolodny and Kaplan (1973) also showed that uranium is present in solution in the pore waters at concentrations in the range 4–120 $\mu g\, l^{-1}$ (mainly in the 4+ state) and that this uranium had an average $^{234}U/^{238}U$ ratio close to 1·15. Uranium concentrations in the pore waters of the sediments of the South West African shelf have been found to lie in the range from 1–650 $\mu g\, l^{-1}$ (Baturin, 1971d). The high concentrations found in both of these areas (compared with values of $\sim 3\ \mu g\, l^{-1}$ for open-ocean water) have been interpreted as being a reflection of the diagenetic mobilization of uranium within the sediments; this is supported by the uranium isotopic ratios reported by Kolodny and Kaplan (1973) which were similar in both the sediments and their associated pore waters.

To summarize, the above data demonstrate that the high uranium concentrations of anoxic sediments are the result of a number of processes.

These include: the formation of uranium(IV) under conditions of relatively low redox potential and its subsequent precipitation, the sorption of uranium onto organic material and the formation of humic complexes.

33.7.6. GEOCHEMISTRY OF HALOGENS IN NEAR-SHORE SEDIMENTS

It has been pointed out above that the concentrations of iodine and bromine provide a useful means of discriminating between the oxic and anoxic sediments in Oslo Fiord. In this instance, iodine is significantly more concentrated in the oxic sediments, whereas bromine is enriched in the anoxic sediments. The relatively high concentrations of halogens in some modern marine sediments were first noted by Vinogradov (1939), and subsequent work (Shishkina and Pavlova, 1965; Price *et al.*, 1970; Bojanowski and Paslawska, 1970) has shown that near-shore sediments are significantly enriched in iodine relative to deep-sea ones (c.f. Bennett and Manuel, 1968). Data for the iodine contents of a variety of modern sediments are shown in Table 33.25. These should be compared with the estimate of 0·05 ppm for iodine in pelagic

TABLE 33.25

Iodine concentrations in some marine sediments[a]

Area	I (ppm)	Source
A. Near-shore sediments		
Russian shelf	1–297	Vinogradov (1939)
Barents Sea shelf	60–828	Price *et al.* (1970)[g]
Baltic Sea	101[b]	Shishkina and Pavlova (1965)
Black Sea	33[c]	Shishkina and Pavlova (1965)
Oslo Fiord, oxic sediments	64–1301	Doff (1969)[g]
Oslo Fiord, anoxic sediments	0–522	Doff (1969)[g]
S.W. African shelf	96–1990	Price and Calvert (1973)
Gulf of California	5–1370	Price and Calvert (unpublished)
B. Pelagic sediments		
Pacific red clays	29[d]	Shishkina and Pavlova (1965)
Oceanic calcareous oozes	39[e]	Shishkina and Pavlova (1965)
Atlantic calcareous oozes	37[f]	Bennett and Manuel (1968)

[a] Values expressed as ppm, on a dry weight basis.
[b] Mean of 2 grey clays.
[c] Mean of 18 grey clays.
[d] Mean of 4 samples.
[e] Mean of 11 samples.
[f] Mean of 4 samples.
[g] The values reported by Doff (1969), Price *et al.* (1970) and Price and Calvert (1973) are corrected for the diluting effect of the salt in the dried examples.

sediments calculated by Turekian and Wedepohl (1961) on the assumption that this element would be present only in the included salts. Doff (1969) has pointed out that the abundances of halogens in modern sediments are in the order I > Br > Cl, i.e. the reverse of their order of abundances in sea water.

Vinogradov (1939) has suggested that the iodine and bromine contents of sediments are related to the concentration of organic matter present, and that these elements are not associated with any other sediment component. This is confirmed by the data for sediments from the Barents Sea (Fig. 33.19), and for oxic sediments of Oslo Fiord (Doff 1969). The general relationship between iodine and texture in marine sediments (also suggested by Vinogradov (1939)), is apparently masked in these instances by the relationship between iodine and organic carbon (Price *et al.*, 1970).

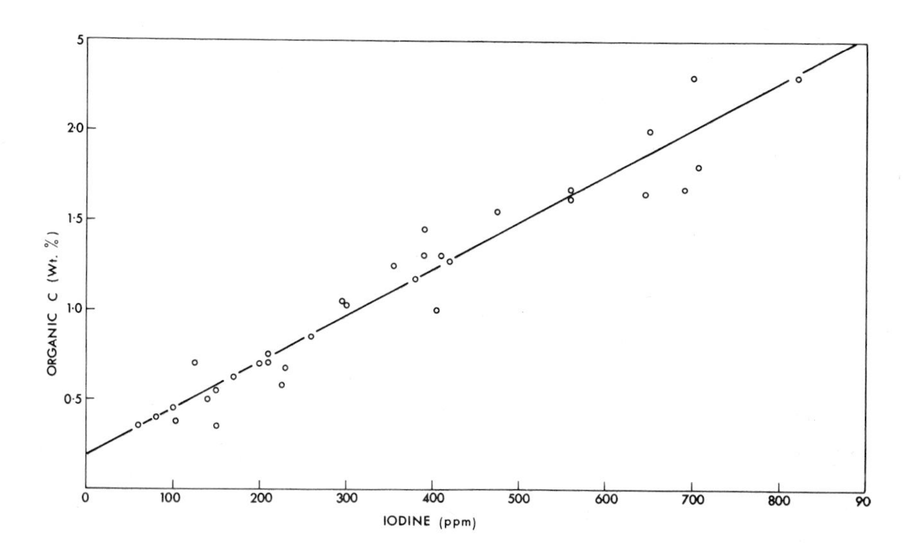

FIG. 33.19. Correlation between iodine and organic carbon in some surface sediments from the southwestern Barents Sea (from Price *et al.*, 1970).

The correlation between iodine and organic carbon shown in Fig. 33.19 is based on analyses of surface (0–1 cm) oxic sediments. In contrast, no such correlation is observed for anoxic surface sediments (Doff, 1969; Price and Calvert, 1973). Price and Calvert (1973) have shown that the I/C ratios of sediments across the South West African shelf varies in a manner which parallels the oxygen content of the bottom waters; a similar relationship appears to exist in the Gulf of California. The range of I/C ratios appears to be from 250–350 × 10^{-4} for oxic sediments to less than 30 × 10^{-4} for anoxic sediments.

The relatively high iodine contents of near-shore sediments (see Table 33.25) are thought to arise from the iodine contributed by organic processes (see above). Marine organisms are known to concentrate iodine from sea water (Shaw, 1962); thus, concentrations in marine plankton have been found to be approximately 300 ppm (Vinogradov, 1953), brown algae have been shown to contain up to 8000 ppm (Vinogradov, 1953), and for marine animals a range of 1–5000 ppm has been reported (Bowen, 1966). These concentrations represent large enrichments with respect to its sea water concentration (0·06 mg kg^{-1}), but they are insufficient to account for its concentrations in sediments. Hence, the primary source of the iodine in sediments is probably not marine organic matter itself, but sea water from which the iodine is sorbed onto degraded organic material at the sediment surface (Price and Calvert, 1973).

The very different I/C ratios of oxic and anoxic surface sediments implies that if the sorption of iodine onto organic material is the controlling process, then it is influenced by the environmental conditions under which the sediments are deposited. In marine organisms iodine is taken up as the element under oxidizing conditions as a result of enzymatic processes on cell surfaces; processes which do not take place under anaerobic conditions (Shaw, 1962). Price and Calvert (1973) have suggested that iodide sorption onto decomposing marine organic material is not controlled by vital cell activities, but only by enzymatic reactions on the organic matter which allow large enrichments to occur. Under anoxic conditions little iodide sorption takes place, presumably because of the lack of surface enzymatic reactions, and the iodine concentrations in the sediments are little greater than they are in living organic materials.

In oxic marine sediments the concentration of iodine decreases with depth of burial (Doff, 1969; Price *et al.*, 1970; Bojanowski and Paslawska, 1970). The I/C ratios at depth approach values of $\sim$30 × 10^{-4}, which are similar to those of surface anoxic sediments. Therefore, iodine is released into pore waters during diagenetic reactions within oxic sediments, and in fact many pore waters do contain relatively high concentrations of dissolved iodine (Tageyeva and Tikhomirova, 1962; Shishkina and Pavlova, 1965; Bojanowski and Paslawska, 1970). In contrast, iodine does not appear to be lost from anoxic sediments upon burial (Price and Calvert, 1973), and for this reason the I/C ratios of surface and sub-surface anoxic sediments are very similar (see also Doff, 1969). The release of organically sorbed iodine from oxic sediments may result from both the higher rates of degradation of organic materials under these conditions (Degens *et al.*, 1961, 1963; Prashnowsky *et al.*, 1961), and the type of bonding of the iodine with the organic materials.

33.7.7. SURFACE AND SUB-SURFACE SEDIMENTS

Differences in redox conditions within a sediment are brought about by diagenetic reactions. The redistribution of elements during this process is discussed more fully in Chapter 30. Attention will be directed here to the differences in chemical composition between surface oxic and the underlying anoxic sediments in near-shore environments.

The surface oxic sediments of Oslo Fiord are underlain by pale grey to green anoxic sediments, and Doff (1969) has provided analyses of these two types of sediments in several cores collected in the fiord. There are only minor differences between the concentrations of Ba, Co, Cr, Mo, Nb, Ni, Rb, Sr, Ti, V, Y and Zr in the surface and sub-surface sediments. These elements reside predominantly in the aluminosilicates (with the exception of Sr which is also associated with the carbonate debris), and the lack of vertical variations in their distributions implies that the composition and texture of the detrital fraction is virtually constant over the sample intervals studied. In contrast, As, Br, C, Cu, I, Mn, P, Pb and Zn are significantly enriched in the oxic layer and S is enriched in the underlying anoxic layer. The higher concentrations of halogens in oxic sediments, and their decrease with burial, have been discussed above. Carbon concentrations decrease with burial because of degradation and utilization of organic material by micro-organisms. The higher concentrations of manganese in the surface sediment are readily explained by the remobilization of the element from the sub-surface sediments and its subsequent reprecipitation in the surface layers (Lynn and Bonatti, 1965). The increased Cu, Pb and Zn concentrations in the oxic sediments are also partly a consequence of this process as these elements are liberated during the dissolution of the manganese (and iron) oxides. An additional source of these minor metals may be the degradation of their complexes with organic material, their migration in the pore waters, and their subsequent adsorption. The enhanced concentrations of arsenic and phosphorus may be explained by adsorption onto positively charged iron oxides which may result in the formation of ferri-arsenate and ferri-phosphate complexes (Doff, 1969). The increased concentration of sulphur in the sub-surface anoxic sediment is due to the presence of sulphide phases, mainly pyrite, which are not formed in the upper oxic layer.

Oxic sediments overlying anoxic sediments are not restricted to shallow water, but they may be found in hemi-pelagic environments (see e.g. Bezrukov, 1960; Skornyakova, 1965; Lynn and Bonatti, 1965). Data on the average chemical compositions of surface and sub-surface hemi-pelagic sediments from the eastern Pacific are shown in Table 33.26 (Bonatti *et al.*, 1971). From these it can be seen that the concentrations of Mn, P, Co, La and Ni are higher in the oxic layer, whereas Cr, U and V are enriched in the anoxic layer. The elements Si, Ti, Al, Mg, Na, K, Fe, Sc and Y, which reside predominantly in aluminosilicate lattices, show very minor differences between the two layers. Bonatti *et al.* (1971) relate these differences to the prevailing

TABLE 33.26

Composition of some surface and sub-surface sediments from the eastern marginal Pacific (from Bonatti et al., 1971)[a]

	Surface[b]	Subsurface[c]
SiO_2	26·2	22·1
TiO_2	0·2	0·17
Al_2O_3	5·8	4·5
Fe_2O_3	2·9	2·6
MgO	1·8	1·7
Na_2O	3·2	4·2
K_2O	9·74	0·71
MnO	3·9	0·26
P_2O_5	0·41	0·11
S	0·37	0·34
N	0·140	0·103
$CaCO_3$	38·4	52·16
B	89	64
Ba	3450	2450
Co	31	9
Cr	25	36
La	33	10
Ni	310	154
Sc	8	7
U	0·50	0·83
V	82	112
Y	21	17
Zr	46	31

[a] Analyses by atomic absorption spectrophotometry (Fe, Mn, Ca, Mg, Na, K), X-ray fluorescence spectroscopy (S), optical spectroscopy (Si, Al, Ti, P, B, Ba, Co, Cr, La, Ni, Sr, V, Y, Zr), fission track methods (U) and Kjeldahl technique (N). Values in wt.% (major elements) and ppm (minor elements).
[b] Mean of 4 samples between surface and 5 cm depth.
[c] Mean of 4 samples between 11 and 20 cm depth.

redox conditions in the two layers. Thus, manganese is present in the upper layer as its oxide which is the host for much of the La, Ni and Co. After burial the manganese is reduced to its lower oxidation state and is present in the anoxic layer as Mn^{2+}. The Ni and Co are also released into the pore waters when the oxides dissolve. In contrast, Cr and V are probably present in solution in the oxic layer, but are reduced to the insoluble hydroxide or sulphide in the anoxic layer. The uranium probably passed from the 6+ to the 4+ state upon burial and was fixed in the anoxic sediment.

Bonatti *et al.* (1971) have attempted to explain the distribution of phosphorus in these sediments by invoking a mechanism in which calcium phosphate (the supposed host of phosphorus in the oxidised layer) is dissolved in the presence of H_2S in the anoxic layer. However, it is more likely that the phosphorus is enriched in the oxide layer because of its adsorption by hydrated ferric hydroxides, or because of the formation of ferri-phosphate phases. Phosphorus adsorption by oxide surfaces, and its release under anoxic conditions, is a process of greater importance for lake sediments (Mortimer, 1941; Hutchinson, 1957; Mackereth, 1966). The enrichment of lanthanum in the oxidized layers of the eastern Pacific sediments may be due to the adsorption of this element on dispersed oxide phases, analogous to the enrichment of lanthanides in ferro-manganese nodules (Goldberg *et al.*, 1963; Barnes, 1967; Piper, 1974).

33.8. The Geochemical Balance and Near-shore Sediments

Near-shore sediments, particularly anoxic ones, may play a significant role in the geochemical cycle of the oceans as a result of their relatively high accumulation rates and the enrichments of certain metals in them. Some of these enrichments have been discussed above, especially those of Mo, Cu, Ni, Pb and Zn. In addition, Veeh (1967) and Berner (1972) have drawn attention to the importance of anoxic near-shore sediments as sinks for uranium and sulphur respectively. Silicon, in the form of biogenous siliceous tests, can be added to this list in view of the importance of near-shore accumulation of diatomites to the overall silicon budget in the oceans (Heath, 1973).

33.8.1. Uranium

Recent estimates of the accumulation rates of uranium in the sediments of some near-shore areas are given in Table 33.27. In general, this data confirms the suggestion made by Holland and Kulp (1954) and Veeh (1967) that some near-shore areas have very high uranium accumulation rates. All the areas listed in Table 33.27, apart from the Yucatan shelf, are areas of organic-rich sedimentation where the sediments, and sometimes the waters, are anoxic.

The sediments of the Yucatan shelf (Sackett and Cook, 1969; Mo *et al.*, 1973) are predominantly calcareous. The uranium concentrations in these sediments may reach 2–4 ppm, and these values are consistent with concentrations of uranium in modern corals of 3–4 ppm (Thompson and Livingstone, 1970; see also Chapter 18). Sackett and Cook (1969) have therefore suggested that, in addition to anoxic sediments, modern shallow-water carbonates may also be important sites for the removal of uranium from the ocean. The data in Table 33.27 show that, on the Yucatan shelf at least, the uranium accumulation rate is much lower than it is in many other near-shore anoxic environments.

TABLE 33.27

Accumulation rates of uranium in anoxic near-shore sediments

Area	Accumulation rate ($\mu g\ cm^{-2}\ yr^{-1}$)
Santa Barbara Basin	0·128[a]
Gulf of California	0·53[a]
Black Sea	0·145[a]
Sea of Azov	0·129[a]
Baltic Sea	0·04–0·2[a]
Saanich Inlet, British Columbia	1·500[b]
Cariaco Trench	0·450–0·910[c]
South West African shelf	0·240–0·770[d]
Yucatan shelf	0·06[e]

[a] From Veeh (1967).
[b] From Kolodny and Kaplan (1970c).
[c] Calculated from data in Dorta and Rona (1971).
[d] From Veeh *et al.* (1974).
[e] Calculated from data in Sackett and Cook (1969) and Mo *et al.* (1973).

Veeh (1967) has used 0·1 $\mu g\ cm^{-2}\ yr^{-1}$ as a conservative estimate of the average uranium accumulation rate in anoxic near-shore sediments, and has pointed out that at this rate only 0·4% of the total area of the World Ocean is required to balance the uranium input (0·0004 $\mu g\ cm^{-2}\ yr^{-1}$). Kolodny and Kaplan (1973) have shown that on the basis of their estimated uranium accumulation rate for Saanich Inlet only 0·025% of the total area of the ocean is required. Mo *et al.*, (1973) have pointed out that since the uranium input rate assumed by Veeh (1967) may be too low, the importance of near-shore anoxic sediments as a uranium sink may be over-emphasized. The validity of this objection may be questioned as the data in Table 33.27 show that uranium accumulation rates in anoxic environments are, in fact, much higher than 0·1 $\mu g\ cm^{-2}\ yr^{-1}$. Moreover, Kolodny and Kaplan (1973) have shown that the residence time for uranium in restricted anoxic basins is several orders of magnitude less than it is in the open-ocean, indicating a very rapid removal under these conditions.

33.8.2. MOLYBDENUM

Bertine (1970) and Bertine and Turekian (1973) have drawn attention to the rapid rate at which molybdenum is removed from sea water by incorporation into near-shore anoxic sediments. Table 33.28 presents data for its accumulation rates in a number of areas and shows that the rates of deposition of this element range from 0·3 to 5·8 μg cm^{-2} $year^{-1}$. The corresponding rate for pelagic sediments is 0·0027 μg cm^{-2} $year^{-1}$ (Bertine, 1970).

TABLE 33.28

Accumulation rates of molybdenum in anoxic near-shore sediments

Area	Accumulation rate (μg cm^{-2} yr^{-1})
Black Sea	0·514[a]
Gulf of California	0·300[b]
Saanich Inlet	0·240[c]
Mljet, Adriatic	0·862[d]
South West African shelf	1·783[e]
Cariaco Trench	2·880–5·830[f]

[a] From data given by Pilipchuk and Volkov (1966) and Ross *et al.* (1970).
[b] Bertine (1970).
[c] From data given by Gross (1967) and Gucluer and Gross (1964).
[d] From data given by Seibold (1958) and Seibold *et al.* (1958).
[e] S. E. Calvert and N. B. Price (unpublished).
[f] From data given by Dorta and Rona (1971).

In order to balance the input of molybdenum, the proportion of the ocean that must be underlain by sediments in which molybdenum is accumulating at 0·3 to 5·8 μg cm^{-2} $year^{-1}$ is 4% to 0·2% assuming a supply of 0·012 μg cm^{-2} $year^{-1}$. At the higher of these accumulation rates the area required is of the same order of magnitude as that for near-shore uranium accumulation.

33.8.3. COPPER, NICKEL, LEAD AND ZINC

Enrichments of Cu, Ni, Pb and Zn in near-shore organic-rich sediments have

TABLE 33.29

Accumulation rates of excess Cu, Ni, Pb and Zn in anoxic near-shore sediments

Area	Accumulation rate (μg cm^{-2} yr^{-1})			
	Cu	Ni	Pb	Zn
Gulf of California[a]	0·644	9·715	1·364	
South West African shelf[b]	1·627	2·427	0·369	1·420
Pelagic sediment, Atlantic[c]	0·019	0·011	0·007	

[a] From data given by Calvert (1966) and Calvert (unpublished).
[b] S. E. Calvert and N. B. Price (unpublished).
[c] From Turekian (1965).

been discussed in Section 33.7. Data on the accumulation rates of excess metals are presented in Table 33.29; these data are scanty, partly because enrichments over background detrital contributions for these metals have seldom been estimated. The data show, however, that anoxic near-shore sediments represent a significant sink for Cu, Ni, Pb and Zn when compared with the accumulation rates for pelagic sediments.

33.8.4. SILICON

The importance of near-shore basins as sinks for silicon has been discussed by Calvert (1966). Estimated rates of accumulation are given in Table 33.30. At the present time, most of the accumulation of silica takes place around Antarctica (Calvert, 1968) at rates which are much less than those found in

TABLE 33.30

Accumulation rates of biogenous silica in near-shore sediments

Area	Accumulation rate ($mg\ cm^{-2}\ yr^{-1}$)
Gulf of California	50[a]
Sea of Okhotsk	1[b]
Saanich Inlet, British Columbia	50[c]
Bering Sea	1[d]
South West African shelf	20[e]
Total input rate per unit area of ocean	0·1[f]

[a] From Calvert (1966).
[b] From data given by Bezrukov (1955).
[c] From data given by Gucluer and Gross (1964).
[d] From data given by Lisitsyn (1966).
[e] S. E. Calvert and N. B. Price (unpublished).
[f] From stream supply given by Livingstone (1963).

near-shore areas. However, poorly-siliceous near-shore sediments which accumulate very rapidly may also be a sink for silicon even though the presence of biogenous silica in them may be masked by detrital materials (Turekian, 1971; Heath, 1973). Supporting data for this process are required in order to resolve some of the conflicts in the oceanic silicon balance (Edmond, 1973; Burton and Liss, 1973).

33.8.5. SULPHUR

Berner (1972) has listed accumulation rates of total sulphur in several near-shore areas (Table 33.31). These rates should be compared with a total supply rate to the World Ocean of 0·34 $mg\ cm^{-2}\ yr^{-1}$. He has emphasized that in the present day absence of evaporite formation the removal of sulphur takes place in anoxic sediments by pyrite formation. However, in order to bring the sulphur budget into balance it is necessary to invoke higher rates of

TABLE 33.31

Accumulation rates of sulphur in near-shore sediments[a]

Area	Accumulation rate (mg cm^{-2} yr^{-1})
Black Sea	0·05–0·85
Santa Barbara Basin	0·08–1·03
Gulf of California	0·05–0·08
Long Island Sound	0·4

[a] From Berner (1972). The sulphur is present entirely as sulphides.

accumulation of sulphur in normal types of near-shore sediments (which actually accumulate under oxygenated conditions, but in which sulphate reduction takes place below the oxic surface layer).

ACKNOWLEDGEMENTS

I wish to thank D. H. Doff and P. L. Wright for permission to use their data on the geochemistry of the sediments of Oslo Fiord and the Barents Sea respectively, and A. H. Stride for his critical comments on portions of the manuscript.

REFERENCES

Allen, E. T., Crenshaw, J. L., Johnston, J. and Larsen, E. S. (1912). *Amer. J. Sci.* **33**, 169.
Allen, J. A. (1960). *Nature, Lond.* **185**, 336.
Altschuler, Z. S., Cisney, E. A. and Barlow, I. H. (1952). *Amer. Min.* **63**, 1230.
Altschuler, Z. S., Clarke, R. S. and Young, E. J. (1958). *U.S. Geol. Surv., Prof. Paper*, No. 314-D.
Altschuler, Z. S., Berman, S. and Cuttitta, F. (1967). *U.S. Geol. Surv., Prof. Paper*, No. 575-B.
Ames, L. L. (1959). *Econ. Geol.* **54**, 829.
Ames, L. L. (1960). *Econ. Geol.* **55**, 354.
d'Anglejan, B. F. (1967). *Mar. Geol.* **5**, 15.
d'Anglejan, B. F. (1968). *Can. J. Earth Sci.* **5**, 81.
Arrhenius, G. (1963). *In* "The Sea" (M. N. Hill, ed.), Vol. 3, p. 655. Interscience, New York.
Arrhenius, G. and Bonatti, E. (1965). *In* "Progress in Oceanography" (M. Sears, ed.), Vol. 3, p. 7. Pergamon Press, London.

Balashov, Y. A. and Kazakov, G. A. (1968). *Dokl. Akad. Nauk SSSR*, **179**, 440 (Eng. Trans. 181).
Barnes, S. S. (1967). PhD. Thesis, Univ. California, San Diego.
Barnes, S. S. and Dymond, J. R. (1967). *Nature, Lond.* **213**, 1218.
Baturin, G. N. (1969a). *Dokl. Akad Nauk SSSR*, **189**, 1359. (Eng. Trans. 227).
Baturin, G. N. (1969b). *Okeanològiya*, **9**, 1031. (Eng. Trans. 828).
Baturin, G. N. (1971a). *In* "Geology of the East Atlantic Continental Margin" (F. M. Delany, ed.), Inst. Geol. Sci. Report 70/13, 87.
Baturin, G. N. (1971b). *Nature, Lond.* **232**, 61.
Baturin, G. N. (1971c). *Okeanologiya*, **11**, 444. (Eng. Trans. 372).
Baturin, G. N. (1971d). *Dokl. Akad Nauk SSSR*, **198**, 1186. (Eng. Trans. 224).
Baturin, G. N., Kochenov, A. V. and Shimkus, K. M. (1967). *Geokhimiya*, **1**, 41.
Baturin, G. N., Kochenov, A. V. and Petelin, V. P. (1970). *Litol. Polez. Iskop.* No. 3, 15. (Eng. Trans. 266).
Baturin, G. N., Kochenov, A. V. and Senin, Y. M. (1971). *Geokhimiya*, 456. (Eng. Trans. *Geochem. Int.* **8**, 281).
Belderson, R. N., Kenyon, N. H. and Stride, A. H. (1971). *In* "Geology of the East Atlantic Continental Margin", (F. M. Delany, ed.), Inst. Geol. Sci. Rep. 70/14, 159.
Baturin, G. N., Merkulova, K. I. and Chalov, P. I. (1972). *Mar. Geol.* **13**, 37.
Bell, D. L. and Goodell, H. G. (1967). *Sedimentology*, **9**, 169.
Belova, I. V. (1970). *Dokl. Akad. Nauk SSSR*, **193**, 433 (Eng. Trans. 210).
Bender, M. L., Ku, T. L. and Broecker, W. S. (1966). *Science, N.Y.* **151**, 2302.
Bennett, J. H. and Manuel, O. K. (1968). *J. Geophys. Res.* **73**, 2302.
Bentor, Y. K. and Kastner, M. (1965). *J. Sediment. Petrol.* **35**, 155.
Berner, R. A. (1962). *Science, N.Y.* **137**, 669.
Berner, R. A. (1964a). *Mar. Geol.* **1**, 117.
Berner, R. A. (1964b). *J. Geol.* **72**, 293.
Berner, R. A. (1964c). *Geochim. Cosmochim. Acta*, **28**, 1497.
Berner, R. A. (1967). *Amer. J. Sci.* **265**, 773.
Berner, R. A. (1970a). *Amer. J. Sci.* **268**, 1.
Berner, R. A. (1970b). *Nature, Lond.* **227**, 700.
Berner, R. A. (1972). *In* "Changing Chemistry of the Oceans" (D. Dyrssen and D. Jagner, eds.), Nobel Symposium 20. John Wiley, New York.
Berrang, P. G. and Grill, E. V. (1974). *Mar. Chem.* **2**, 125.
Bertine, K. K. (1970). PhD. Thesis, Yale University.
Bertine, K. K. (1972). *Mar. Chem.* **1**, 43.
Bertine, K. K. and Turekian, K. K. (1973). *Geochim. Cosmochim. Acta*, **37**, 1415.
Bezrukov, P. L. (1955). *Dokl. Akad. Nauk SSSR*, **103**, 473.
Bezrukov, P. L. (1960). 21st Int. Geol. Congr. Copenhagen. Rep. Part X, 39.
Birch, G. F. (1971). *SANCOR Mar. Geol. Prog. Bull.* **4**, 134.
Black, W. A. P. and Mitchell, R. L. (1952). *J. Mar. Biol. Ass. U.K.* **30**, 575.
Bojanowski, R. and Paslawska, S. (1970). *Acta Geophys. Pol.* **18**, 277.
Bonatti, E., Fisher, D. E., Joensuu, O. and Rydell, H. S. (1971). *Geochim. Cosmochim. Acta*, **35**, 189.
Borchert, H. (1965). *In* "Chemical Oceanography" (J. P. Riley and G. Skirrow, eds.), 1st Ed., Vol. 2, 159. Academic Press, London and New York.
Bowen, H. J. M. (1966). "Trace Elements in Biochemistry", Academic Press, London and New York.
Bramlette, M. N. (1961). *In* "Oceanography" (M. Sears, ed.). AAAS Publ. 67, p. 345.
Breger, I. A. and Deul, M. (1956). *U.S. Geol. Surv., Prof. Paper*, **300**.
Brindley, G. W. (1951). *Min. Mag.* **29**, 502.

Brindley, G. W. and Youell, R. F. (1953). *Min. Mag.* **30**, 57.
Brongersma-Sanders, M. (1951). *Proc. 3rd World Petrol. Cong. Sec. 1*, 401.
Brooks, R. R., Presley, B. J. and Kaplan, I. R. (1968). *Geochim. Cosmochim. Acta*, **32**, 392.
Buchanan, J. Y. (1891). *Trans. Roy. Soc. Edinburgh*, **36**, 459.
Burst, J. F. (1958). *Amer. Min.* **43**, 481.
Burton, J. D. and Liss, P. S. (1973). *Geochim. Cosmochim. Acta*, **37**, 1761.
Buser, W. and Grütter, A. (1956). *Schwiez. Mineral. Petrog. Mitt.* **36**, 49.
Byrne, J. V. and Emery, K. O. (1960). *Bull. Geol. Soc. Amer.* **71**, 983.
Caillière, S. and Martin, L. (1972). *C.R. Acad. Sci., Paris*, **274**, 2273.
Calvert, S. E. (1966). *Bull. Geol. Soc. Amer.* **77**, 569.
Calvert, S. E. (1968). *Nature, Lond.* **219**, 919.
Calvert, S. E. and Price, N. B. (1970a). *Nature, Lond.* **227**, 593.
Calvert, S. E. and Price, N. B. (1970b). *Contrib. Mineral. Petrol.* **29**, 215.
Calvert, S. E. and Price, N. B. (1971a). *In* "Geology of the East Atlantic Continental Margin" (F. M. Delany, ed.), Inst. Geol. Sci. Rep. 70/16, 171.
Calvert, S. E. and Price, N. B. (1971b). *Deep-Sea Res.* **18**, 505.
Calvert, S. E. and Price, N. B. (1972). *Earth Planet. Sci. Lett.* **16**, 245.
Carrigy, M. A. (1956). *J. Sediment. Petrol.* **26**, 228.
Carroll, D. (1958). *Geochim. Cosmochim. Acta*, **14**, 1.
Cayeux, L. (1897). *Ann. Soc. Geol. Nord, Mem.* **4**, 589.
Cayeux, L. (1935). "Les Roches Sedimentaires de France-Roches Carbonatees". Masson et Cie, Paris.
Chave, K. E. (1967). *A.G.I. Council Educ. Geol. Sci. Short Rev.* **7**, 200.
Cheney, E. S. and Vredenburg, L. D. (1968). *J. Sediment. Petrol.* **38**, 1363.
Chester, R. and Elderfield, H. (1968). *Geochim. Cosmoschim. Acta*, **32**, 1128.
Chester, R. and Hughes, M. J. (1967). *Chem. Geol.* **2**, 249.
Ciresse, P. (1965). *C. R. Acad. Sci., Paris*, **260**, 5597.
Clark, A. H. (1966). *Neues Jahrb. Mineral., Monatsh.* **10**, 300.
Cloud, P. E. (1955). *Bull. Amer. Ass. Petrol. Geol.* **39**, 484.
Copenhagen, W. J. (1934). *Invest. Rep. No. 3 Fish. Mar. Biol. Surv., Union S. Africa*, 1.
Copenhagen, W. J. (1953). *Invest. Rep. No. 14, Div. Fish. Union S. Africa*, 1.
Coryell, C. D., Chase, J. W. and Winchester, J. W. (1963). *J. Geophys. Res.* **68**, 559.
Crecelius, E. A. (1969). *Trans. Amer. Geophys. Union*, **50**, 208.
Cronan, D. S. (1967). PhD. Thesis, Univ. of London.
Curray, J. R. (1960). *In* "Recent Sediments North-west Gulf of Mexico" (F. P. Shepard, F. B. Phleger and T. H. Van Andel, eds.), p. 221. Amer. Ass. Petrol. Geol., Tulsa.
Curray, J. R. (1961). *Bull. Geol. Soc. Amer.* **72**, 1707.
Curray, J. R. (1965). *In* "The Quaternary of the United States" (H. E. Wright and D. G. Frey, eds.), p. 922. Princeton Univ. Press.
Curtis, C. D. (1966). *In* "Advances in Organic Geochemistry" (G. D. Hobson and M. C. Louis, eds.), p. 330. Pergamon Press, Oxford.
Curtis, C. D. and Spears, D. A. (1968). *Econ. Geol.* **63**, 257.
Debyser, J. (1957). *Rev. Inst. Fr. Petrole Ann. Combust. Liquides*, **12**, 3.
Degens, E. T. (1967). *In* "Diageneses of Sediments" (G. Larsen and G. V. Chilingan, ed.), p. 343, Elsevier, Amsterdam.
Degens, E. T., Prashnowsky, A., Emery, K. O. and Pimenta, J. (1961). *Neues Jahrb. Geol. Palaeontol. Monatsh.* 413.
Degens, E. T., Emery, K. O. and Reuter, J. H. (1963). *Neues Jahrb. Geol. Palaeontol. Monatsh.* 231.
Deuser, W. G. (1970). *Science, N.Y.* **168**, 1575.

Deuser, W. G. (1971). *Deep-Sea Res.* **18**, 995.
Dietz, R. S., Emery, K. O. and Shepard, F. P. (1942). *Bull. Geol. Soc. Amer.* **53**, 815.
Doetler, C. (1926). "*Handbuch der Mineral-Chemie*", p. 526.
Doff, D. H. (1969). PhD Thesis, Univ. of Edinburgh, p. 245.
Dorta, C. C. and Rona, E. (1971). *Bull. Mar. Sci.* **21**, 745.
Doyle, R. W. (1968). *Amer. J. Sci.* **266**, 980.
Drever, J. L. (1971). *Science, N.Y.* **172**, 1334.
Duchart, P., Calvert, S. E. and Price, N. B. (1973). *Limnol. Oceanogr.* **18**, 605.
Edelman, C. H. (1933). "Petrologische Provincies in het Nederlandse Kwartain." Centen, Amsterdam.
Edmond, J. M. (1973). *Nature, Lond.* **241**, 391.
Ehlmann, A. J., Hullings, N. C. and Glover, E. D. (1963). *J. Sediment. Petrol.* **33**, 87.
Emery, K. O. (1960). "The Sea off Southern California", p. 366. John Wiley, New York.
Emery, K. O. (1968). *Bull. Amer. Ass. Petrol. Geol.* **52**, 445.
Emery, K. O. and Bray, E. E. (1962). *Bull. Amer. Ass. Petrol. Geol.* **46**, 1839.
Emery, K. O. and Rittenberg, S. C. (1952). *Bull. Amer. Ass. Petrol. Geol.* **36**, 735.
Erd, R. C., Evans, H. T. and Richter, D. H. (1957). *Amer. Min.* **42**, 309.
Fleischer, M. (1955). *Econ. Geol. 50th Anniv. Vol. Pt. 2*, 970.
Fleming, R. H. (1940). *Proc. 6th Pacific Sci. Cong.* **3**, 535.
Fomina, L. S. (1966). *Dokl. Akad. Nauk SSSR*, **170**, 1181 (Eng. Trans. 221).
Fomina, L. S. and Volkov, I. I. (1969). *Dokl. Akad. Nauk SSSR*, **185**, 188. (Eng. Trans. 158).
Frondel, C., Marvin, U. and Ito, J. (1960). *Amer. Min.* **45**, 1167.
Galliher, E. W. (1933). *J. Sediment. Petrol.* **3**, 51.
Galliher, E. W. (1935). *Bull. Geol. Soc. Amer.* **46**, 1351.
Galliher, E. W. (1939). *In* "Recent Marine Sediments" (P. D. Trask, ed.), p. 513. Amer. Ass. Petrol. Geol. p. 513.
Giresse, P. (1965). *C.R. Acad. Sci. Paris*, **260**, 2550.
Glagoleva, M. A. (1961). *Dokl. Akad. Nauk SSSR*, **136**, 195 (Eng. Trans. 1).
Glagoleva, M. A. (1970). *Dokl. Akad. Nauk SSSR*, **193**, 184. (Eng. Trans. 203).
Glasby, G. P. (1972). *Marine Geol.* **13**, 57.
Goldberg, E. D. (1954). *J. Geol.* **62**, 249.
Goldberg, E. D. and Arrhenius, G. O. S. (1958). *Geochim. Cosmochim. Acta*, **13**, 153.
Goldberg, E. D., Koide, M., Schmitt, R. A. and Smith, R. H. (1963). *J. Geophys. Res.* **68**, 4209.
Goldschmidt, V. M. (1937). *J. Chem. Soc.*, 655.
Goldschmidt, V. D. (1954). "Geochemistry", p. 730. Clarendon Press, Oxford.
Goldsmith, J. R. and Graf, D. L. (1960). *J. Geol.* **68**, 324.
Gorshkova, T. I. (1957). *Tr. Vses. Gidrobiol. Obshchest.* **8**, 68.
Gorsline, D. S. (1963). *J. Geol.* **71**, 422.
Gorsline, D. S. and Milligan, D. B. (1963). *Deep-Sea Res.* **10**, 259.
Grill, E. V., Murray, J. W. and Macdonald, R. D. (1968). *Nature, Lond.* **219**, 358.
Gripenberg, S. (1934). *Havforskningsinst. Skr.* **96**, 1.
Gross, M. G. (1967). *In* "Estuaries" (G. H. Lauff, ed.), p. 273. Amer. Ass. Adv. Sci.
Grout, F. F. (1925). *Bull. Geol. Soc. Amer.* **36**, 393.
Groves, A. W. (1951). "Silicate Analysis". Allen and Unwin, London.
Grütter, A. and Buser, W. (1957). *Chimia*, **11**, 132.
Gucluer, S. M. and Gross, M. G. (1964). *Limnol. Oceanogr.* **9**, 359.
Gulbrandsen, R. A. (1966). *Geochim. Cosmochim. Acta*, **30**, 769.

Gulbrandsen, R. A. (1969). *Econ. Geol.* **64**, 365.
Gulbrandsen, R. A., Kramer, J. R., Beatty, L. B. and Maya, R. E. (1966). *Amer. Min.* **51**, 819.
Hartmannn, M. (1964). *Meyniana*, **14**, 3.
Heath, G. R. (1973). *In* "Geologic History of the Oceans" (W. W. Hay, ed.). Soc. Econ. Palaeontol. Mineral. Spec. Publ.
Hegemann, F. (1949). *Heidelberg. Beitr. Mineral. Petrogr.* **1**, 690.
Hinze, C. and Meischner, D. (1968). *Mar. Geol.* **6**, 53.
Hirst, D. M. (1962a). *Geochim. Cosmochim. Acta*, **26**, 309.
Hirst, D. M. (1962b). *Geochim. Cosmochim. Acta*, **26**, 1147.
Hirst, D. M. (1973). *In* "The Black Sea: Its Geology, Chemistry and Biology". (E. T. Degens and D. A. Ross, eds.). Amer. Ass. Petrol. Geol. Mem. (in press).
Hirst, D. M. and Nichols, G. D. (1958). *J. Sediment. Petrol.* **28**, 468.
Høgdahl, O. T., Melsom, S. and Bowen, V. T. (1968). *Adv. Chem. Ser.* **73**, p. 308.
Holland, H. D. and Kulp, J. L. (1954). *Geochim. Cosmochim. Acta*, **5**, 197.
Hower, J. (1961). *Amer. Min.* **46**, 313.
Hülsemann, J. (1966). *J. Sediment. Petrol.* **36**, 622.
Hutchinson, G. E. (1957). "A Treatise on Limnology". John Wiley, New York.
Ignatius, H., Niemistö, L. and Voipio, A. (1971). *Eripainos Geologilehdestä*, *No3*, 43.
Irving, H. and Williams, R. J. P. (1953). *J. Chem. Soc.* 3192.
James, H. L. (1966). *U.S. Geol. Surv., Prof. Paper*, **60**.
Jelgersma, S. (1961). *Maastright, van Aelst, Geol. Stichting Ser.* C-IV.
Jones, G. E. and Starkey, R. L. (1957). *J. Appl. Microbiol.* **5**, 111.
Jones, L. H. P. and Milne, A. A. (1956). *Min. Mag.* **31**, 283.
Kaplan, I. R., Rafter, T. A. and Hulston, J. R. (1960). *N.Z. J. Sci.* **3**, 338.
Kaplan, I. R., Emery, K. O. and Rittenberg, S. C. (1963). *Geochim. Cosmochim. Acta*, **27**, 297.
Kaplan, I. R. and Rittenberg, S. C. (1964). *J. Gen. Microbiol.* **34**, 195.
Kazakov, A. V. (1937). *USSR Sci. Inst. Fertilisers and Fungicides Trans.* **142**, 95.
Kee, N. S. and Bloomfield, C. (1961). *Geochim. Cosmochim. Acta*, **24**, 206.
Kemp, A. L. W. and Thode, H. G. (1968). *Geochim. Cosmochim. Acta*, **32**, 71.
Kenyon, N. H. and Stride, A. H. (1970). *Sedimentology*, **14**, 159.
Kinsman, D. J. J. (1969). *J. Sediment. Petrol.* **39**, 486.
Klenova, M. V. (1938). *Dokl. Akad. Nauk SSSR*, **19**, 8.
Klenova, M. V. (1960). "Geology of the Barents Sea". Izv. Akad. Nauk SSSR, Moscow.
Kochenov, A. V., Baturin, G. N., Kovaleva, S. A., Yemelyanov, Y. M. and Shimkus, K. M. (1965). *Geokhimiya*, **3**, 302.
Koczy, F. F., Tomic, E. and Hecht, F. (1957). *Geochim. Cosmochim. Acta*, **11**, 86.
Koide, M., Soutar, A. and Goldberg, E. D. (1972). *Earth Planet Sci. Lett.* **14**, 442.
Kolodny, Y. (1969). *Nature, Lond.* **224**, 1017.
Kolodny, Y. and Kaplan, I. R. (1970a). *J. Sediment. Petrol.* **40**, 954.
Kolodny, Y. and Kaplan, I. R. (1970b). *Geochim. Cosmochim. Acta*, **34**, 3.
Kolodny, Y. and Kaplan, I. R. (1970c). *In* "International Symposium Hydrogeochemistry and Biogeochemistry." Tokyo.
Kolodny, Y. and Kaplan, I. R. (1973). *In*"Proc. Symposium on Hydrochemistry and Biogeochemistry" (E. Ingerson, ed.), Vol. I, p. 418. The Clarke Co., Washington, D.C.
Kolyadin, L. B., Mikolayev, D. S., Grashchenko, S. M., Kuznestov, Y. V. and Lazarev, K. F. (1960). *Dokl. Akad. Nauk. SSSR*, **132**, 915 (Eng. Trans. 456).
Korolev, D. F. (1958). *Geokhimiya*, **4**, 452.

Krauskopf, K. B. (1955). *Econ. Geol. 50th Aniv. Vol. Part 1*, 411.
Krauskopf, K. B. (1956). *Geochim. Cosmochim. Acta*, **9**, 1.
Krauskopf, K. B. (1967). "Introduction to Geochemistry". McGraw-Hill, New York.
Krinsley, D. (1960). *Micropaleontology*, **6**, 297.
Kriss, A. E. (1963). "Marine Microbiology". Oliver and Boyd, Edinburgh.
Krynine, P. D. (1948). *J. Geol.* **56**, 130.
Ku, T. L. and Broecker, W. S. (1969). *Deep-Sea Res.* **16**, 625.
Ku, T. L. and Glasby, G. P. (1972). *Geochim. Cosmochim. Acta*, **36**, 699.
Landergren, S. (1964). *Rep. Swed. Deep-Sea Exped. 1947–1948*, **10**, IV, 57.
Lees, A. and Buller, A. T. (1972). *Mar. Geol.* **13**, 67.
Lees, G. H. (1937). *Bull. Amer. Ass. Petrol. Geol.* **21**, 1579.
Lehr, J. R., McClellan, C. H., Smith, J. P. and Frazier, A. W. (1967). *Colloq. Int. sur Phosphates Min. Solides*, 29.
Li, Y. H., Bischoff, J. and Mathieu, G. (1969). *Earth Planet. Sci. Lett.* **7**, 265.
Lisitsyn, A. P. (1966). "Recent Sedimentation in the Bering Sea". Izdatel 'stvo Nauka, Moscow.
Livingstone, D. A. (1963). *U.S. Geol. Surv., Prof. Paper* **440-G**.
Loring, D. H. and Nota, D. J. G. (1968). *J. Fish. Res. Bd. Can.* **25**, 2327.
Lowenstam, H. A. (1972). *Chem. Geol.* **9**, 153.
Lubchenko, I. Y. (1970). *Dokl. Akad. Nauk SSSR*, **193**, 445 (Eng. Trans. 221).
Lutenauer, J. L. and Pilkey, O. H. (1967). *Mar. Geol.* **5**, 315.
Lynn, D. C. and Bonatti, E. (1965). *Mar. Geol.* **3**, 457.
Mackereth, F. J. H. (1966). *Phil. Trans. Roy. Soc., Lond.* **250B**, 165.
McClellan, G. H. and Lehr, J. A. (1969). *Amer. Min.* **54**, 1374.
McConnell, D. (1952). *Bull. Soc. Fr. Min. Cristallogr.* **75**, 428.
McConnell, D. (1970). *Amer. Min.* **55**, 1659.
MacGregor, M., Lee, G. W. and Wilson, G. V. (1920). *Geol. Surv. Gt. Brit. Spec. Rep. Min. Reserves*, **11**, 240.
McKelvey, V. E. (1967). *U.S. Geol. Surv. Bull.* **1252-D**.
McKelvey, V. E., Swanson, R. W. and Sheldon, R. P. (1953). *Int. Geol. Cong. 19th Algiers Comp. Rend. Sec. 11 pt. 11*, 45.
Manghnani, M. H. and Hower, J. (1964). *Amer. Min.* **49**, 586.
Manheim, F. T. (1961a). *Geochim. Cosmochim. Acta*, **25**, 52.
Manheim, F. T. (1961b). *Stockholm Contrib. Geol.* **8**, 27.
Manheim, F. T. (1965). *Narragansett Mar. Lab., Occas. Publ.* **3**, 217.
Manskaya, S. M. and Drozdova, T. V. (1968). "Geochemistry of Organic Substances". Pergamon Press, London.
Marchant, J. M. (1928). *Spec. Rep. No. 5 Fish Mar. Biol. Surv., Union South Africa.*
Marchig, V. (1972). "*Meteor*" *Forschungsergeb.* Ser. C. No. 11.
Mero, J. L. (1965). "The Mineral Resources of the Sea". Elsevier, Amsterdam.
Meyer, F. H., Riggs, O. L., McGlasson, R. L. and Sudbury, J. D. (1958). *Corrosion*, **14**, 69.
Miller, L. P. (1950). *Contrib. Boyce Thomson Inst.* **16**, 85.
Milliman, J. D. and Emery, K. O. (1968). *Science, N.Y.* **162**, 1121.
Minguzzi, C. and Talluri, A. (1951). *Atti Soc. Toscana Sci. Nat. Pisa Proc. Verb. Mem. Ser.* **A.58**, 89.
Mo., T., Suttle, A. D. and Sackett, W. H. (1973). *Geochim. Cosmochim. Acta*, **37**, 35.
Mohr, P. A. (1959). *Contrib. Geophys. Observ. Univ. Coll., Addis Ababa*, 1.
Moore, J. R. (1963). *J. Sediment. Petrol.* **33**, 511.
Moore, J. R. (1968). *Bull. Brit. Mus. (Natur. Hist.)* **2**, 1.

Mortimer, C. H. (1941). *J. Ecol.* **29**, 280; *J. Ecol.* **30**, 147.
Müller, G. (1967). *J. Sediment. Petrol.* **37**, 957.
Murray, J. (1900). *Scott. Geog. Mag.* **16**, 673.
Murray, J. and Irvine, R. (1893). *Trans. Roy. Soc., Edinburgh*, **37**, 481.
Murray, J. and Renard, A. F. (1891). "Scientific Report *Challenger* Expedition", Vol. 3. HMSO, London.
Murty, P. S. N., Reddy, C. V. G. and Varedachari, V. V. R. (1968). *Proc. Nat. Inst. of Sci., India*, **34**B, 134.
Nelson, B. W. and Roy, R. (1954). *In* "Clays and Clay Minerals". (A. Swineford and N. V. Plummer, eds.), Proc. 2nd Conf., p. 334. Pergamon Press, Oxford.
Nicholls, G. D., Curl, H. and Bowen, V. T. (1959). *Limnol. Oceanogr.* **4**, 472.
Nissenbaum, A., Baedecker, M. J. and Kaplan, I. R. (1972). *In* "Advances in Organic Geochemistry". (H. R. von Gaertner and H. Wehner, eds.) Pergamon Press, Oxford.
Noddack, I. and Noddack, W. (1931). *Z. Phys. Chem.* **154A**, 207.
Norris, R. M. (1964). *N.Z. Dep. Sci. Ind. Res., Bull.* **159**, 39.
Nota, D. J. G. and Loring, D. H. (1964). *Mar. Geol.* **2**, 198.
Oborn, E. T. and Hem, J. D. (1961). *U.S. Geol. Surv. Water Supply Paper*, **1459H**.
Odum, H. T. (1957). *Publ. Inst. Mar. Sci.* **4**, 38.
Ostroumov, E. A. (1953). *Trudy Inst. Okeanol. Akad. Nauk SSSR*, **7**, 70.
Ostroumov, E. A., Volkov, I. I. and Fomina, L. S. (1961). *Tr. Inst. Okeanol. Akad. Nauk SSSR*, **50**, 93.
Parker, R. J. (1971). *SANCOR Mar. Geol. Prog. Bull.* **2**, 94.
Peck, L. C. and Smith, V. C. (1964). *Talanta*, **11**, 1343.
Pettijohn, F. J. (1957). "Sedimentary Rocks". Harper Bros., New York.
Pevear, D. R. (1966). *Econ. Geol.* **61**, 251.
Pevear, D. R. (1967). *Econ. Geol.* **62**, 562.
Pevear, D. R. and Pilkey, O. H. (1966). *Bull. Geol. Soc. Amer.* **77**, 849.
Pilipchuck, M. F. and Volkov, I. I. (1966). *Dokl. Akad. Nauk SSSR*, **167**, 1 (Eng. Trans. 152).
Pilipchuck, M. F. and Sevast'yanov, V. F. (1968). *Dokl. Akad. Nauk SSSR*, **179**, 1. (Eng. Trans. 194).
Pilipchuck, M. F. and Volkov, I. I. (1968). *Litol. Polez. Iskop.* **4**, 5. (Eng. Trans. 389).
Piper, D. Z. (1971). *Geochim. Cosmochim. Acta*, **35**, 531.
Piper, D. Z. (1974). *Geochim. Cosmochim. Acta*, **38**, 1007.
Porrenga, D. H. (1966). *In* "Clays and Clay Minerals", Proc. 145th Nat. Con.. (S. W. Bailey, ed.). Pergamon Press, Oxford.
Porrenga, D. H. (1967). *Mar. Geol.* **5**, 495.
Prashnowsky, A., Degens, E. T., Emery, K. O. and Pimenta, J. (1961). *Neues. Jahrb. Geol. Palaeontol. Monatsh.* 400.
Pratt, R. M. (1963). *Deep-Sea Res.* **10**, 245
Pratt, R. M. and McFarlin, P. F. (1966). *Science, N.Y.* **151**, 1080.
Pratt, R. M. and Thompson, S. L. (1962). *Woods Hole Oceanog. Inst. Rep.* **62**.
Pratt, W. L. (1963). *In* "Essays in Marine Geology in Honor of K. O. Emery". (T. Clements, ed.), p. 201. Univ. of Southern Calif. Press.
Price, N. B. (1967). *Mar. Geol.* **5**, 511.
Price, N. B. and Calvert, S. E. (1970). *Mar. Geol.* **9**, 145.
Price, N. B. and Calvert, S. E. (1973). *Geochim. Cosmochim. Acta*, **37**, 2149.
Price, N. B. and Wright, P. L. (1971). *In* "The Geology of the East Atlantic Continental Margin (F. M. Delany, ed.). I.G.S. Rep. 70/14, p. 17.
Price, N. B., Calvert, S. E. and Jones, P. G. W. (1970). *J. Mar. Res.* **28**, 22.

Richards, F. A. (1960). *Deep-Sea Res.* **7**, 163.
Richards, F. A. (1965). *In* "Chemical Oceanography". (J. P. Riley and G. Skirrow, eds.), 1st Ed., Vol. I, p. 611. Academic Press, London and New York.
Richards, F. A. (1970). *In* "Organic Matter in Natural Waters". (D. W. Hood, ed.), p. 399. Inst. Mar. Sci. Univ. Alaska, Occas. Publ. 1.
Richards, F. A. (1971). *In* "Impingement of Man on the Oceans". (D. W. Hood, ed.), p. 201. John Wiley, New York.
Richards, F. A. and Vaccaro, R. F. (1956). *Deep-Sea Res.* **3**, 214.
Richards, F. A. and Benson, B. B. (1961). *Deep-Sea Res.* **7**, 254.
Richards, F. A., Cline, J. D., Broenkow, W. W. and Atkinson, L. P. (1965). *Limnol. Oceanogr. (Suppl.)*, **10**, 185.
Rickard, D. T. (1969a). *Stockholm Contrib. Geol.* **20**, 49.
Rickard, D. T. (1969b). *Stockholm Contrib. Geol.* **20**, 67.
Rittenberg, S. C., Emery, K. O. and Orr, W. L. (1955). *Deep-Sea Res.* **3**, 23.
Roberts, W. M. B., Walker, A. L. and Buchanan, A. S. (1969). *Miner. Deposita*, **4**, 18.
Rohrlich, V., Price, N. B. and Calvert, S. E. (1969). *J. Sediment. Petrol.* **39**, 624.
Rooney, T. P. and Kerr, P. F. (1967). *Bull. Geol. Soc. Amer.* **78**, 731.
Ross, D. A., Degens, E. T. and MacIlvaine, J. (1970). *Science, N.Y.* **170**, 163.
Sackett, W. M. and Cook, G. (1969). *Trans. Gulf Coast Ass. Geol. Soc.* **19**, 233.
Schoen, R. (1964). *J. Sediment. Petrol.* **34**, 855.
Scholl, D. W. (1963). *Bull. Amer. Ass. Petrol. Geol.* **47**, 1581.
Seibold, E. (1958). *Geol. Rundsch.* **47**, 100.
Seibold, E., Müller, G. and Fesser, H. (1958). *Erdöl Kohle*, **11**, 296.
Sevast'yanov, V. F. (1967). *Dokl. Akad. Nauk SSSR*, **176**, 191 (Eng. Trans. 180).
Sevast'yanov, V. F. and Volkov, I. I. (1966). *Dokl. Akad. Nauk SSSR*, **166**, 701.
Sevast'yanov, V. F. and Volkov, I. I. (1967). *Tr. Inst. Okeanol.* **83**, 137.
Shaw, T. I. (1962). *In* "Physiology and Biochemistry of Algae". (R. A. Lewin, ed.), p. 247. Academic Press, New York and London.
Sheldon, R. P. (1964). *U.S. Geol. Surv., Prof. Paper* **501C**
Shepard, F. P. (1932). *Bull. Geol. Soc. Amer.* **43**, 1017.
Shepard, F. P. (1963). *In* "Essays in Marine Geology in Honor of K. O. Emery", (T. Clements, ed.), p. 1. Univ. Southern Calif. Press.
Shepard, F. P., Phleger, F. B. and Van Andel, Tj. H. (1960). "Recent Sediments, Northwest Gulf of Mexico". Amer. Ass. Petrol. Geol., Tulsa.
Shishkina, O. V. and Pavlova, G. A. (1965). *Geokhimiya*, **6**, 739 (Eng. Trans. 559.).
Shterenberg, L. E., Gorshkova, T. I. and Naktinas, E. M. (1968). *Litol. Polez. Iskop.* **4**, 63. (Eng. Trans. 438).
Silverman, S. R., Fuyat, R. K. and Weiser, J. D. (1952). *Amer. Min.* **37**, 211.
Simpson, D. R. (1964). *Amer. Min.* **49**, 363.
Simpson, D. R. (1967). *Amer. Min.* **52**, 896.
Skopintsev, B. A., Romenskaya, N. N. and Smirnov, E. V. (1966). *Okeanologiya*, **6**, 799. (Eng. Trans. 653).
Skornyakova, N. S. (1965). *Int. Geol. Rev.* **7**, 2161.
Smulikowski, K. (1954). *Polsk. Akad. Nauk Kom. Geol. Arch-Mineral*, **18**, 21.
Solov'yev, A. V. (1960). *Int. Geol. Rev.* **4**, 17.
Sorokin, Y. I. (1964). *J. Cons. Perm. Int. Explor. Mer.* **29**, 41.
Spencer, D. W. and Brewer, P. G. (1971). *J. Geophys. Res.* **76**, 5877.
Spencer, D. W., Degens, E. T. and Kulbicki, G. (1968). *In* "Origin and Distribution of the Elements" (L. H. Ahrens, ed.), p. 981. Pergamon Press, Oxford.
Spencer, D. W., Brewer, P. G. and Sachs, P. L. (1972). *Geochim. Cosmochim. Acta*, **36**, 71.

Starik, I. E. and Kolyadin, L. B. (1957). *Geokhimiya*, 3, 204.
Stetson, T. R. (1961). *Woods Hole Oceanog. Inst. Rep.* **61–35**, 34 pp.
Straczec, J. A., Horen, A., Ross, M. and Warshaw, C. W. (1960). *Amer. Min.* **45**, 1174.
Strakhov, N. M. (1959). *Eclogae Geol. Helv.* **51**, 753.
Strakhov, N. M. (1959). *Eclogae. Geol. Helv.* **51**, 761.
Strakhov, N. M. (1961). "Recent Sediments of Seas and Oceans", Izv. Akad. Nauk SSSR.
Strakhov, N. M. (1962). "Principles of Lithogenesis", Oliver and Boyd, Edinburgh.
Strakhov, N. M. (1966). *Int. Geol. Rev.* **8**, 1172.
Strakhov, N. M. and Nesterova, I. L. (1968). *Geochem. Int.* **5**, 644.
Stride, A. H. (1963). *Quart. J. Geol. Soc. London*, **111**, 175.
Strøm, K. M. (1936). *Skifter Norsk Viden-Akad Oslo*, *I*, **7**, 1.
Strøm, K. M. (1948). *Nature, Lond.* **162**, 922.
Summerhayes, C. P. (1967). *N.Z. J. Mar. Fresh. Res.* **1**, 627.
Summerhayes, C. P. (1971). *PhD Thesis, University of London*, 282.
Summerhayes, C. P., Birch, G. and Rogers, J. (1971). *S. African Nat. Comm. Oceanog. Res. Tech. Rep.* **3**.
Summerhayes, C. P., Nutter, A. H. and Tooms, J. S. (1972). *Sediment. Geol.* **8**, 3.
Swaine, D. J. (1962). *Commonwealth Agric. Bureau Tech. Comm.* **52**.
Swanson, V. E. (1961). *U.S. Geol. Surv., Prof. Paper*, **356-C**.
Swanson, V. E., Frost, I. C., Rader, L. F. and Huffman, C. (1966). *U.S. Geol. Surv., Prof. Paper*, **50-C**.
Swift, D. J. P., Stanley, D. J. and Curray, J. R. (1971). *J. Geol.* **79**, 322.
Szalay, A. (1958). *Int. Conf. Peaceful Uses Atom. Energy, 2nd Proc. Geneva*, **2**, 182.
Szilagyi, M. (1967). *Geokhimiya*, 1489 (Eng. Trans. 1165).
Tageyeva, N. V. and Tikhomirova, M. N. (1962). "Geochemistry of Pore Solutions in Diagenesis of Marine Sediment". Izv. Akad. Nauk SSSR, Moscow.
Takahashi, J. I. and Yagi, T. (1929). *Econ. Geol.* **24**, 838.
Taylor, S. R. (1965). *In* "Physics and Chemistry of the Earth" (L. H. Ahrens, F. Press, S. B. Runcorn and H. C. Urey, eds.), Vol. 6, p. 133. Pergamon Press, Oxford.
Thode, H. G., Kleerekoper, H. and McElcheran, D. (1951). *Research*, **4**, 581.
Thompson, G. and Livingstone, H. D. (1970). *Earth Planet. Sci. Lett.* **8**, 439.
Thompson, T. G. and Chow, T. J. (1955). *Deep-Sea Res.* **3** (*Suppl*), 20.
Thurber, D. L. (1962). *J. Geophys. Res.* **67**, 4518.
Tooms, J. S. (1967). *Hydrospace*, **1**, 40.
Tooms, J. S. and Summerhayes, C. P. (1968). *Nature, Lond.* **218**, 1241.
Tooms, J. S., Summerhayes, C. P. and Cronan, D. S. (1969). *Oceanogr. Mar. Biol. Ann. Rev.* **1**, 49.
Trask, P. D. (1953). *Paper Phys. Oceanog. Meteorol, Mass. Inst. Technol.* **12**, 51.
Triplehorn, D. M. (1966). *Sedimentology*, **6**, 247.
Tuddenham, W. M. and Lyon, R. J. P. (1960). *Anal. Chem.* **32**, 1630.
Turekian, K. K. (1964). *Trans. N.Y. Acad. Sci.* **26**, 312.
Turekian, K. K. (1965). *In* "Chemical Oceanography" (J. P. Riley and G. Skirrow, eds.), 1st Ed., Vol. 2, p. 81. Academic Press, London and New York.
Turekian, K. K. (1971). *In* "Impingement of Man on the Oceans" (D. W. Hood, ed.), p. 9. John Wiley, New York.
Turekian, K. K. and Wedepohl, K. H. (1961). *Bull. Geol. Soc. Amer.* **72**, 175.
Van Andel, Tj, H. (1959). *J. Sediment. Petrol.* **29**, 153.
Van Andel, Tj, H. (1964). *In* "Marine Geology of the Gulf of California" (Van Andel, Tj. H. and Shor, G. G., eds.), Amer. Ass. Petrol. Geol. Mem., 3, p. 216.

Van Andel, Tj. H. and Poole, D. M. (1960). *J. Sediment. Petrol.* **30**, 91.
Van Andel, Tj. H. and Postma, H. (1954). *Kon. Ned. Akad. Wetensch. Versl.* **20**, 254.
Van Andel, Tj. H. and Shor, G. G. (1964). *In* "Marine Geology of the Gulf of California", Amer. Assoc. Petrol. Geol. Memoir., **3**, 408 pp.
Van Andel, Tj. H. and Veevers, J. J. (1967). *Bur. Min. Resources, Geol. Geophys. Bull.* **83**, 173 pp.
Van Straaten, L. M. J. U. (1954). *Leidse Geol. Meded.* **19**, 1.
Veeh, H. H. (1967). *Earth Planet. Sci. Lett.* 3, 145.
Veeh, H. H., Bennett, W. C. and Soutar, A. (1973). *Science, N.Y.* **181**, 845.
Veeh, H. H., Calvert, S. E. and Price, N. B. (1974). *Mar. Chem.* **2**, 189.
Vine, J. D. (1962). *U.S. Geol. Surv., Prof. Paper,* **256-D**.
Vine, J. D. and Tourtelot, E. B. (1970). *Econ. Geol.* **65**, 253.
Vinogradov, A. P. (1939). *Tr. Biogeokhim. Lab. Akad. Nauk SSSR*, **5**, 19. (Eng. Trans. 33).
Vinogradov, A. P. (1953). "The Elementary Chemical Composition of Marine Organisms". Seas Foundation, New Haven, Conn.
Vinogradova, Z. A. and Koval'skiy, V. V. (1962). *Doll. Akad. Nauk SSSR*, **147**, 1458. (Eng. Trans. 217).
Volkov, I. I. (1961). *Tr. Inst. Okeanol. Akad. Nauk SSSR*, **50**, 68.
Von der Borch, C. C. (1970). *J. Geol. Soc. Aust.* **16**, 755.
Wasmund, E. (1930). *Geol. Foren Stockholm Foerh.* **52**, 315.
Wedepohl, K. H. (1964). *Geochim. Cosmochim. Acta*, **28**, 305.
Wedepohl, K. H. (1971). *In* "Physics and Chemistry of the Earth", Vol. 8, (L. H. Ahrens, F. Press, S. R. Runcorn and H. C. Urey, eds.), p. 307. Pergamon Press, Oxford.
White, S. M. (1970). *J. Sediment. Petrol.* **40**, 38.
Wildeman, T. R. and Haskin, L. (1965). *J. Geophys. Res.* **70**, 2905.
Winterhalter, B. (1966). *Geotek. Julk,* **69** 1.
Winterhalter, B. and Siivola, J. (1967). *Compt. Rend. Soc. Geol. Finlande*, **39**, 161.
Wright, P. L. (1972). PhD Thesis, Univ. of Edinburgh, 266 pp.
Yoder, H. S. and Eugster, H. P. (1955). *Geochim. Cosmochim. Acta*, **8**, 225.
Zen, E. (1959). *J. Sediment. Petrol.* **29**, 513.
Zobell, C. E. (1946). "Marine Microbiology". Chronica Botanica, Waltham, Mass.

Chapter 34

The Geochemistry of Deep-sea Sediments

R. CHESTER
Department of Oceanography, The University, Liverpool, England

and

S. R. ASTON
Department of Environmental Sciences, The University, Lancaster, England

34.1. INTRODUCTION

Deep-sea sediments, which cover more than fifty per cent of the surface of the earth, are very different from those deposited in the near-shore or the continental environments. The first major survey of the deep-sea sediments of the World Ocean was made by scientists on the famous *Challenger Expedition* (1873–1876) who laid down many of the foundations of marine geology (see for example, Murray and Renard, 1891; Chester, 1972a). Subsequent oceanographic expeditions involved in the collection of deep-sea sediments have been listed in the reviews compiled by Revelle (1944), Sverdrup *et al.* (1946), Kuenen (1950), El Wakeel and Riley (1961), Ericson *et al.* (1961), Wüst (1964), Riley and Chester (1971) and Deacon (1971), and have been described in many research papers.

The thickness of the sediment cover over deep-sea areas varies locally with topography, and also from one ocean to another; for example, in the Atlantic it averages >1 km and in the Pacific <1 km, with a World Ocean average of $\sim 0{\cdot}5$ km. Until recently, little of the total sediment thickness had been sampled, and although the advent of the piston corer allowed cores of up to ~ 50 m to be taken, it was not until the Deep Sea Drilling Project (DSDP) was initiated in 1968 that complete sections of the sedimentary column were obtained. None the less, prior to the DSDP there had been a fairly extensive sampling of the upper (0–~ 50 m) layer of deep-sea sediments, particularly in the Atlantic and Pacific Oceans. Although this coverage was by no means complete it did provide sufficient material to permit some of the overall geological, mineralogical and geochemical trends in deep-sea sediments to be established. The geological and mineralogical trends have been described in previous chapters, and in the present chapter the geochemistry of the sediments will be discussed and related to the general patterns of deep-sea sediment deposition.

The classification of deep-sea sediments has involved many difficulties, and has produced a plethora of often mutually inconsistent terms which can be found scattered through the literature. One of the classic examples of the confused terminology is the use of the descriptive category "red clay", since many sediments described in this manner are not red and are sometimes not even clays. Traditionally, a fundamental distinction has been made between non-biogenous deep-sea sediments, i.e. those which contain $< \sim 30\%$ of organic skeletal remains, and biogenous deep-sea deposits, i.e. those which contain $> \sim 30\%$ of organic skeletal remains and which are usually termed oozes. It has long been recognized that although the components of deep-sea sediments can occur in various proportions, and thus can produce a variety of sediment types, there are at least two genetically different kinds of inorganic deep-sea sediments. A number of terms have been used to describe these two sediment types, the most common of which are *pelagic* and *non-pelagic*. These sediment types are difficult to categorize precisely, and recent attempts to define them have involved two of the most important parameters of the land-derived material in deep-sea sediments, i.e. particle size, and rate of deposition.

The particle size spectrum of the land-derived material in deep-sea sediments is dominated by the <2 μm size class which usually constitutes 60–70% of these solids—see Table 34.1. This land-derived material has been

TABLE 34.1

*The average content (wt. %) of the <2 μm fraction in some sediments from the World Ocean**

Type	Location	No. of samples	<2 μm fraction (%(wt))
Deep-sea sediment	Atlantic	23	58
	Pacific	22	61
	Indian	52	64
Continental shelf sediment	Atlantic coast U.S.	12	2
	Gulf of Mexico	11	27
	Gulf of California	15	19
	Sahul Shelf, N.W. Australia	9	72
Suspended river sediment	33 U.S. rivers	1026	37

* Data from Griffin *et al.* (1968).

brought to the oceans by a variety of transport processes which include river run-off, wind, and ice transport. The material is redistributed within the oceans themselves by several mechanisms, and it is convenient to distinguish between that land-derived material brought to deep-sea areas in the upper water layers by oceanic currents and that carried by bottom processes such as turbidity currents. The former material has usually remained in suspension in sea water for relatively long periods, whereas the latter has been deposited initially on the continental shelf areas which have acted as intermediate sediment traps. Turekian (1967) used this fundamental distinction to make a broad definition of pelagic and non-pelagic deep-sea sediments. He proposed that, in general, the term pelagic (or eupelagic) should be applied to those deep-sea sediments, the land-derived mineral component of which, has been deposited from a dilute suspension in which the particles have had a long residence time and from which they have settled slowly from the upper layers of the overlying water. In contrast, non-pelagic (or hemi-pelagic) deep-sea sediments are considered to be those that contain land-derived components which have not been in suspension for long periods in the upper water layers, but which have usually been deposited by sea bottom processes such as turbidity currents.

This kind of approach has been used by Davies and Laughton (1972) (see also Chapter 24) who made a detailed assessment of the sedimentary processes which operate in the North Atlantic. These authors were able to distinguish between three kinds of sedimentary transport mechanisms.
(i) Turbidity currents which transport material, which has a wide variety of particle sizes, from the continental shelves to deep-sea regions. These turbidity currents are influential in the formation of abyssal plains by smoothing out irregularities on the sea floor.
(ii) Deep-ocean and bottom currents, which transport mainly fine-grained material along the deep-sea floor, and subsequently redeposit it as piles and long ridges which are elongated in the direction of the bottom current.
(iii) Surface and near-surface currents which transport the living biomass and fine-grained land-derived mineral matter; such currents also control the distribution of ice-rafted sediments. In the absence of effective bottom currents, particles from the upper water layers will settle out and blanket the bottom topography with a cover of sediment. It is sediments of this type which Davies and Laughton (1972) refer to as pelagic, and define as "those laid down in deep-water under quiet current conditions"; these sediments are thus to be contrasted with those resulting from bottom current transport. Their definition is based, therefore, on the process of sedimentation rather than on the sediment composition, and thus biogenous sediments are included in the pelagic category.

Pelagic deep-sea sediments are deposited from material which has been

TABLE 34.2
*The concentration of total suspended particulate material in some surface waters from the World Ocean**

Oceanic area	No. of samples	Concentration (μg l^{-1}) Range	Average	Time of year
Eastern margins of North Atlantic	25	121–1168	467	April, July
Open-ocean South Atlantic	23	67–448	150	April, July
Open-ocean Indian Ocean	30	43–110	66	April, June
China Sea	22	29–338	127	May, June
Open-ocean (average)	75	29–448	110	—
Near-shore localities (average)	37	84–3641	1050	—

* Data from Chester and Stoner (1972).

in suspension for relatively long periods in the upper water column. There have been various estimates of the concentration of this material (see e.g. Lisitzin, 1960; Gordeyev, 1963; Jacobs and Ewing, 1969; Ewing and Thorndike, 1966; Svirenko, 1970; Bogdanov *et al.*, 1970), which range between <100–$>1{,}000\ \mu g\ l^{-1}$. Chester and Stoner (1972) have reported the concentrations of total particulate material, i.e. inorganic + organic components, in some surface (0–$\sim$5 m) sea waters of the World Ocean, and their findings are summarized in Table 34.2. It can be seen from this table that, although the concentrations include the surface biomass, the particulate material in the upper layers of sea water is present at a very low "open-ocean" average of $\sim 100\ \mu g\ l^{-1}$. However, the particulate material is not evenly distributed throughout the water column, and may be concentrated within specific layers. For example, nepheloid layers, which contain a relatively high concentration of suspended material, can extend for several hundred metres above the sea floor; such layers are an important factor in the deep-sea bottom movement of sedimentary material (see e.g. Jacobs and Ewing, 1969).

Investigations carried out both on the particulate material in sea water and on bottom sediments suggest that in deep-sea areas land-derived material is present in the form of a dilute suspension at an average concentration of $<100\ \mu g\ l^{-1}$. The suspended matter in the suspension has an average particle size of $<2\ \mu m$ and has a characteristically slow rate of deposition which is reflected in the slow accumulation of land-derived particles in those deep-sea sediments to which it makes a significant contribution. There are various ways of dating marine sediments most of which are based on magnetic reversals, fossil assemblages or decay of natural radio-nuclides. Ku *et al.* (1968) compared sedimentation rates measured by palaeomagnetic, ionium/thorium (i.e. $^{230}Th/^{232}Th$) and ^{14}C techniques, and concluded that there is a relatively good agreement between them. However, they did point out that their results were slightly at variance with those found by Goldberg and his co-workers who used an ionium/thorium technique. According to Goldberg (1968) these discrepancies result largely from differences in the interpretation of both the effects of the presumed shortening of open-barrel sediment cores and the decay curves of the $^{230}Th/^{232}Th$ ratio with depth in the cores. Some examples of the sedimentation rates of deep-sea sediments are listed in Table 34.3, in which the values are given on a carbonate-free basis and are thus estimates of the rates of accumulation in the sediments of the land-derived material and any authigenic components. The oceanic average sedimentation rates reported by both Ku *et al.* (1968) and Goldberg and his co-workers are shown in Table 34.3. Despite the differences between the two sets of results it may be concluded that the average rate of accumulation of the land-derived fractions of deep-sea sediments in the World Ocean

TABLE 34.3. *Rates of sedimentation of non-carbonate material in deep-sea sediments from the World Ocean*

Oceanic region	Rate ($mm/10^3$ yr) (a)	(b)
South Pacific	~0·45	~1·0
North Pacific	~1·5	~5·8
South Atlantic	~1·9	~6·0
North Atlantic	~1·8	~5·7
Indian Ocean	—	~4·4

[a] Data from Goldberg and Koide (1962), Goldberg *et al.* (1964) and Goldberg and Griffin (1964).
[b] Data from Ku *et al.* (1968), see text.

is of the order of a few millimetres (usually ≲ 10 mm) per 1,000 y. However, in some regions (e.g. those adjacent to the continents, and portions of the Mid-Atlantic Ridge) which have locally controlled sedimentation regimes, the land-derived fractions of the sediments have accumulation rates ≳ 10 mm per 1,000 y.

The problems involved in establishing definitive parameters for pelagic deep-sea sediments have been discussed by Arrhenius (1963). He defended the use of the term pelagic and, like Bramlette (1961), he suggested that such sediments should be defined on the basis of a maximum value for the rate of deposition of their land-derived mineral component. He concluded that this value should be of the order of a few millimetres per 1000 y. It may be concluded, therefore, that it is useful to make a general distinction between pelagic deep-sea sediments, which have been deposited at rates of ≲ 10 mm per 1000 y from material suspended in the upper water layers, and non-pelagic varieties, which have been transported by bottom processes and have accumulated at rates of ≳ 10 mm per 100 y.* However, this somewhat over-simplified classification will be continually modified and refined as more sophisticated oceanic sedimentation models are produced (see Chapter 24).

34.2. THE DISTRIBUTION OF THE CHEMICAL ELEMENTS IN DEEP-SEA SEDIMENTS

34.2.1. INTRODUCTION

All of the elements found in the crust of the earth are probably present in deep-sea sediments, although not all of them have yet been detected. The

* Many authors have simply used the term "deep-sea clay" to describe any non-biogenous deep-sea sediment; this terminology is adhered to in the present chapter when such work is quoted.

TABLE 34.4

*The concentrations of some trace elements in deep-sea sediments (values in ppm)**

Element	Deep-sea carbonate	Deep-sea clay
Li	5	57
Be	X	2·6
B	55	230
Sc	2	19
V	20	120
Cr	11	90
Mn	1000	6700
Fe	9000	65000
Co	7	74
Ni	30	225
Cu	30	250
Zn	35	165
Ga	13	20
Ge	0·2	2
As	1	13
Se	0·17	0·17
Rb	10	110
Sr	2000	18
Y	42	9
Zr	20	150
Nb	4·6	14
Mo	3	27
Ag	X	0·11
Cd	0·23	0·21
In	0·02	0·08
Sn	X	1·5
Sb	0·15	1·0
Cs	0·4	6
Ba	190	2300
La	10	115
Ce	35	345
Pr	3·3	33
Nd	14	140
Sm	3·8	38
Eu	0·6	6
Gd	3·8	38
Tb	0·6	6
Dy	2·7	27
Ho	0·8	7·5
Er	1·5	15
Tm	0·1	1·2
Yb	1·5	15
Lu	0·5	4·5
Hf	0·41	4·1

continued overleaf

TABLE 34.4—*continued*

Element	Deep-sea carbonate	Deep-sea clay
Re	0·004	0·001
Hg	0·46	0·32
Tl	0·16	0·8
Pb	9	80
Th	—	5
U	—	1

* Data from Turekian and Wedepohl (1961); Aston *et al.* (1972 b, c); data for In and Re from A. D. Matthews (unpublished); Two types of deep-sea sediments are listed: (i) pelagic clay, which is essentially free from $CaCO_3$; (ii) carbonates, which in the purest sampled form contained ~10% clay. For some elements only order of magnitude estimates could be made; these are indicated by the symbol X.

concentrations of a selection of these elements are given in Table 34.4. In this table the sediments are sub-divided into "clays" and "carbonates", which are quantitatively the two most important types of deep-sea sediment. The distinction between them is a fundamental genetic one because the principal components in each type have different origins. The deep-sea clays consist predominantly of fine-grained land-derived (lithogenous) particles which have been deposited at a slow rate, but may also contain authigenic (hydrogenous) components, such as ferro-manganese phases, which are rich in certain trace elements. In contrast, deep-sea carbonates have a much faster rate of deposition, and contain a significant proportion of calcareous shell debris which is usually relatively impoverished in trace elements, other than strontium.

There are a large number of elements listed in Table 34.4, and as analytical techniques are developed and improved more and more of those elements present at even lower concentrations will be determined. However, many of the initial studies on the geochemistry of deep-sea sediments (see e.g. Revelle, 1944; Young, 1954; Goldberg and Arrhenius, 1958; El Wakeel and Riley, 1961; Landergren, 1964) were limited to the major elements and to a relatively small range of trace elements. This early work revealed that there are very large variations between the concentrations of the major elements of deep-sea sediments of different lithologies. However, with the exception of iron and manganese, the major elements in deep-sea clay, carbonate and siliceous sediments have similar concentration ranges to those in their near-shore and continental counterparts. In contrast, the early investigations showed that certain deep-sea sediments have trace element assemblages which are characteristically different from those of sediments deposited in continental and near-shore environments. In particular, deep-sea clays from the present day sediment surface are considerably enriched in some trace elements, e.g. Ni, Co, Cu, Pb and Zn.

The overall geochemistry of deep-sea sediments is discussed in the following sections, and for convenience the major and the trace elements are considered separately.

34.2.2 MAJOR ELEMENTS

34.2.2.1. *Introduction*

The relative proportions of the component minerals of deep-sea sediments exert a fundamental control on the chemical composition of the sediments, particularly with respect to their contents of the major elements. This is apparent from Table 34.5 in which the major element compositions of three of the principal types of deep-sea sediments are listed. The differences between the elemental compositions of the three types of sediment are largely a result of variations in the proportions of the principal minerals in them. For example, calcium is concentrated in the carbonates, aluminium in the clays and silicon in the siliceous sediments. However, most of the major elements in deep-sea sediments have more than one source, and often more than one host mineral. The distributions of the elements among the most important minerals in the sediments are summarized in Table 34.6, and each element is considered below individually.

34.2.2.2. *Silicon*

Silicon is the second most abundant element found in the earth's crust, and is an important constituent of many igneous, metamorphic and sedimentary rocks. With the exception of quartz (SiO_2), most of the various igneous and metamorphic minerals which contain silicon are susceptible to chemical weathering (see Chapter 25). As a consequence of this, silicon is brought to the oceans in three principal forms: (i) silicon which has been leached into solution, or which is in a colloidal state; (ii) silicon held within minerals which are resistant to, or which have escaped, chemical weathering; and (iii) silicon contained in minerals which are weathering products.

34.2.2.2.1. *Silicon in solution.* The principal sources of dissolved silicon to the oceans are thought to be river run-off ($\sim 4{\cdot}3 \times 10^{14}$ g SiO_2 yr^{-1}; Livingstone, 1963) and glacial weathering ($5\text{–}8 \times 10^{14}$ g SiO_2 yr^{-1}; Schutz and Turekian, 1965). Sea water contains only a relatively small amount of silicon in solution (usually <4 mg l^{-1}), and it is apparent that large quantities of the silicon brought to the oceans annually are removed from sea water. However, there is a good deal of controversy over the mechanisms which bring about this removal (see Section 11.3.4). The budget of silica in

TABLE 34.5

Average chemical composition of pelagic sediments (wt. % oxides)*

	Original compositions				Compositions on carbonate, water and organic C-free basis			
	Calcareous	Lithogenous clay	Siliceous	Oceanic average	Calcareous	Lithogenous	Siliceous	Oceanic average†
SiO_2	26·96	55·34	63·91	42·72	59·86	60·44	70·61	61·52
TiO_2	0·38	0·84	0·65	0·59	0·94	0·92	0·72	0·90
Al_2O_3	7·97	17·84	13·30	12·29	18·34	19·06	14·72	18·12
Fe_2O_3	3·00	7·04	5·66	4·89	9·77	8·93	7·08	9·09
FeO	0·87	1·13	0·67	0·94	—	—	—	—
MnO	0·33	0·48	0·50	0·41	0·75	0·52	0·55	0·64
CaO	0·30	0·93	0·75	0·60	1·04	1·01	0·83	1·00
MgO	1·29	3·42	1·95	2·18	3·05	3·73	2·17	3·19
Na_2O	0·80	1·53	0·94	1·10	2·50	1·67	0·93	1·98
K_2O	1·48	3·26	1·90	2·10	3·29	3·56	2·09	3·23
P_2O_5	0·15	0·14	0·27	0·16	0·46	0·16	0·30	0·33
H_2O	3·91	6·54	7·13	5·35	—	—	—	—
$CaCO_3$	50·09	0·79	1·09	24·87	—	—	—	—
$MgCO_3$	2·16	0·83	1·04	1·51	—	—	—	—
Org. C.	0·31	0·24	0·22	0·27	—	—	—	—
Org. N.	—	0·016	0·016	0·015	—	—	—	—
Total	100·0	100·0	100·0	100·0	100·0	100·0	100·0	100·0
Total Fe_2O_3	3·89	8·23	6·42	—	—	—	—	—

* Data based on the analyses of 25 pelagic sediments from all the major oceans (El Wakeel and Riley, 1961).

† Weighted mean calculated on an areal basis. Percentage of deep-ocean floor covered by sediments, 48·7% calcareous, 37·8% lithogenous, 13·5% siliceous.

TABLE 34.6*

*The distribution of major elements among the minerals of deep-sea sediments**

	Lithogenous minerals			Hydrogenous minerals		Biogenous minerals	Cosmogenous minerals	
Element	Resistant minerals	Weathering residues	Igneous, metamorphic and sedimentary minerals; and pyroclastic material	Primary	Secondary			Average content in deep-sea sediments (wt. %, oxide)†
Si	Quartz	**Clay Minerals**	MICAS, FELDSPARS, AMPHIBOLES, PYROXENES, PYROCLASTIC MATERIAL	—	**Zeolites, Montmorillonite**	**Opal**	—	SiO_2, 42·7
Al	—	**Clay Minerals**	MICAS, FELDSPARS, AMPHIBOLES, PYROXENES, PYROCLASTIC MATERIAL	—	**Zeolites, Montmorillonite**	—	—	Al_2O_3, 12·3
Ti	Rutile, Anatase	**Clay Minerals**	MICAS, FELDSPARS, AMPHIBOLES, PYROXENES, PYROCLASTIC MATERIAL	ANATASE(?) FERRO-MANGANESE NODULES	**Zeolites, Montmorillonite**	—	—	TiO_2, 0·59
Na, K	—	**Clay Minerals**	MICAS, FELDSPARS, AMPHIBOLES, PYROXENES, PYROCLASTIC MATERIAL	—	**Zeolites, Montmorillonite**	—	—	Na_2O, 1·1 K_2O, 2·1

TABLE 34.6—*continued*
*The distribution of major elements among the minerals of deep-sea sediments**

	Lithogenous minerals			Hydrogenous minerals		Biogenous minerals	Cosmogenous minerals	
Element	Resistant minerals	Weathering residues	Igneous, metamorphic and sedimentary minerals; and pyroclastic material	Primary	Secondary			Average content in deep-sea sediments (wt. %, oxide)†
P	—	**Clay Minerals**	MICAS, FELDSPARS, AMPHIBOLES, PYROXENES, PYROCLASTIC MATERIAL	PHOSPHATES	**Zeolites, Montmorillonite**	**Phosphates**	—	P_2O_5, 0·16
Ca	—	**Clay Minerals**	MICAS, FELDSPARS, AMPHIBOLES, PYROXENES, CALCITE, DOLOMITE, PYROCLASTIC MATERIAL	Carbonates, Phosphates	**Zeolites, Montmorillonite**	**Carbonates**	—	CaO 0·75
Mg	—	**Clay Minerals**	MICAS, FELDSPARS, AMPHIBOLES, PYROXENES, CALCITE, DOLOMITE, PYROCLASTIC MATERIAL	Carbonates	**Zeolites, Montmorillonite**	**Carbonates**	—	MgO, 2·18

TABLE 34.6—*continued*

Fe	Haematite, Goethite	**Clay Minerals**	MICAS, FELDSPARS, AMPHIBOLES, PYROXENES, PYROCLASTIC MATERIAL, **Iron oxides** (as coatings on other minerals, and as discrete grains)	FERRO-MANGANESE NODULES, GOETHITE	**Zeolites, Montmorillonite**	Associated with calcareous shell debris as oxide coatings.	Cosmic spherules	Fe_2O_3, 24·89
Mn	—	**Clay Minerals**	MICAS, FELDSPARS, AMPHIBOLES, PYROXENES, PYROCLASTIC MATERIAL, **Manganese oxides** (as coatings on other minerals, and as discrete grains).	**Ferro-manganese nodules, Manganese oxides**	ZEOLITES, MONTMORILLONITE	Associated with calcareous shell debris as oxide coatings.	—	MnO, 0·41

* In this table the minerals are classified in terms of a number of sedimentary components according to the scheme outlined in Table 34.8 Some of these components, e.g. ferro-manganese nodules, contain more than one mineral, and for these the individual minerals are not specified. Only the principal host minerals are listed and their quantitative importance is indicated by the following typescripts, e.g. **Quartz** > CLAY MINERALS > Carbonates. From Riley and Chester (1972).

† From El Wakeel and Riley (1961).

the oceans has been discussed by various authors including, Bien *et al.* (1958), Siever and Scott (1963), Schutz and Turekian (1965), Harriss (1966), Gregor (1968), Calvert (1968) and Burton and Liss (1968). According to Calvert (1968) the most important removal mechanism for silica from sea water is a biological one in which opaline siliceous shells are formed by diatoms and radiolarians. In some regions, e.g. around Antarctica, the shells are deposited to form siliceous sediments. However, Burton and Liss (1968) have criticized several aspects of Calvert's silica budget. For example, Calvert has proposed that Antarctica is the principal area for the removal of dissolved silicon from the World Ocean, whereas Burton and Liss (1968) consider that the data of Schutz and Turekian (1965) suggest that, in fact, Antarctica is a major source of silicon for the rest of the oceans. Burton and Liss (1968) concluded that there is a significant imbalance between the input of silica to the World Ocean and its biological removal, and suggested that non-biological removal mechanisms are important in the oceanic silica budget. One mechanism which they suggested for the non-biological removal of silicon is its reaction with suspended material in estuaries during an early stage of the mixing of fresh and saline waters (see also Bien *et al.*, 1958). Another non-biological mechanism which was postulated was the "reverse-weathering" reaction between dissolved silicon and degraded aluminosilicates (see also Sillén, 1961; Mackenzie and Garrels, 1966; and Chapter 5). Silicon is also incorporated into authigenic silicates such as quartz overgrowths (Arrhenius, 1963), and aluminosilicates, e.g. orthoclase, palygorskite and sepiolite (Hathaway and Sachs, 1965; Heezen *et al.*, 1965; Bonatti and Joensuu, 1967). Secondary authigenic minerals, i.e. those which are formed by the alteration of pre-existing minerals, such as the zeolites, may also remove silicon from sea water during their formation. Chert horizons have recently been discovered at depth in the deep-sea sediment columns by DSDP drillings (see e.g. Ewing *et al.*, 1969; Peterson *et al.*, 1970; Gartner, 1970; Calvert, 1971).

There is no doubt that biological processes do remove very considerable amounts of silica from sea water; estimates range from $\sim 1{\cdot}9 \times 10^{14}$ g SiO_2 yr^{-1} (Burton and Liss, 1968) to $\sim 3{\cdot}6 \times 10^{14}$ g SiO_2 yr^{-1} (Calvert, 1968), i.e. between $\sim 44\%$ and $\sim 83\%$ of the silica brought to the oceans by river run-off. Four types of organisms contribute significant quantities of siliceous skeletal remains to deep-sea sediments. These are, in order of decreasing importance, diatoms, radiolarians, sponges and silicoflagellates (Riedel, 1959). Silica is precipitated biologically as opaline silica (an amorphous substance, $SiO_2.nH_2O$), and according to Riedel (1959) there are three main factors which control the contributions made by opal-secreting organisms to marine sediments. These are: (i) the rate of production of siliceous organisms in the overlying waters; (ii) the degree of dilution of siliceous remains by terrigenous and volcanic materials and by calcareous organisms; and, (iii) the extent of the dissolution of siliceous skeletons. Factors which affect the dissolution of silica at both the sediment surface, and during diagenesis at depth, have been discussed by Arrhenius (1963) and by Siever and Scott (1963), and controls on the preservation of siliceous shell populations are described in Chapter 29.

Lisitzin (1972) has investigated the distribution of opaline silica in both surface sea water and marine sediments. The opaline silica concentrations in the former vary between ~0·1 and ~1000 μg SiO_2 l^{-1}. His studies of the distribution of opaline silica in marine sediments have shown that there are three major belts of recent biogenous silica deposition. (i) A polar belt which almost encircles the globe in the Southern Hemisphere. This belt, which accounts for >75% of all silica accumulation in the World Ocean (Lisitzin, 1966), is between ~900 km and ~2000 km wide and its northern boundary coincides with the Antarctic Convergence. The principal siliceous organisms in the sediments of this belt are diatoms. (ii) A northern belt which is found in the Pacific Ocean, in the Sea of Okhotsk and in the Bering and Japan Seas, but which is absent from the Atlantic and Indian Oceans. (iii) An equatorial, or near-equatorial, belt which is well defined in the Pacific and Indian Oceans, but which is less distinct in the Atlantic Ocean.

34.2.2.2.2. *Silicon in resistant minerals.* Quartz, which is an important constituent of many deep-sea sediments, is the most important weathering-resistant silicon-containing mineral. Almost all of that found in such sediments is detrital and originates from the weathering of terrestrial rocks, or is formed during sub-aerial, or submarine, volcanic processes, and is transported to deep-sea areas by wind, water or ice mechanisms Investigations of the origin and the distribution of quartz in deep-sea sediments have been made by many workers. For example, the importance of wind-transported quartz has been discussed by Murray and Renard (1891), Radczeweski (1939), Rex and Goldberg (1958), Goldberg and Griffin (1964), Chester and Hughes (1969), Beltagy *et al.* (1972), Chester *et al.* (1972) and Aston *et al.* (1973). Volcanically-derived quartz has been described by Peterson and Goldberg (1962), Griffin and Goldberg (1963) and Goldberg (1964); water-transported quartz has been identified by Arrhenius (1963). The overall distribution of this mineral has been described in Chapter 26.

Silicon is an important constituent of many other minerals which resist continental chemical weathering, either partially or completely, or which are produced by submarine volcanic processes. These minerals include the feldspars, amphiboles, pyroxenes, micas and olivine. However, with the exception of the feldspars, these minerals usually make up only a very small fraction of deep-sea sediments; the distribution of feldspars in the sediments has been discussed by, for example, Peterson and Golderg (1962), Arrhenius (1963), Goldberg *et al.* (1963) and Biscaye (1964); see also Chapter 26.

34.2.2.2.3. *Silicon in weathering residues.* By far the most quantitatively important silicon-containing weathering products brought to the oceans from the land areas are the clay minerals, the predominant members of which belong to the illite, montmorillonite, chlorite and kaolinite groups, or are mixed-layer varieties. The distributions of these clay mineral species in deep-sea sediments are now known in some detail and have been summarized in Chapter 26 (see also for example, Yeroshchev-Shak, 1961; Biscaye, 1964; Griffin *et al.*, 1968). Some of the clays in deep-sea sediments are undoubtedly detrital in origin. Thus, kaolinite, chlorite and illite exhibit definite latitudinal distribution patterns in the sediments, which are related to those found in continental soils, atmospheric dusts and marine particulate material (see e.g. Griffin *et al.*, 1968; Chester *et al.*, 1972; Chester *et al.*, 1974; Behairy *et al.*, 1975; see also Chapter 26). In contrast, much of the montmorillonite in deep-sea sediments has an authigenic (hydrogenous) origin and has been formed by the alteration of volcanic debris (see e.g. Griffin *et al.*, 1968; Chester *et al.*, 1974; see also Chapters 26 and 27).

34.2.2.3. *Aluminium*

Aluminium is the third most common element in the earth's crust and, like silicon, it is an important constituent of all igneous, metamorphic and sedimentary rocks. However, its geochemistry in deep-sea sediments differs significantly from that of silicon in several ways. The two most important of which are: (i) the absence from marine sediments of significant amounts of Al-bearing minerals which are resistant to chemical weathering; c.f., quartz which has a major role in the distribution of silicon in the sediments; and, (ii) the fact that aluminium is not incorporated into marine sediments by biogenous shell sedimentation.

The distribution of aluminium in deep-sea sediments is almost exclusively controlled by the input of detrital aluminosilicates to the oceans, either from the continents or from submarine volcanic activity. Quantitatively, the clay minerals and the feldspars are the most important aluminosilicates found in deep-sea sediments, but aluminium also occurs in many other minerals which are present in most deep-sea sediments in only small amounts. Because aluminium in deep-sea sediments is largely located in detrital minerals it has been used as an indicator of the amount of terrigenous debris in the sediments (see e.g. Arrhenius, 1952; Landergren, 1964). This relationship has recently been utilized by Boström *et al.* (1969) to identify chemically specialized metalliferous sediments found at some active centres of sea floor spreading. These sediments are thought to have originated, at least in part, by precipitation from hydrothermal solutions and therefore contain only relatively small amounts of terrigenous material. For this reason, Boström *et al.* (1969) were able to use the ratio Al/Al + Fe + Mn to identify the principal regions in which these sediments occur, and to delineate the geographical extent to which the hydrothermal processes have influenced the geochemistry of the adjacent deep-sea sediments.

34.2.2.4. *Titanium*

Titanium may be present in a number of the components of deep-sea sediments, including terrigenous material, basaltic debris, authigenic (hydrogenous) precipitates and biological phases.

Titanium occurs as a minor constituent in many aluminosilicates, e.g. the clays, feldspars, micas etc. as well as in the form of its polymorphic oxides, anatase and rutile. Continentally-derived minerals such as these account for most of the titanium found in deep-sea sediments. However, Goldberg and Arrhenius (1958) have pointed out that in those sediments containing relatively high contents of titanium (i.e. $> \sim 0{\cdot}70\%$ Ti) it is almost invariably associated with basaltic debris; for example, the titanium contents of a series of Pacific Ocean deep-sea sediment cores decreased from $2{\cdot}5\%$ Ti adjacent to a volcanic source to $0{\cdot}78\%$ Ti some 350 km distant. Similar distributional trends have also been reported for Atlantic deep-sea sediments by Correns (1937, 1954). This association of basaltic material with titanium has also been used, in conjunction with the distributions of other elements, by Boström *et al.* (1972, 1973) to estimate the amount of basaltic detritus in deep-sea sediments. These authors concluded that although oceanic basaltic sources of titanium may be important locally, most of this element in Indo-Pacific deep-sea sediments has a continental origin (see also Horowitz, 1974).

Titanium is found in ferro-manganese nodules, in which it has an average concentration of $\sim 0{\cdot}60\%$ Ti (Chester, 1965a). Goldberg (1954) found a covariance between Fe and Ti in some nodules, and El Wakeel and Riley (1961) found a similar relationship between the two elements in deep-sea sediments from the major oceans. The reasons for this association are not clear, but Goldberg (1954) has suggested that some of the titanium in nodules and deep-sea sediments has been scavenged from sea water by hydrous ferric oxides, (see also Cronan, 1972). There is also some evidence that authigenic titanium-bearing minerals may precipitate from interstitial waters, and Correns (1954) has linked the formation of authigenic anatase to the post-depositional migration of titanium in the deep-sea sediment/interstitial water complex.

Titanium has frequently been found in marine organisms (see e.g. Nicholls *et al.*, 1959; Boström *et al.*, 1974). Martin and Knauer (1973) have reported that phytoplankton usually contain $\lesssim 150$ ppm Ti, although siliceous frustules can have titanium concentrations of up to ~ 1000 ppm (see also

Griel and Robinson, 1952). However, Goldberg (1954) has found that, in general, Fe : Ti ratios in marine organisms are similar to those in deep-sea sediments, and has concluded that most of the titanium in the organisms was probably acquired by the ingestion of particles which had scavenged this element from sea water. Furthermore, Goldberg and Arrhenius (1958) did not find abnormally high Ti : Al ratios in equatorial Pacific deep-sea sediments having a large biogenous component, and these authors concluded that the biosphere is not an important contributor of titanium to the sediments.

It may be concluded, therefore, that most of the titanium in deep-sea sediments, is located in terrigenous material and, less importantly, in basaltic debris; authigenic minerals and biogenous components contribute only minor amounts.

34.2.2.5. *Phosphorus*

Phosphorus occurs in the terrigenous, submarine basaltic, authigenic and biogenous components of marine sediments. Small amounts of the element are found in structural positions in many crustal minerals, and it is transported to deep-sea areas within these minerals by water, wind and ice.

A number of authors have pointed out that there is an association between phosphorus and biogenous material in deep-sea sediments. For example, El Wakeel and Riley (1961) found that, on average, deep-sea carbonates contain about twice as much phosphorus as do deep-sea clays. However, there was no apparent association between phosphorus and the total carbonate content of the sediments, and this element is not concentrated in calcareous shell material. The major host mineral for biogenous phosphate is skeletal apatite, which can constitute a few per cent (by wt.) of some deep-sea sediments. On the deep-ocean floor biogenous apatite undergoes dissolution, and in the slowly accumulating pelagic sediments only the most resistant phosphate structures such as shark's teeth and the ear bones of whales are preserved (Arrhenius, 1963). In sediments which accumulate at a faster rate, e.g. some deep-sea carbonates, structures such as fish scales may be preserved. Apatite is an effective adsorbent for the lanthanides and thorium; since these elements form very insoluble phosphates this tends to inhibit the dissolution of the skeletal apatite. According to Arrhenius *et al.* (1957) skeletal fish debris in deep-sea sediments is frequently relatively rich in these elements. However, it seems likely that they were adsorbed from sea water after the death of the organisms and that their phosphates remain preserved as microscopic crystals even after the skeletal apatite has completely dissolved.

There are four principal types of marine authigenic phosphates; phosphatic nodules, phosphatic pellets, phosphorite-rock and phosphorite minerals such as francolite and the lanthanide phosphates. The nodules, pellets and phosphorite-rock are largely confined to near-shore environments and have been extensively reviewed in Chapter 33. Francolite ($Ca_5(PO_4, CO_3, OH)_3F$) is the only authigenic phosphate mineral to have a widespread occurrence in deep-sea sediments: it occurs particularly in deposits on seamounts and other topographic highs.

Phosphorus occurs in ferro-manganese nodules which, on average, are about three times richer in the element than are deep-sea sediments. The phosphorus contained in the nodules is probably associated with manganese oxide phases (Arrhenius, 1952), an association which has also been demonstrated for deep-sea sediments from the Pacific Ocean (Revelle, 1944).

34.2.2.6. *Sodium and potassium*

Deep-sea and near-shore sediments contain similar concentrations of sodium and potassium, and the bulk of both these elements is associated with the clay minerals (Welby, 1958; Heier and Adams, 1964). However, other minerals in deep-sea sediments, which are quantitatively much less important than the clays, may contain relatively high concentrations of sodium (e.g. sodic feldspars) and potassium (e.g. phillipsite).

The alkali elements may be associated with the clay minerals in two principal ways; in lattice structures, and in surface and inter-sheet positions. It is important to distinguish between these two types of elemental associations because the surface-held and inter-sheet sodium and potassium can be extensively involved in diagenetic reactions in both fluvial and marine environments. A voluminous literature exists on the changes which can affect sodium and potassium when river-transported clays are introduced into sea water, and also during sediment diagenesis (see Chapters 5 and 27 for reviews of these processes). Various authors have stressed that reactions between the sodium and potassium associated with the clay minerals and the elements present in sea water may play a large part in controlling the composition of ocean water (see e.g. Chapter 5; Sillén, 1963, 1965, 1967; Mackenzie and Garrels, 1966).

In general, the content of potassium in deep-sea sediments exceeds that of sodium (see Table 34.5). However, El Wakeel and Riley (1961) have observed that two highly calcareous deep-sea sediments contained more sodium than potassium; they interpreted this as resulting from the biological removal of sodium from sea water by certain calcareous organisms.

34.2.2.7. *Calcium and magnesium*

The distribution of calcium, and to a lesser extent that of magnesium, in deep-sea sediments is largely controlled by biological processes, and biogenous carbonates (chiefly calcite and aragonite) may constitute $>90\%$ of some of these sediments. Calcareous oozes, which contain $>30\%$ of skeletal carbonate, cover almost half of the deep-ocean floor, and are quantitatively the most important type of deep-sea sediment. The principal classes of carbonate-secreting organisms in the upper layers of deep-sea waters are the foraminifera, coccolithophorids and pteropods. The sedimentation, preservation and distribution of these calcareous organisms have been discussed in detail in Chapter 29.

Both calcium and magnesium are found in many crustal minerals (e.g. the clay minerals, feldspars etc.) and are transported to the oceans in terrestrial detritus. Volcanic minerals may also contribute significant amounts of both elements to sediments; thus, El Wakeel and Riley (1961) reported average concentrations of 6·26% CaO and 8·34% MgO in two volcanic clays from the Pacific; these concentrations may be compared to 0·93% CaO and 3·42% MgO for lithogenous, i.e. non-volcanic, deep-sea clays. In this connection Goldberg and Arrhenius (1958) have pointed out that pyroxenes are an important contributor of calcium to some Pacific deep-sea sediments. The zeolite phillipsite also contains relatively large concentrations of this element.

Some deep-sea sediments contain authigenic calcium and magnesium minerals (e.g. calcite, aragonite, dolomite, magnesite, ferroan magnesite and hydromagnesite). Bonatti (1966) has described the precipitation of a number of carbonate minerals from hydrothermal volcanic emanations, and Leir (1959) has suggested that the dolomite in the sediments of the Peru-Chile Trench is authigenic. However, authigenic calcium and magnesium carbonates are not common in the marine environment, and play only a very minor role in the deep-sea sediment geochemistry of these elements. Lithified carbonates, such as those described by Thompson *et al.* (1968), are a specialised type of deep-sea carbonate and their significance is not fully understood (see Chapter 27).

34.2.2.8. *Iron and manganese*

The geochemistry of both iron and manganese in deep-sea sediments is intimately related to those of a number of trace elements, and is discussed in detail in Section 34.2.3. For this reason the factors controlling the distribution of these two elements will only be summarized briefly at this stage.

34.2.2.8.1. *Iron.* The iron in deep-sea sediments may have several origins.

(i) It may have been transported to the oceans in association with terrigenous solids, particularly the clay minerals which form a large fraction of the land-derived material in deep-sea sediments. According to Carroll (1958) iron may be associated with the clay minerals in a number of ways, viz. (a) as an essential constituent within the crystal lattice; (b) as a minor constituent within the crystal lattice; and, (c) as iron oxide coatings on the surfaces of the clay particles. These oxide coatings, which can be formed in the river, estuarine, or marine environments, are geochemically important because they can act as "scavengers" for trace elements, and subsequently may be transported to deep-sea areas together with the clays. Iron is also transported from the continents in association with minerals such as feldspars, olivine, augite, hornblende, magnetite and ilmenite (see e.g. Goldberg and Griffin, 1964). Some of the iron in deep-sea sediments is located in submarine volcanic debris, and El Wakeel and Riley (1961) found relatively large amounts of ferrous iron ($\sim 6\%$) in two Pacific cores which contained unaltered volcanic material; most deep-sea sediments contain $< 1\%$ FeO. In addition to inorganic solids from terrestrial weathering and volcanic sources, a very small proportion of the iron in deep-sea sediments is present in cosmic spherules which, as the name implies, have a cosmic origin (see Section 34.2.3.3.1.).

(ii) It may have been deposited through the agency of a biological mechanism, and the importance of biologically associated iron in deep-sea sediments has been stressed by various authors (see e.g. Clarke and Wheeler, 1922; Bradley and Bramlette, 1942; Revelle, 1944; El Wakeel and Riley, 1961). Although iron can be associated with organisms in several ways much of the biologically-derived iron which is preserved in deep-sea sediments occurs as coatings on skeletal parts, particularly those of the foraminifera. For example, El Wakeel and Riley (1961) reported that the acid-insoluble residues of certain highly calcareous Atlantic deep-sea sediments have total iron contents ($\sim 18\%$ as Fe_2O_3) which are considerably higher than those of lithogenous clays (average; 7% as Fe_2O_3). These authors concluded that this "excess" iron had been brought to the sediment surface by the deposition of globigerina, or other organisms, and had remained behind as an insoluble residue after the dissolution of the calcareous parts of the shells. Correns (1941), who investigated the extraction of iron from sea water by foraminifera, has suggested that the iron may have been initially acquired by feeding on diatoms which can concentrate this element (Harvey, 1939) and subsequently retained. However, Pettersson (1945) concluded that iron associated with shell material may have been removed from sea water by inorganic processes after the deposition of the tests.

(iii) It may have originated through processes involving the authigenic removal of iron from sea water. Such mechanisms have received considerable attention in recent years, particularly with regard to the formation of ferro-manganese nodules (see Chapter 28) and the chemically specialized metalliferous deep-sea sediments (see Section 34.2.3.3.2).

34.2.2.8.2. *Manganese.* Manganese is one of the few major elements which are concentrated in deep-sea relative to near-shore sediments; for example, some Pacific deep-sea sediments contain an order of magnitude more manganese than do near-shore ones. Geochemical balance calculations (see e.g. Horne and Adams, 1966) have shown that there is a considerable imbalance between the amount of manganese supplied to sea water from rock weathering and that present in deep-sea sediments. A voluminous literature exists on all aspects of this "excess" manganese, and its origin has been the subject of much speculation which started as far back as the last century (see e.g. von Gumbel, 1878; Murray and Renard, 1891; Murray and Irvine, 1895).

Manganese is present in a number of components of deep-sea sediments. It is transported from the continents in a particulate form both in association with weathered minerals (e.g. the clay minerals, feldspars etc.) and as discrete particles of hydrous manganese oxides which, because of their small size, may be transported for long distances by oceanic current systems. Chester and Messiha-Hanna (1970) studied the partition of manganese among the components of a series of North Atlantic deep-sea sediments and showed that only about 30% of the manganese in the sediments was held in lattice positions within continentally-derived minerals. The remainder of it was present as hydrous manganese or ferro-manganese oxides, and had been removed from solution during the history of the sediment components. According to Turekian (1965a, 1967) much of this removal occurred in the river environment, and the precipitates were subsequently transported to deep-sea areas by oceanic currents (see Section 34.2.3.5.). However, much of the manganese in deep-sea sediments is associated with ferro-manganese nodules, phases which have precipitated from material dissolved in sea water, and the removal of manganese from sea water into the sediment complex is an important process in the oceanic budget of this element. For example, Chester *et al.* (1973) have shown that the hydrogenous (authigenic) phases of some Atlantic deep-sea clays contain ~8% Mn and approximate to the composition of ferro-manganese nodules. They have suggested that this provides evidence that the accumulation of manganese in nodules and deep-sea sediments are related processes. Manganese may be incorporated into deep-sea sediments by biological processes, e.g. as coatings on foraminiferal shells (Turekian, 1965a), by organisms which remove manganese from sea water by metabolic processes (Graham, 1959); and also by sedimentation of organic bubble-formed aggregates (Riley *et al.*, 1964, 1965). During recent years considerable attention has been devoted to problems involved in the generation of ferro-manganese nodules and metalliferous deep-sea sediments. Because these phases have been reviewed in Chapter 28 and Section 34.3 they are not discussed in this section.

One of the major problems relating to the deep-sea geochemistry of manganese is the ultimate source of the element, and the oceanic budget of manganese has been discussed by numerous authors (see e.g. Pettersson, 1945; Goldberg, 1954; Goldberg and Arrhenius, 1958; Lynn and Bonatti, 1965; Bonatti and Nayudu, 1956; Boström, 1967; Boström *et al.*, 1969; Bender *et al.*, 1971; Calvert and Price, 1972; Glasby, 1973; Hart, 1973; Elderfield, 1975). Three principal sources have been suggested for it viz. river run-off, post-depositional migration from sub-surface sediments and submarine volcanic activity. Many authors (see e.g. Wedepohl, 1960; Goldberg and Arrhenius, 1958; Horne and Adams, 1966) have concluded that river run-off cannot supply all of the manganese which is found in deep-sea sediments. A post-depositional, or diagenetic, supply mechanism was proposed by Murray and Irvine (1895) to account for the enrichment of manganese in the upper oxic layers of some deep-sea sediments. In recent years the processes involved in the chemical diagenesis of manganese within the sediment/interstitial water complex have been studied in some detail (see e.g. Chapter 30; Lynn and Bonatti, 1965; Li *et al.*, 1969; Bender, 1971; Calvert and Price, 1972). Our knowledge of the processes involved in the diagenesis of manganese has been considerably advanced by these investigations, and a number of theoretical models describing the post-depositional migration of manganese have been proposed (see e.g. Spencer and Brewer, 1971; Calvert and Price, 1972). Although the diagenetic reactions are still by no means fully understood it is apparent that post-depositional processes are important in controlling the distribution of manganese in many near-shore (see e.g. Manheim, 1965) and some deep-sea (see e.g. Lynn and Bonatti, 1965; Li *et al.*, 1969) sediments. However, diagenetic processes are largely controlled by the redox conditions in the sediment/interstitial water complex, and many deep-sea sediments have oxic layers which extend to considerable depths below the sediment surface. Bender (1971) has concluded that manganese is unlikely to diffuse through these extensive oxic layers rapidly enough to supply manganese at the rate at which it is accumulating at the surfaces of many deep-sea sediments. In general, manganese (and other elements) may be introduced into sea water by volcanic processes in three ways: (i) by the violent leaching of newly extruded lava by sea water; (ii) by the debouching of hydro-thermal, possibly mantle-derived, solutions directly onto the sea floor; and, (iii) by the low temperature weathering of volcanic rock which is exposed to sea water over relatively long periods. These processes are discussed in detail in Section 34.2.3.3.2. Much of the early speculation on the importance of volcanically-derived manganese was hindered by a lack of real evidence about the type of process, or processes, involved in the liberation of elements from volcanic material (see e.g. Pettersson, 1945). The picture changed considerably, however, with the discovery of deep-sea sediments rich in iron and manganese on the crests of some active ridge centres of sea floor spreading (see e.g. Boström *et al*., 1969). Following this discovery models have been proposed to explain both the hydrothermal debouching (see e.g. Boström and Pettersson, 1969), and the volcanic rock weathering of manganese and other elements (see e.g. Corliss, 1971; Hart, 1973). These models are described in details in Section 34.2.3.3.2., and their overall importance to deep-sea sediment trace element budgets is discussed in Section 34.2.3.5. At this stage, it may be concluded that manganese in solution is supplied to deep-sea sediments by river run-off, post-depositional migration and volcanic processes, and that the relative importance of each source varies locally.

34.2.3 TRACE ELEMENTS

34.2.3.1. *Introduction*

One of the characteristic features of certain types of deep-sea sediments is that they contain enhanced concentrations of a number of trace elements. The distribution of such elements in these sediments is now known in some detail, and a number of fundamental trends have been established; these are summarized below, and representative analyses of a number of deep-sea sediment types are given in Table 34.7.

(i) On average, deep-sea carbonates are impoverished in most trace elements relative to both near-shore muds and deep-sea clays.

(ii) Some trace elements, e.g. Cr, V, Ga, have similar concentrations in both near-shore muds and deep-sea clays.

(iii) Other trace elements, e.g. Mn, Cu, Ni, Co, Pb, are concentrated in deep-sea clays relative to near-shore muds, and are further enriched in Pacific compared to Atlantic deep-sea clays.

(iv) Those trace elements which are enriched in deep-sea clays often have their highest concentrations in sediments deposited farthest from the land masses under conditions of relatively slow accumulation.

(v) Chemically specialized, or so called metal-rich, deep-sea sediments found on the crests of some active ocean ridges contain higher concentrations of certain elements, e.g. Fe, Mn, Cu, Ni, Co, Zn, than do "normal" deep-sea clays.

(vi) Ferro-manganese nodules have relatively very high concentrations of those trace elements, e.g. Mn, Cu, Ni, Co, Pb, which are enhanced in deep-sea clays, but only small concentrations of those trace elements, e.g. Cr, Ga, which are not enhanced in the clays.

In recent years a great deal of research has been carried out with the aim of establishing the sources, modes of transport and mechanisms of incorporation of trace elements into deep-sea sediments. Many fundamental questions still remain unanswered, but important advances have been made. In the present chapter these advances are reviewed in terms of the overall geochemistry and mineralogy of deep-sea sediments.

TABLE 34.7

The average distributions of trace elements in some deep-sea sediments (in ppm)

Trace element	Near-shore muds[a]	Deep-sea carbonate[b]	Atlantic deep-sea clay[c]	Pacific deep-sea clay[d]	Active ridge sediment[e]	Ferro-manganese nodules[f]
Cr	100	11	86	77	55	10
V	130	20	140	130	450	590
Ga	19	13	21	19	—	17
Cu	48	30	130	570	730	3300
Ni	55	30	79	293	430	5700
Co	13	7	38	116	105	3400
Pb	20	9	45	162	—	1500
Zn	95	35	130	—	380	3500
Mn	850	1000	4000	12500	60000	220000
Fe	69900	9000	82000	65000	180000	140580

[a] Data from Wedepohl (1960).
[b] Data from Turekian and Wedepohl (1961).
[c] Data from Wedepohl (1960) and Turekian and Imbrie (1967).
[d] Data from Goldberg and Arrhenius (1958), El Wakeel and Riley (1961) and Landergren (1964).
[e] Data for East Pacific Rise, given on a carbonate-free basis, from Boström and Peterson (1969).
[f] Data from Chester (1965).

TABLE 34.8

*The principal minerals in the various components of deep-sea sediments**

Lithogenous components	Hydrogenous components.† Primary	Secondary	Biogenous components	Cosmogenous components
Clay minerals	*Ferro-manganese nodules*	*Montmorillonite* (inc. nontronite)	*Carbonates*	*Cosmic spherules*
Quartz	*Iron oxides*	*Zeolites*	*Opal*	
Feldspars	*Manganese oxides*		Phosphates	
Micas	Francolite			
Amphiboles	Barite			
Pyroxenes	Celestite			
Pyroclastic material	Calcite			
Rutile	Aragonite			
Anatase	Dolomite			
Haematite	Quartz-overgrowths			
Goethite	Anatase (?)			
Calcite				
Dolomite				
Iron oxides				
Manganese oxides				

* The quantitatively important minerals are italicized.
† For classification of hydrogenous components, see text.

The chemical compositions of deep-sea sediments are governed by a number of complex inter-relating controls. Riley and Chester (1971) have suggested that in order to provide a framework in which to discuss deep-sea sediment geochemistry these controls can be divided into a number of categories. These are: the relative proportions of the minerals constituting deep-sea sediments; the pathways by which the elements are introduced into the marine environment; the mechanisms by which the elements are incorporated into the sediments; and, the overall pattern of sedimentation in a particular oceanic area. Each of these controls is discussed individually below, although it must be remembered that they are not mutually independent but rather combine together to exert an overall influence on the geochemistry of the sediments.

34.2.3.2. *The relative proportions of the minerals constituting deep-sea sediments*

The relative proportions of the mineral, and multi-mineral, components will obviously exert a fundamental control on the chemical compositions of deep-sea sediments, particularly since some components are considerably richer in trace elements than are others. There have been a number of attempts to classify the components of deep-sea sediments, and although none of these is entirely satisfactory the scheme suggested by Goldberg (1954) provides a useful framework within which the various components can be conveniently discussed. In this classification the components are divided into five broad categories according to the geosphere in which they originated. *Lithogenous* components are defined as those arising from land erosion, submarine volcanoes, or from underwater weathering, where the solid phase undergoes no major change during its residence in sea water. *Hydrogenous* components are those which result from the formation of solid matter in the sea by inorganic reactions, i.e. non-biological processes. Chester and Hughes (1967) divided the hydrogenous components of deep-sea sediments into two broad genetic groups: (i) primary material, which is formed directly from sea water; and, (ii) secondary material, which results from the submarine alteration of pre-existing minerals. This classification has been extensively modified and discussed in Chapter 27. *Biogenous* components are produced in the biosphere, and include inorganic shell material and organic matter. *Cosmogenous* components are those derived from extra-terrestrial sources. Some of the more important mineral, and multi-mineral, components of deep-sea sediments are classified according to Goldberg's scheme in Table 34.8. To a large extent the relative proportions of these components control the overall trace element compositions of deep-sea sediments. However, the formation of these components is intimately related to the geochemical history of the trace elements themselves, and this history must be examined before any attempt is made to explain elemental variations. *Pore waters* are the fifth category.

34.2.3.3. *The pathways by which trace elements are introduced into the marine environment*

It is useful to make a fundamental distinction between those trace elements which are introduced into the water overlying a sediment as solid or colloidal phases, and those which are introduced as dissolved material

34.2.3.3.1. *Trace elements associated with solid phases.* Material directly entering the oceans in a solid form comprises the lithogenous and cosmogenous categories of the classification scheme outlined by Goldberg (1954).

Lithogenous material originating from continental weathering consists of both inorganic and organic components, and makes by far the most important contribution to the total amount of solids brought to the oceans. It is transported to deep-sea areas by a variety of agencies which include water, wind and ice.

Rivers supply the largest proportion of lithogenous (land-derived) solids to the oceans (Kuenen, 1965). However, since the amount of river run-off varies from one oceanic area to another, this type of lithogenous material is not evenly distributed in sea water. The areas of the major oceans, and of the complementary land masses draining into them, are given in Table 34.9, and

TABLE 34.9
*Oceanic areas and complementary land areas draining into them (units: 10^6 km^2)**

Ocean	Area	Land area drained	Percentage
Atlantic	98	67	68·5
Indian	65	17	26·0
Antarctic	32	14	44·0
Pacific	165	18	11·0

* Data from Goldberg (1967).

details of individual river discharges are listed in Table 34.10. Of the various continents, Asia produces by far the highest annual river-transported sediment yield with $\sim 600 \times 10^3$ kg for every km^2 of land surface, with Africa, Europe and Australia each producing less than one tenth of this (Holeman, 1968). The total discharge of river sediment to the oceans has been estimated to be $\sim 20 \times 10^{12}$ kg yr^{-1}, of which $\sim 80\%$ originates in Asia which constitutes only ~25% of the total land area draining into the World Ocean (Holeman, 1968). However, river-transported sediment has a wide range of particle sizes, and only a small proportion of material ⩾4 μm in size reaches the deep-sea areas directly. One reason for this is that many of the world's major sediment-transporting rivers flow into marginal seas rather than into "open" coastal areas, with the result that much of the sediment is deposited before it can reach deep-sea areas. Further, even in "open" coastal areas much of the river-transported sediment is initially deposited on the continental shelf and slope regions which act as intermediate sediment traps. Subse-

TABLE 34.10

*River discharge of suspended sediment into the oceans**

A. *River discharge of suspended sediment from the continents*

Continent	Area draining to ocean (10^6 km^2)	Annual suspended sediment discharge (g × 10^{13})	Annual suspended sediment discharge (kg km^{-2} of land surface)
Asia	26·6	1600	600 × 10^3
North America	20·4	200	96 × 10^3
South America	19·2	120	62 × 10^3
Africa	19·7	54	27 × 10^3
Europe	9·2	32	35 × 10^3
Australia	5·1	23	45 × 10^3

B. *Discharge of suspended sediment from some major rivers*

River	Location	Drainage area (10^3 km^2)	Annual suspended sediment discharge (g × 10^{13})
Yellow (Hwang Ho)	China	666	208
Ganges	India	945	160
Brahmaputra	Bangladesh	658	80
Yangtze	China	1920	55
Indus	Pakistan	957	48
Amazon	Brazil	5710	40
Mississippi	U.S.A.	3180	34
Irrawaddy	Burma	425	33
Mekong	Thailand	786	19
Red	N. Vietnam	117	41
Nile	Egypt	2944	12
Congo (Zaire)	Africa	3968	7
Niger	Nigeria	1100	0·50
St. Lawrence	Canada	1275	0·40

* Data from Holeman (1968).

quently, these shelf and slope sediments may be transported to deep-sea areas by continental margin processes such as sliding, slumping and turbidity currents.

The overall chemical composition of river-transported sediment is controlled by a number of factors. These include: (i) the geological nature of the catchment area deposits; (ii) the drainage conditions; (iii) the intensity of local weathering processes; and, (iv) anthropogenic contamination, which can be significant in certain regions. Konovalov and Ivanova (1968) have investigated some of the geological and environmental factors which control the discharge of both soluble and particulate trace elements from rivers of the U.S.S.R. into marine basins. The results of their investigations can be summarized as follows (i) Discharges of V, Mn, Co, Ni, Cu, Zn, Ag and Mo are controlled by the chemical properties of the elements, by local physiographic catchment conditions (e.g. water flow, relief etc.), and by the petrographic nature of the rocks in the river basins. (ii) Mountain rivers tend to transport more suspended material than do lowland rivers, and also to carry trace elements mainly in dissolved forms. (iii) The dissolved supply of some trace elements remains constant among various hydrological regimes in the river basins, indicating that chemical properties rather than physiographical

TABLE 34.11

*The average trace element composition of suspended sediment from rivers with both mineralized and unmineralized catchment zones in Cornwall (U.K.) (in ppm)**

River	Trace element	Mineralized tributaries	Unmineralized tributaries
Red	Mn	1020	657
	Cu	706	563
	Zn	1590	358
	Pb	775	133
	Fe	57000	39000
Carnon	Mn	1054	964
	Cu	3000	416
	Zn	1839	391
	Pb	540	402
	Fe	67000	48000
Gannel	Mn	3165	1000
	Cu	313	46
	Zn	3995	255
	Pb	2670	89
	Fe	46000	42000

* Data from Aston *et al.* (1974).

conditions control their discharges; however, the detrital supply of trace elements appears to be considerably more dependent on physiographical parameters.

Factors such as those listed above can produce large variations in the trace element compositions of river-transported sediment even between rivers which are geographically close. For example, Aston *et al.* (1974) have analysed suspended material from three rivers in Cornwall (England), each of which has feeder tributaries from both mineralized and unmineralized catchment areas. The results of their survey are given in Table 34.11, and show that some trace elements, e.g. Mn, Cu, Zn, Pb, are considerably enhanced in material derived from mineralized zones. The effects of pollution on the trace element composition of river-transported solids has been demonstrated by Turekian and Scott (1967). These authors determined the trace element compositions of solids from 17 rivers on the eastern and Gulf of Mexico seaboards of the United States, and found the following average concentrations (in ppm): Cr 194; Ag 2.2; Ni 126; Co 36; Mn 4199. For solids from the Susquehannah River (Penn.), however, the following averages were obtained (in ppm): Cr 290; Ag 15; Ni >1000; Co >500; Mn >12000. The authors concluded that the enhanced concentrations in the solids from the Susquehannah River resulted, at least in part, from industrial pollution.

In addition to the more obvious effects of industrial pollution, Turekian and Scott (1967) have highlighted another important trend in the trace element compositions of river particulates. United States rivers to the west of the Mississippi have suspended solids with concentrations of Mn, Ni, Co and Cr which are essentially similar to those of the average shale and have probably acquired the elements from weathering with little subsequent chemical modification. In contrast, the rivers of the eastern United States, and possibly the Rio Maipo (Chile) which drains the Andes, have higher trace element concentrations which, according to Turekian and Scott (1967), are indicative of other elemental sources in addition to natural chemical weathering. The results of this study are summarized in Table 34.12, together with that of Martin and Meybeck (1975).

An estimate of the trace element composition of that fraction of river-transported sediment which is deposited on the shelf areas can be obtained from the average compositions of near-shore and estuarine muds. There are inherent difficulties in attempting to establish the average concentration of trace elements in these sediments because of the wide range of depositional environments in near-shore areas (see Chapter 33). Furthermore, sediments in some areas will be significantly more affected by pollution than will those in others. For these reasons a range of trace element concentrations in both "unpolluted" and "polluted" oxic and anoxic shelf and estuarine muds is given in Table 34.13. However, because of the wide range of trace element

TABLE 34.12

The average trace element composition of suspended sediment from a number of rivers (in ppm)

A. Data from Turekian and Scott (1967)

River	Mn	Ni	Co	Cr	Ag	Mo
3 U.S. Rivers west of the Mississippi	596	25	15	73	0·4	9
Rhone River (France)	820	60	29	150	0·7	14
14 U.S. Rivers east of the Mississippi	4971	150	40	220	2·6	23
Rio Maipo (Chille)	2400	40	76	68	1·0	44
Shale*	850	68	19	90	0·1	2·6

B. Data from Martin and Meybeck (1976)

River	Al	Fe	Mn	Ni	Co	Ga	Cu	Cr	Mo	Pb	Zn	V
Amazon	115000	53800	390	58	15	19	375	116	0·7	105	426	232
Congo	100000	51500	1200	74	20	25	350	141	4	455	400	163
Ganges	77000	37000	1060	—	14	—	—	71	—	—	—	—
Mekong	112000	55700	940	99	20	28	107	102	2	113	300	—
Orinoco	113000	57500	740	30	10	28	73	70	—	76	119	127
Average	86200	43800	700	65	16	25	226	100	2·2	187	310	97

* Data from Turekian and Wedepohl (1961).

TABLE 34.13

The distribution of trace elements in some estuarine and shelf sediments (in ppm except Fe which is as % wt.)

Sediment type	Location	Mn	Fe	Cu	Ni	Co	Ga	Cr	V	Ba	Sr	Sn	Zn	Pb
"Unpolluted" sediments*	Solway Firth, U.K.[a]	360	—	10	38	16	15	35				7		
	Gulf of Paria[b]	2000	—	19	31	12		100	130	750	250			20
	Saanich Inlet, Canada[c]													
	oxic sediments	370	—	38	33	9		86	110				88	20
	anoxic sediments	400	—	45	26	8		35	37				80	tr
	Baltic Sea[d] (anoxic)	4030	—	78	43	22		90	130	750	130	6	110	25
"Polluted" sediments*	Swansea Bay, U.K.[e]	—	—	81									128	126
	Severn Estuary, U.K.[f]	—	—										470	163
	Firth of Clyde, U.K.[g]	1118	—	37	50	34		64				19	165	86
	Clyde Estuary, U.K.[h]	1600	—	225	69	60		624				85	1680	528
	Sorfjord, Norway[i]	—	—	2424					1617			303	20000	11000
	Severn Estuary, U.K.[j]	1820	4·5	38	36	7	17	71	86	250	400	101	280	119
Average: detrital near-shore sediments[10]		866	3·1	17	39	8	14	87	91	416	194	—	—	17
Average: anoxic near-shore sediments[10]		26850	5·3	86	65	41	—	101	203	522	447	—	377	67
Average: near-shore mud[k]		850	6·9	48	55	13	19	100	130	750	250		95	20

[a] Perkins *et al.* (1973).
[b] Hirst (1962).
[c] Gross (1967).
[d] Manheim (1961).
[e] Bloxham *et al.* (1972).
[f] Butterworth *et al.* (1972).
[g] Mackay *et al.* (1972).
[h] Skei *et al.* (1972).
[i] Chester and Stoner (1975a).
[j] Chapter 33.
[k] Wedepohl (1960).

* "Unpolluted" sediments are those which so far as is known have not been subjected to pollution. "Polluted" sediments are those which have been reported as probably suffering from some pollution effects.

concentrations in near-shore sediments it is impossible in our present state of knowledge to assess fully the trace element differentiation which occurs before river-transported sediment is finally deposited in the marine environment. It will be necessary to carry out much more work on individual river systems before the problem can be solved.

Perhaps the most important "physical" process affecting river-transported sediment on its introduction to the marine system is particle size separation. The process itself is very complex, and involves a number of physico-chemical reactions (see e.g. Whitehouse and McCarter, 1958; Welder, 1959; Postma, 1967). The processes involved in particle size separation have been considered by Turekian (1967) who divided river-transported particles into two very broad categories. Coarser particles, (usually ~ 2–~ 4 μm in size) tend in general to be deposited adjacent to the continents and may subsequently be redistributed to deep-sea areas by sea bottom processes such as turbidity currents etc. These particles, which include significant amounts of relatively coarse lithogenous minerals such as quartz, contain comparatively low concentrations of trace elements, e.g. Ni, Co, Pb, etc. In contrast, the finer particles, (<2 μm in size) are composed mainly of clay minerals (which may be coated with iron and manganese oxides), together with discrete grains of iron and manganese hydrous oxides. These particles have a relatively large surface area and will have acquired a high content of trace elements in the river environment by adsorption and co-precipitation reactions. Because of their size these particles may escape deposition on the shelf regions and be transported directly to deep-sea areas by surface and intermediate oceanic current systems. The overall effect of this process of differential transport is the separation of river-transported particulates having a "low" trace element content from those with a "high" trace element component. The effect of this process is reflected in the differences between the trace element contents of near-shore, non-pelagic and pelagic sediments. Evidence for the partitioning of trace elements between "coarse" and "fine" particles of suspended river solids has been provided by Per hac (1972) who separated and analysed two size classes of suspended particles, one of which was $>0{\cdot}15$ μm and the other $<0{\cdot}15$ μm, from two streams in Tennessee (U.S.). Some of his data are given in Table 34.14, and it can be seen that the trace elements Co, Cu, Ni and Pb are considerably enhanced in the small-sized particles.

There are a number of complex inter-related chemical processes which affect river-borne solids when they are transported from fresh water through brackish to saline water. Desorption is one of the most important of these processes and may have a direct influence on the trace element chemistry of river-transported solids (see Chapter 27). The process has been studied by Kharkar *et al.* (1968) who concluded that trace elements which have been

TABLE 34.14

*The particle size distribution of some trace elements in solid material from two Tennessee (U.S.) streams (in ppm)**

Trace element	Particles >0·15 μm	Particles <0·15 μm
Mn	1628	1205
Fe	21 500	25 700
Zn	947	1031
Cu	2500	3375
Co	55	199
Ni	65	355
Pb	278	1638

* Data from Per hac (1972).

adsorbed from solution by clay minerals suspended in river water are desorbed, to a greater or lesser extent, on contact with sea water. For example, they found that the supply of cobalt desorbed from river-transported clay particles on contact with sea water exceeds that originally dissolved in the river water by a factor of two. In contrast, cobalt was considerably less efficiently desorbed from iron and manganese oxides than it was from clay mineral particles. However, the overall effects of desorption are not well understood, and there is little data on the subsequent fate of desorbed trace elements. It may well be that elements desorbed from clay mineral surfaces in estuarine waters are re-adsorbed onto river-transported, or freshly precipitated, iron and manganese hydrous oxides, or are bound up with organic complexes.

Most of the lithogenous solids transported by rivers are either rock fragments or mineral weathering residues. However, in addition to these components, the solids also contain phases which have been precipitated from weathering solutions, e.g. hydrous oxides of manganese and iron. These oxides are particularly important in marine geochemistry because they can adsorb trace elements from river waters and subsequently transport them to deep-sea areas.

In most oceanic regions the input of lithogenous material by wind (i.e. aeolian) transport is very small relative to that from rivers. However, in certain areas, e.g. the North Atlantic underlying the path of the North East Trades, aeolian solids are the principal source of lithogenous sedimentary material. Chester (1972b) and Aston *et al.* (1973) have given data on the dust-loadings of soil-sized aeolian material from the lower troposphere over various areas of the World Ocean, and their findings are summarized in Table 34.15. The interacting processes which control the quantity and nature of the soil-sized dust in the lower marine troposphere include: (i) the strength and circulation patterns of the principal wind systems; (ii) the nature of the soils in the continental source areas; and (iii) the local effectiveness of the mechanisms by which the dust is removed from the atmosphere, i.e. gravitational fall-out and rain-scrubbing. The highest natural dust-loadings are found over marine areas adjacent to the major deserts which act as reservoirs of loose easily lifted surface soils. These major deserts include the West African, the Arabian, the Mediterranean, the Australian, the North and South American and the Asian arid and semi-arid regions.

TABLE 34.15

*Atmosphere soil-sized dust-loadings in the lower marine troposphere**

Oceanic area	Principal wind system	Dust-loading (μg m^{-3} of air)
North Atlantic	Westerlies	10^{-3} – 10^{0}
	North East Trades	10^{0} – 10^{2}
South Atlantic	South East Trades	10^{-2} – 10^{-1}
Northern Indian	Northeast Monsoon	10^{-1} 10^{1}
	Southwest Monsoon	
Southern Indian	Northwest Monsoon	10^{-2}
	South East Trades	
China Sea	Southwest Monsoon	10^{-1}

* Data from Aston *et al.* (1973). The order of magnitude dust-loadings are based on collection by a mesh technique and are lower limit estimates.

Lower tropospheric marine dust is composed of a variety of materials of various origins and includes: continental weathering products, e.g. quartz and the clay minerals; biological material, e.g. spores and fresh water and marine organisms; sea salts; extra-terrestrial solids, e.g. cosmic spherules; and, man-made (anthropogenic) contaminants. According to Junge (1972) most of the Sahara Desert dust collected over marine areas lies in the size range 0·5–20 μm, and, in general, all soil-derived dusts will have a similar size range with the majority of particles being >1 μm (Cawse and Peirson, 1972). In contrast, industrial processes involving combustion yield particles which are usually <1 μm in size. (Cawse and Peirson, 1972). Although anthropogenic particles can make a significant contribution to the overall trace element content of present-day aeolian dust populations, their effect is almost entirely confined to sediments laid down during the last few hundred years. Chester and Stoner (1973, 1974) have collected and analyzed soil-sized dusts from the lower troposphere over various oceanic areas. Although the collection technique employed recovered mainly soil-sized particles, i.e. $\lesssim 1$ μm, it was found that in some regions the dusts contained a significant anthropogenic fraction of elements such as Pb, Sn and Zn (see Table 34.16).

TABLE 34.16
*The average trace element composition of soil-sized dusts from the lower marine troposphere (in ppm; except Fe which is in % wt.)**

Element	Average concentration[a]	Enrichment factor[b]
Fe	5·2	—
Sr	101	0·35
Co	9	0·45
Cr	85	1·2
V	145	1·4
Ba	487	1·4
Mn	1312	1·5
Ni	91	1·5
Ga	21	1·6
Cu	157	3·4
Zn	683	13
Sn	30	19
Pb	465	49

* Data from Chester and Stoner (1973).
[a] Average of 51 soil-sized dusts from the Atlantic and Indian Oceans, China Sea and various coastal localities.
[b] The enrichment factor is based on the concentration ratio of the elements to iron in the dusts, divided by the ratio of the same elements in average crustal material. To a first approximation the enrichment factor should be close to unity for those elements which have a predominantly natural source from crustal material.

Because of their anthropogenic fraction the present day soil-sized dusts cannot be used to compute the average input of particulate atmospheric trace elements to the oceans over geological time. However, Chester and Stoner (1974) found that in some regions there was an input of zonal dusts from non-industrialized catchment areas which mask the effects of any anthropogenic components which are present in the global background dusts. One such region is that which underlies the North East Trade Winds in the Atlantic Ocean, and which receives the major proportion of its dust from the adjacent Sahara Desert. The trace element composition of this dust should approximate to that of "natural" soil-sized aeolian dust. The average trace element composition of the Atlantic North East Trade dusts is listed in Table 34.17, and it can be seen that most of the trace elements (Zn, Pb and Sn are exceptions) are present in concentrations which are similar to their crustal abundances, i.e. they have predominantly "natural" sources. The trace element compositions of "natural" marine soil-sized aeolian dusts will vary locally in the same manner as do those of their parent soils. However, the Atlantic North East Trade dusts can be used to give, at least to a first approximation, an estimate of the composition of aeolian material brought to the oceans before the onset of anthropogenic contamination. Comparative data for the average compositions of these dusts and of near-shore and deep-sea sediments are given in Table 34.17.

TABLE 34.17

The average trace element composition of soil-sized dusts from the Atlantic North East Trades, near-shore sediments and Atlantic deep-sea clays (in ppm except Fe which is as % wt.)

Element	N.E. Trade dusts; enrichment factor in parentheses[a]		Near-shore sediment[b]	Atlantic deep-sea clay[c]
Fe	5·8		6·9	8·2
Mn	1813	(2·0)	850	4000
Cu	79	(2·0)	48	130
Ni	43	(0·8)	55	79
Co	11	(0·5)	13	38
Ga	22	(1·6)	19	21
Cr	67	(0·8)	100	86
V	103	(0·9)	130	140
Ba	484	(1·4)	750	700
Sr	136	(0·4)	<250	120
Sn	4	(2·4)	—	—
Zn	338	(5·1)	95	130
Pb	140	(12)	20	45

[a] Data from Chester and Stoner (1973); for explanation of enrichment factor see legend to Table 34.16.
[b] Data from Wedepohl (1960).
[c] Data from Wedepohl (1960), and Turekian and Imbrie (1967).

Ice-transported lithogenous material has a significant effect on deep-sea sedimentation only in the polar regions. Other factors such as rock type etc. being equal, the kind of material which is weathered from the earth's crust depends on local environmental conditions, and in this respect the polar latitudes have several characteristic features which have been discussed by Lisitzin (1962). The most important of these are: (i) the lack of river drainage, and therefore of river transported debris, as a result of the ice and snow cover which often extends to sea level; and, (ii) the fact that chemical weathering, which may even be locally absent, is usually less intense than it is in temperate and tropical latitudes because continental water is largely held in the form of snow and ice. The overall result is that in polar latitudes the predominant form of weathering is the mechanical action of glacial scour and rock pulverization, and the principal transport mechanisms are movement by glaciers on land and ice-rafting in the oceans.

One of the major products of glacial weathering is "rock flour" which is debris consisting of undecomposed grains of rocks and minerals ground to a fine (usually clay sized) powder. In addition to "rock flour", ice also transports much large material ranging in size from silt to boulders. Yeveyev (1959) has estimated that $\sim 0{\cdot}69 \text{ km}^3 \text{ yr}^{-1}$ of rock flour is eroded from Antarctica, and on the assumption that the density of the material is $2{\cdot}7 \text{ g cm}^{-3}$ and that Antarctica constitutes $\sim 90\%$ of the present day ice caps, Garrels and Mackenzie (1971) have calculated that $\sim 20 \times 10^{14}$ g of material is contributed annually to the oceans by the world's glaciers. This is to be contrasted with the annual contribution of river-transported solids, i.e. $\sim 183 \times 10^{14}$ g. However, it must be remembered that the supply of glacial material has varied considerably in the past, particularly during glacial-interglacial periods.

Ice-rafting can result in the long distance transport of rocks and rock fragments (see e.g. Lisitzin, 1960; Kimo *et al.*, 1956; Needham, 1962). However, the characteristic marine deposits of polar regions are the glacial marine sediments, which are particularly predominant around Antarctica. According to Ericson *et al.* (1961) these sediments can be distinguished from turbidite deposits by the complete absence of particle size sorting. Stetson and Upton (1937) investigated a series of sediments from the Ross Sea, Antarctica, and concluded that their compositions were dominated by glacially-transported lithogenous material which was poorly sorted and chemically unaltered. The sediments resembled glacial tills deposited on land. Similar conclusions were drawn by Angino (1966), Angino and Andrew (1968) and by Edwards and Goodell (1969) for sediments from the Amundsen, Bellinghausen, Ross and Weddell Seas, respectively.

The trace element contents of Antarctic glacial marine sediments vary considerably from one area to another, probably as a result of differences in the elemental chemistry of the source rocks (see e.g. Angino 1966; Angino and Andrews, 1968). The average concentrations in several series of glacial marine sediments are given in Table 34.18. For various reasons, including the lack of chemical alteration and the poor sorting of their lithogenous components, glacial marine sediments are unique deep-sea deposits. This is confirmed by their trace element assemblages which, on average, resemble those of near-shore muds more closely than those of deep-sea clays (see Table 34.22).

TABLE 34.18

The concentrations of trace elements in some Antarctic glacial marine sediments (ppm except Fe and Al which are given as % wt.)

Location	Trace element						
	Al	Fe	Cr	Ni	Co	V	Cu
Ross Sea*	6·8	3·4	94	40	30	160	90
Amundsen Sea*	7·5	4·1	90	40	30	240	185
Bellinghausen Sea*	7·6	3·5	70	30	20	185	120
Weddell Sea*	6·8	2·1	104	46	20	158	70
Average: Antarctic glacial marine sediment†	7·3	3·7	89	37	27	200	137

* Data from Angino and Andrew (1968).
† Data from Angino (1966).

Volcanic (volcanogenic) material is an ubiquitous component of deep-sea sediments. However, although its concentration varies considerably from one area to another, it is only rarely the principal sediment component. Volcanic, or pyroclastic, material includes ash, scoria, lava, pumice and glass, and also alteration products of these such as palagonite, phillipsite, montmorillonite etc.

Some volcanic material, e.g. ash, is transported to deep-sea areas via the atmosphere. According to Ninkovich *et al.* (1964) the transportation and deposition of volcanic ash depends on a number of factors which include: (i) the initial force of the volcanic eruption; (ii) the angle of ejection; (iii) the height reached by the ash particles; (iv) their composition and density: and, (v) the velocity and turbulence of the transporting winds. It was shown above that continental soil-sized marine atmospheric dust has its highest concentration at low latitudes adjacent to arid or semi-arid regions. In contrast, the distribution of volcanic dust is independent of latitude. As a consequence of this, ash horizons are found in many localities, e.g. the Mediterranean (Mellis, 1954), the Gulf of Alaska (Nayndu, 1964), the North Atlantic (Bramlette and Brady, 1940), and the Equatorial Pacific (Worzel, 1959). For an historical assessment of the importance of volcanic dust in the atmosphere the reader is referred to Lamb (1970).

Volcanic ash consists of glass shards together with smaller amounts of mineral and crystal fragments. Of these, it is the shards which are transported over the greatest distances by the wind. Unlike ash, other volcanic material can only be transported for relatively shorter distances through the atmosphere (sub-aerial volcanoes), or through the hydrosphere (submarine volcanoes). However, pumice (a general term for highly vesicular solidified lava of any chemical and mineralogical composition) may be transported over relatively long distances in the hydrosphere because of its porous nature.

Volcanic glass is a common form of pyroclastic material in deep-sea sediments, and its composition depends on the type of parent vulcanism. For example, Peterson and Goldberg (1962) identified two types of volcanic glass in South Pacific pelagic sediments; a clear colourless glass associated with acidic vulcanism, and a dark reddish-brown highly altered glass which results from basic vulcanism. These authors have also identified a series of volcanic feldspars in the sediments, and estimated from their distribution that the dissemination distance for volcanic material >32 μm in size is of the order of a few hundred kilometres, whereas that for material 4–8 μm in size can be as much as 2000 km.

Basalts are the predominant type of volcanic rock found on the ocean floor, although other types, e.g. trachytes, occur in subordinate amounts (Nicholls, 1965a). Oceanic basalts are divided into two principal groups, or suites: the tholeiitic basalts, and the alkali basalts; each of which can be divided into a number of sub-groups. Both tholeiitic and alkalic basalts are common on oceanic islands, but basalts dredged from the deep-sea floor are largely tholeiitic in character and range in composition from "typically" tholeiitic, which are not obviously associated with mid-ocean ridges, to high alumina basalts, which do appear to have a definite association with the ridge systems (Nicholls, 1965b).

Bonatti (1965) has studied the emplacement of basalts on the deep-sea floor, and has concluded that at least two types of submarine eruptions are involved in the production of these rocks. In one type, the lava appears to have flowed quietly onto the ocean floor, and any interaction with sea water is prevented by the instantaneous formation of a thin insulating crust of glass at the surface of the flow. These lavas appear to be associated primarily with hilly bottom topography, and they are thought to result from fissure-type eruptions. In the second type of eruption, (hyaloclastic) there is considerable chemical and physical interaction between the hot lava and sea water during the outflow, and much of the lava undergoes fragmentation (i.e. it is shattered and pulverized). The resulting lava debris is rapidly hydrated at the prevailing high temperature, and is susceptible to alteration, forming mainly palagonitic glass, which may itself subsequently be altered to smectites (e.g. montmorillonite) or zeolites (e.g. phillipsite).

The concentrations of trace elements in a number of oceanic basalts are given in Table 34.19. However, the basalts, particularly those having a hyaloclastic origin, can undergo very considerable mineralogical and chemical alteration. For example, McKelvey and Fleet (1974) and Fleet and Kempe (1974), who have studied the geochemistry of a series of ancient deep-sea volcanic and detrital sediments from the Indian Ocean, have shown that there are differences between the trace element concentrations in relatively "unaltered" and relatively "altered" pyroclastic debris—see Table 34.20.

TABLE 34.19

The trace element compositions of some oceanic basalts (in ppm)

	Oceanic alkali basalts[a]	Oceanic tholeiitic basalts[b]	Basalts; Mid-Atlantic Ridge[c]	Basalts; Mid-Indian Ocean Ridge[c]	Basalt; Tonga Island Arc[d]
Fe	82600	76800	63616	68282	54008
Mn	1084	1239	—	—	—
Cu	36	77	87	90	51
Ni	51	97	123	242	25
Co	25	32	41	73	30
Ga	22	17	18	20	13
Cr	67	297	292	347	75
V	252	292	289	340	230
Ba	498	14	12	—	14
Sr	815	130	123	131	115

[a] Data from Engel *et al.* (1965), and Nicholls (1965).
[b] Data from Melson and Thompson (1971).
[c] Data from Cann (1969).
[d] Data from Bryan *et al.* (1972).

The leaching of trace elements from basaltic material, and hydrothermal sources of trace elements, are discussed in a subsequent section.

Several kinds of material in deep-sea sediments have been considered, at least tentatively, to have a cosmic origin. Such material includes "iron" spherules, "stony" spherules, micro-tektites and cosmic-ray produced radioactive or stable nuclides.

Small black magnetic, or "iron", spherules have been found in deep-sea sediments from all the major oceans (see e.g. Murray and Renard, 1891; Brunn *et al.*, 1955; Hunter and Parkin, 1960; Crozier, 1960; Grjebine, 1965; Millard and Finkelman, 1970). These spherules, which have a density of ~6 and a size of <100 μm, have a metallic nucleus surrounded by a shell usually consisting of magnetite. Murray and Renard (1891) have suggested that the shell was formed by the oxidation of the outer layers of molten droplets (ablated from a meteor) during their passage through the atmosphere. This type of spherule is generally considered to have a cosmic origin. "stony" spherules, which have a density of ~3 and a size range of ~15–~250 μm, have a silicate groundmass. They were initially considered to have originated from "stony" meteorites, but it is now thought that many of them may have been formed in the terrestrial environment. In addition to the two major groups of spherules, i.e. "irons" and "stones", Hunter and Parkin (1960) have tentatively distinguished a third group, the "stony irons" in deep-sea sediments.

Tektites are small greenish to jet black natural glass bodies which are found in large numbers over a few restricted areas of the earth's surface. Micro-tektites have been found in 30–40 cm thick bands at depths of 60–1000 cm in the deep-sea sediments in parts of the Indian Ocean. These micro-tektites are probably associated with the Australasian strewn field (Glass, 1972). Micro-tektites have also been found in deep-sea sediments from the equatorial Atlantic Ocean; these are thought to be derived from the Ivory Coast strewn field (Glass, 1968, 1969).

TABLE 34.20

Trace element compositions of volcanic and detrital deep-sea sediments from the Indian Ocean (in ppm)

	Basaltic pyroclastic debris[a]	Altered basaltic pyroclastic debris[b]	Detrital clay[c]
Fe	72988	95253	46000
Mn	968	880	2100
Cu	142	118	105
Ni	85	39	96
Co	59	67	70
Cr	163	91	88
Zn	79	138	<88

[a] Data from McKelvey and Fleet (1974); coarse vitric lappilli tuff and vitric ash; D.S.D.P. Site 253.
[b] Data from McKelvey and Fleet (1974); medium and fine-grained vitric ash containing secondary clay minerals and zeolite cement; D.S.D.P. Site 253.
[c] Data from Fleet and Kempe (1974); average for detrital clays; D.S.D.P. Sites 250, 256, 257 and 258.

Over the past few years a good deal of attention has been directed towards the identification and measurement of cosmogenic nuclides, i.e. cosmic-ray produced radioactive or stable nuclides in deep-sea sediments and polar ice. The nuclides studied have included; ^{36}Cl, ^{26}Al, ^{10}Be, ^{64}Fe, ^{3}He, ^{4}He, ^{40}Ar, ^{36}Ar (see e.g. Schaffer *et al*., 1964; Merrihue, 1964; Lal and Venkatavaradan, 1966; Amin *et al*., 1966; Tilles, 1966; McCorkell *et al*., 1967; Arrhenius, 1967; Wasson *et al*., 1968; Tanaka *et al*., 1968).

Many of the investigations carried out to assess the effect of cosmic material on the trace element geochemistry of deep-sea sediments have concentrated on the "iron" spherules, even though they form only a very minor fraction of the total sediment. Petersson (1959) suggested that the inflow of these cosmic spherules could account for the relatively high content of nickel in some deep-sea sediments. Grjebine *et al*. (1964) investigated the

TABLE 34.21

Elemental compositions of some cosmic spherules and microtektites

A. Cosmic spherules*

	Element (in ppm)							
	Fe	Ni	Co	Mn	Al	Si	Ti	Cr
Spherule 1								
Inner core	216000	775000	11900	800	<200	500	400	<200
Outer core	410000	202000	3800	900	7700	27000	3000	<300
Magnetite	710000	14000	2500	<200	2300	1400	<200	600
Rim	640000	19300	2700	300	3300	2000	200	700
Spherule 2 (total)	646000	16200	3800	<200	<200	<300	<200	< 200
Spherule 3 (total)	682000	7200	2500	<200	<200	<200	900	2000
Spherule 12 (total)	640000	4100	2000	<200	<600	600	<200	<200

B. Microtektites

		wt. %
Range for 121 Australasian microtektites†	SiO_2	55·0 –81·0
	TiO_2	0·3 – 2·09
	Al_2O_3	7·4 –23·4
	FeO	2·78– 9·68
	MnO	0·03– 0·15
	MgO	1·2 –12·7
	CaO	0·5 – 6·18
	Na_2O	0·32– 2·8
	K_2O	0·19– 3·7
	P_2O_5	0·01– 0·17
		ppm
Composite of 5 Australasian microtektites†	Cr	81
	Mn	820
	Co	10
	Ba	530
	Fe	36226

* Data from Schmidt and Keil (1966).
Spherule 1 was probably produced during the surface ablation of a meteorite on entry to the earth's atmosphere and has undergone fractionation. Spherules 2, 3 and 12 probably have a similar origin to that of Spherule 1, but may have been more weathered and oxidized. They are texturally homogenous.
† Data from Glass (1972).

TABLE 34.22

The average trace element compositions of various solid materials brought to the oceans together with data for some marine sediments (in ppm)

Trace element	River transported sediment[a]	Ice transported sediment[b]	Wind transported dust[c]	Oceanic alkali basalt[d]	Oceanic tholeiitic basalt[e]	Altered basaltic debris[f]	Cosmic spherules[g]	Near-shore mud[h]	Atlantic deep-sea clay[i]	Pacific deep-sea clay[j]
Cr	100	90	67	67	297	91	<200 700	100	86	77
V	97	186	103	252	292	—	—	130	140	130
Ga	25	—	22	22	17	—	—	19	21	19
Cu	226	116	79	36	77	118	—	48	130	570
Ni	65	39	43	51	97	39	4100 775000	55	79	293
Co	16	25	11	25	32	67	2000 11900	13	38	116
Pb	187	<50	140	—	—	—	—	20	45	162
Zn	310	<200	338	—	—	—	—	95	130	—
Mn	700	1143	1813	1084	1239	880	<200–810	850	4000	12500
Fe	43800	33000	58000	82600	76800	92253	216000–710010	69000	82000	65000

[a] Data from Table 34.12.
[b] Data from Table 34.18.
[c] Data from Atlantic N.E. Trade wind soil-sized dusts from Table 34.17.
[d,e] Data from Engels *et al.* (1965).
[f] Data from Table 34.20.
[g] Data from Table 34.21. The range is for total spherules and some fractionated portions.
[h] Data from Wedepohl (1960).
[i] Data from Wedepohl (1960), and Turekian and Imbrie (1966).
[j] Data from Goldberg and Arrhenius (1958), El Wakeel and Riley (1961) and Landergren (1964).

distribution of magnetic spherules in three sediment cores from the Mediterranean Sea and concluded that the iron and manganese contents of the sediments are independent of the number of spherules present, but that there is a relationship between the concentration of nickel and the abundance of the spherules. In agreement with this Del Monte (1972) has suggested that black spherules have contributed an appreciable proportion of the nickel in Adriatic near-shore sediments. In contrast, Laevastu and Mellis (1955) concluded that the "iron" spherules do not appreciably influence the nickel content of deep-sea sediments (see also Goldberg, 1954). Chester and Hughes (1966) investigated the distribution of cosmic spherules in Pacific deep-sea sediments and postulated that even in those sediments which contain the highest number of spherules reported in the literature, i.e. ~5000 per kg of dry sediment, nickel of cosmic origin would amount to only ~2% of the total nickel present. In this context, it should be noted that Yabuki and Shima (1972) could find no correlation between the abundances of cosmic spherules and the nickel concentrations in a series of Pacific deep-sea clays. Other elements found in cosmic spherules which have been studied in relation to deep-sea sediment geochemistry include iridium and osmium (Barker and Anders, 1968). Millard and Finkleman (1970) have concluded that 10–50% of the iridium in Pacific deep-sea sediments may have a cosmic origin. Some analyses of "iron" spherules and micro-tektites are given in Table 34.21.

The trace element compositions of the various kinds of solid material brought to the oceans are summarized in Table 34.22. These concentrations are those of the "raw" solids. However, it has been shown above that "raw" material can undergo considerable trace element differentiation, both physical and chemical, before being deposited as marine sediments. In considering the processes of chemical differentiation which affect lithogenous material it is necessary to distinguish between two kinds of trace elements. (i) Those held in lattice positions within lithogenous minerals; and, (ii) those held in surface and inter-sheet positions, i.e. trace elements which have been removed from solution. For lithogenous material in the sediments this distinction is necessary, because although the former elements have a continental origin, the latter can be acquired from solution in either the continental, estuarine or marine environments. To establish the distribution of trace elements which are held in lattice positions in lithogenous material Chester and Hughes (1967) outlined a chemical technique designed to separate these elements from others present in deep-sea sediments. The technique, which involves leaching the sediments with a combined acid-reducing agent solution, can be applied to those marine sediments, and suspended materials, in which the sole non-lithogenous components are ferro-manganese minerals, carbonate minerals and adsorbed trace elements, i.e. authigenic (hydrogenous) silicates and biogenous opal are excluded.

The non-lithogenous material of deep-sea sediments which are being deposited at the present day bottom of the North Atlantic consists primarily of ferro-manganese nodules and carbonate minerals, together with only minor amounts of authigenic monøtmorillonites, zeolites, and biogenous opal. Because of this, the residues obtained by leaching the sediments with the acid-reducing agent consist of continentally-derived minerals in which the trace elements are held in lattice positions. Chester and Messiha-Hanna (1970) used this technique to study the partition of a series of trace elements between the "lattice-held" and the "non-lattice-held" fractions of a number of North Atlantic deep-sea sediments. The average trace element contents of the "lattice-held" fractions and total samples of the sediments are given in Table 34.23, together with those of the average near-shore mud. Near-shore muds may be regarded as representative of an early stage in the adjustment of lithogenous material to the marine environment, and for this reason their compositions should give an indication of those of deep-sea sediments in the absence of, or prior to, the onset of processes which produce trace element enrichment. A number of important conclusions can be drawn from the data in Table 34.23. It was shown above that certain elements, e.g. Mn, Ni, Co, are enriched in deep-sea relative to near-shore sediments, whereas others, e.g. Cr, V, are not. However, it is apparent from the table that the concentrations of Mn, Cu, Ni, Co, Cr and V are reasonably similar in both the total samples of near shore- sediments and the "lattice-held" fractions of deep-sea sediments. This involves two important geochemical implications. (i) Those elements, e.g. Cr and V, which have a similar abundance in near-shore muds and deep-sea clays are largely lithogenous in origin, i.e. they are transported from the continents in lattice-positions within crustal minerals. This was confirmed by Chester and Messiha-Hanna (1970) who have shown that $\sim 70\%$ of the Cr and V in North Atlantic deep-sea sediments is in "lattice-held" positions. (ii) Those elements, e.g. Mn, Ni and Co, which are enhanced in deep-sea clays have their "excess" concentrations in "non-lattice-held" positions, i.e. they have been removed from solution at some time during the history of the sedimentary components. For example, Chester and Messiha-Hanna (1970) have shown that, on average, $<50\%$ of the Co, Ni and Mn in deep-sea sediments is held in lattice positions. Further, they observed that the "non-lattice-held" contributions of these elements vary geographically, e.g. the fraction of "lattice-held" nickel decreases from $\sim 66\%$ near to the continents to $\sim 30\%$ in the area of the Mid Atlantic Ridge system. Table 34.23 also includes data on the partition of the trace elements in some samples of river-transported sediment. From these it can be seen that although the suspended sediment has considerably more Mn and Co than does the average near-shore sediment, the "lattice-held" concentrations of these elements are similar to those in deep-sea sediments.

TABLE 34.23

The average concentrations of lattice-held trace elements in various sedimentary materials (in ppm)

Trace element	Average; Atlantic deep-sea clay[a]	Average; lattice-held fractions of 38 N. Atlantic deep-sea sediments[b]	Average; 12 suspended sediments from Mersey and Weaver rivers (U.K.)[c]	Average; lattice-held fractions of 12 suspended sediments from Mersey and Weaver rivers (U.K.)[d]	Average; near-shore muds[e]
Mn	4000	582	9920	1040	850
Fe	82000	66621	97000	77000	69000
Cu	130	67	476	205	48
Ni	79	63	36	37	55
Co	38	12	26	10	13
Cr	86	72	148	170	100
V	140	120	94	106	130

[a] Data from Wedepohl (1960), and Turekian and Imbrie (1967).
[b] Data from Chester and Messiha-Hanna (1970).
[c,d] Data from J. H. Stoner (unpublished material).
[e] Data from Wedepohl (1960).

It may be concluded, therefore, that there are considerable variations in the trace element contents of solid materials brought to the oceans. These variations are most pronounced for river-transported sediment in which they may result from a combination of effects related to geographical, seasonal and anthropogenic factors. In general, volcanic debris is only a quantitatively important component of deep-sea sediments in local restricted regions, e.g. some parts of the South Pacific, and the trace element input from ice and wind-transported material cannot account for the enrichment of deep-sea clays with trace elements (see Table 34.22). In contrast, river-transported sediment could account, at least in part, for these enrichments if differentiation resulted in small "trace-element-rich" particles being selectively transported to deep sea areas, and in larger "trace-element-poor particles" being deposited either as near-shore sediments, or as deep-sea ones adjacent to the shelf regions.

This differential transport is considered in detail in a subsequent section (34.2.3.5).

34.2.3.3.2. *Elements in solution.* Elements which are incorporated into deep-sea sediments from solution may have been introduced into the sea by river run-off, glacial weathering, post-depositional migration or submarine weathering. The inputs from these processes have been considered by various authors, and have also been discussed in Chapter 7.

River run-off is by far the most important single process bringing dissolved material to sea water. Garrels and Mackenzie (1971) have estimated that a total of $\sim 2{\cdot}5 \times 10^{16}$ g of material are added to the oceans each year and that $\sim 90\%$ of this, i.e. $\sim 2{\cdot}25 \times 10^{16}$ g yr^{-1}, is transported by rivers with $\sim 0{\cdot}42 \times 10^{16}$ g yr^{-1} being in a dissolved form. That is, dissolved river-transported components constitute $\sim 17\%$ of the total material brought to the oceans, and river particulates $\sim 72\%$. However, the overall average dissolved composition of river water cannot be evaluated with a great deal of certainty because regional variations in the petrology of the rocks from the drainage areas of different rivers, and seasonal variations in drainage characteristics in the same river, can produce very large differences in the concentrations, and compositions of the dissolved material. Further, anthropogenic contamination can have an important effect on the dissolved elemental load of some rivers—see Table 34.24. An average dissolved trace element composition of river water is given in Table 34.25 but, for the various reasons listed above, this must only be regarded as a world-wide average, and should not be related to any particular river system.

It was shown above that there are various complex processes which can affect both the dissolved and the particulate trace elements in estuarine waters. These processes have been considered in Chapter 7, in which the

TABLE 34.24

The concentrations of some trace elements in rivers and estuaries with mineralized and anthropogenic inputs ($\mu g\ l^{-1}$)

Trace element	Rivers draining mineralized zones[a]	Estuary with an anthropogenic input[b]	Average river water[c]	Average sea water[d]
Mn	9–28	—	~5	~2
Zn	50–130	5–>30	10	5
Cd	1·0–3·4	0·2–~4	—	0·07
Pb	3·7–17·6	<1–~5	3	3×10^{-2}

[a] Data from Abdullah and Royle (1973): Mn, Zn, Cd—River Rheidal, Wales; Pb—River Dovey, Wales.
[b] Data from Abdullah *et al.* (1972), for the Bristol Channel, U.K.
[c] Data from Table 34.25.
[d] Data from Table 34.25; Cd from Chester and Stoner (1974).

importance of adsorption—desorption and precipitation—distribution reactions at the river/ocean boundary have been considered. Some of these reactions lead to the net addition of dissolved trace elements to estuarine waters, and some to their net loss from them. An example of the latter process has been reported by Coonley *et al.* (1971) who found that dissolved iron was depleted with increasing salinity in the waters of the lower reaches of a New Jersey (U.S.) river; they concluded that the depletion was a result of the oxidation of Fe (II) to Fe (III), hydrolysis, flocculation and settling of the resultant particules. Aston and Chester (1973) have shown that both increasing salinity and the presence of suspended sediment particles increase the rate, and extent, of the precipitation of iron in the river-estuarine-marine system. Once they have been formed some of the iron particles may become involved in biological processes, some may sink to the bottom and be incorporated into estuarine sediments (see e.g. Williams and Chan, 1966), and some may be transported by tidal currents and subsequently by current action to deep-sea areas. The latter particles are relatively small with large surface areas, and may scavenge trace elements such as Ni and Co from estuarine and marine waters (see e.g. Goldberg, 1954; Krauskopf, 1956). These particles may play an important role in the transport of trace elements to deep-sea areas where they are subsequently incorporated into the underlying sediments. Reactions which lead to the net addition of trace elements to estuarine waters include desorption. This process has been studied by Kharkar *et al.* (1968) who have shown experimentally that some trace elements, e.g. Co, are desorbed from river-transported clay minerals on contact with waters of high salinity (see also Section 27.3.3.1).

TABLE 34.25

*The concentrations of some trace elements in rivers and sea water**

Trace element	River water ($\mu g\ l^{-1}$)	Sea water ($\mu g\ l^{-1}$ for S = 35‰)
Si	4000	10^3†
Al	400	5†
Ti	3	<1
V	1	1·5
Cr	1	0·6†
Mn	~5	2†
Fe	670	3†
Co	0·2	0·08†
Ni	0·3	2
Cu	5	1†
Zn	10	5
Ga	0·1	3×10^{-2}
As	~1	2·3
Sr	50	$8{\cdot}5 \times 10^3$
Ag	0·3	0·1
Sn	0·04	0·01
Hg	0·07	5×10^{-2}†
Pb	3	3×10^{-2}†

* Data from Riley and Chester (1971).
† Considerable variations occur.

The input of material to the oceans from glacial action has been discussed above (Section 33.4.3.3.1). The total amount of material supplied to the oceans by glaciers has been estimated to be $\sim 0{\cdot}2 \times 10^{16}$ g y^{-1} (Garrels and Mackenzie, 1971). This is about 8% of the total material delivered annually by the world's rivers; however, it is not known how much of the glacially-supplied material is in a dissolved, and how much is in a particulate, state. Schutz and Turekian (1965) have suggested that leaching of glacial flour which enters the Southern Ocean from Antarctica may be responsible for the enrichment of Co, Ni and Ag in waters of moderate depths south of 68°S. According to Burton and Liss (1973) the data obtained by Schutz and Turekian (1965) suggests that there is a net outflow of dissolved silicon from Antarctica to the rest of the World Ocean.

The post-depositional remobilization and migration of trace elements occurs in the interstitial water/sediment complex, and may result in a supply of trace elements to the upper layers of some deep-sea sediments. Post-depositional processes are largely controlled by the redox and pH conditions in the interstitial water/sediment environment. The uppermost layers of the majority of deep-sea sediments are in an oxidized condition, but the thickness of the oxidized zone varies considerably from one type of sediment to another. For example, Lynn and Bonatti (1965) have shown that in Pacific Ocean sediments the thickness of the oxidized layer increases seawards from < 1 cm in near-shore areas to between ~ 8 and ~ 15 cm in marginal areas. In deep-sea areas of the central Pacific the sediments were still in an oxidized state at the full extent of "normal" coring (i.e. to depths of tens of metres). Lynn and Bonatti (1965) found relatively high contents of manganese in the upper layers of several deep-sea cores from the Pacific. In these cores the sediments underlying the manganese-rich zones showed evidence of reducing conditions. The authors concluded that the manganese had been mobilized by reduction in the reduced layers, and had migrated upwards through the interstitial waters by ionic or molecular diffusion, and had re-precipitated in the surface oxidized layers. Van der Weijden *et al.* (1970) found that manganese was enriched in the oxidized upper layers of a number of Atlantic Ocean deep-sea sediment cores relative to the lower more reducing layers. Similar findings have been reported for sediments from the west Pacific (Bezrukov, 1960); the Baltic Sea (Manheim, 1965); the Arctic Sea (Li *et al.*, 1969), and; the Barents and Kara Seas (Brujevicz, 1938).

Li *et al.* (1969) studied the distribution of manganese in both the sediments and the interstitial waters of an Arctic deep-sea core. In the upper layer of the core the conditions were oxidizing, but below a depth of ~ 20 cm the sediments were in a reduced condition. The manganese contents of the sediments decreased with depth, being highest in the upper oxidizing layer. In contrast, the manganese content of the interstitial waters showed a constant decrease with depth down to about one metre from the top of the core. The authors concluded that manganese oxides had been reduced at depth in the sediment, and that the released manganese had migrated upwards through the interstitial water until precipitation occurred in the upper oxidizing layers. Calvert and Price (1972) have also shown that manganese is present in relatively high concentrations in interstitial waters from which manganese oxides are precipitating. In order to explain this type of distribution the authors postulated a model which involved the rapid recycling of manganese between solid and dissolved phases leading to the depletion of the element in sub-surface reduced sediments, and to its enrichment in surface oxidized ores.

The post-depositional mobility of a number of elements has been investigated by Bonatti *et al.* (1971) for a core from the East Pacific; the core had both oxidized and reduced layers. The conclusions of their study may be summarized as follows. (i) Manganese is by far the most mobile of the elements studied, and is enriched by about one order of magnitude in the sediments of the oxidized zone. Nickel and cobalt are also enriched in the oxidized zone, but to a lesser extent than is manganese. (ii) Iron and copper are not significantly mobilized. (iii) Chromium, vanadium and uranium are enriched in the lower, reduced, layer of the core. These elements tend to be insoluble in their lower oxidation states, and are therefore not released into the interstitial waters upon reduction. (iv) Phosphorus and lanthanum are enriched in the upper oxidized layer, but their post-depositional mobility appears to be only indirectly linked to redox reactions. The processes which occur in the interstitial water/sediment complex are reviewed in detail in Chapter 32, and the diagenetic changes which affect marine sediments are discussed in Chapter 30.

Many authors have suggested that certain of the trace elements which are enriched in deep-sea sediments have a mainly submarine volcanic origin (see e.g. Pettersson, 1945; Goldberg and Arrhenius, 1958; Wedepohl, 1960; Arrhenius *et al.*, 1964). However, the importance of volcanically-derived trace elements has given rise to much speculation, and has been the subject of a great deal of controversy. There are three principal ways by which dissolved trace elements can be introduced into sea water from volcanic activity. (i) From the violent leaching of newly extruded lava by sea water; (ii) from the debouching of hydrothermal, possibly mantle-derived, solutions directly onto the sea floor; and, (iii) from the low temperature weathering of volcanic rock which has been exposed to sea water over relatively long periods of time.

Bonatti (1965) has classified submarine volcanic eruptions into two types: a quiet type of effusion in which the interaction between sea water and lava is prevented by the instantaneous formation of a glass crust, and in which pillow structures are often formed, and; a hyaloclastic type of effusion in which there is considerable interaction between sea water and lava which results in the fragmentation of the volcanic material. Bonatti (1965) and Bonatti and Nayudu (1965) have proposed that during hyaloclastic effusions the extruded lava may be altered to material such as palagonite, zeolites and smectites, and that certain elements, e.g. Fe and Mn, may be leached into sea water. Olafsson (1975) has shown that during the eruption of Heimaey (Iceland) in 1973 the concentrations of dissolved Si, Mn, Fe and Zn in sea water increased locally as a result of the leaching of hot lava.

Hart (1970) has considered the chemical effects which may result when basalts are exposed to hydration by sea water at low temperatures over long periods of time, i.e. low temperature weathering (or halmyrolysis). The author investigated the chemical compositions of a series of 112 deep-ocean basalts, and was able to discern systematic chemical trends in their major element concentrations with respect to distance from ridge centres of sea floor spreading. He related these changes to the effects of low temperature weathering, and concluded that for tholeiitic ridge basalts this type of weathering results in the rocks gaining K^+, Fe^{2+}, Mn^{2+}, Na^+ and P from, and losing Si^{4+}, Ca^{2+} and Mg^{2+}, to sea water. In order to assess the importance of the low temperature weathering of basalts in the geochemical budgets of some major elements

Hart estimated their annual contribution per square centimetre of ocean floor from a 1 cm deep layer of basalt, and compared this to the annual dissolved stream supply—see Table 34.26. However, in a later publication Hart (1973) made the assumption that the rate of change of seismic velocity as a function of age represents an integrated value for the rate of alteration of basalts in the

TABLE 34.26

The supply of major elements to sea water from basaltic weathering and rivers[a]

1. Based on the analyses of oceanic basalts with decreasing distance from ridge spreading centres—see text

Element	Annual weathering contribution from 1 cm layer of basalt[b]	Annual river contribution[c]	Weathering depth in cm[d]
SiO_2	+7·4	$+1{\cdot}3 \times 10^5$	17600
TiO_2	−1·6	$+0{\cdot}3 \times 10^2$	18
Fe^{2+}	−3·3	$+0{\cdot}6 \times 10^3$	182
Mn^{2+}	−0·06	$+0{\cdot}7 \times 10^2$	1166
Mg^{2+}	+2·3	$+0{\cdot}4 \times 10^5$	17400
Na	−0·21	$+6{\cdot}3 \times 10^4$	300000
Ca^{2+}	+2·1	$+1{\cdot}5 \times 10^5$	71500
K^+	−0·9	$+2{\cdot}3 \times 10^4$	25500

[a] Data from Hart (1970).
[b] + indicates addition to sea water, − indicates subtraction from sea water. Units; 10^{-9} g cm^{-2} of sea floor.
[c] Data from Turekian (1969). Units; 10^{-9} g cm^{-2} of sea floor.
[d] Indicates the depth of basalt required to be weathered for the weathering contribution from basalt to equal that from rivers.

2. Based on analyses of oceanic basalts with differing seismic velocities—see text[a]

Element	Annual weathering contribution from a 0·66 km layer of basalt[b]	Annual river contribution[c]
SiO_2	+2·20	$+1{\cdot}3 \times 10^5$
Fe^{2+}	+0·45	$+0{\cdot}6 \times 10^3$
Mn^{2+}	+0·006	$+0{\cdot}7 \times 10^2$
Mg^{2+}	+0·92	$+0{\cdot}4 \times 10^5$
Na^+	+0·25(?)	$+6{\cdot}3 \times 10^4$
Ca^{2+}	+1·42	$+1{\cdot}5 \times 10^5$
K^+	−0·45	$+2{\cdot}3 \times 10^4$

[a] Data from Hart (1973).
[b] + indicates addition to sea water, − indicates subtraction from sea water. Units; 10^{-9} g cm^{-2} of sea floor.
[c] Data from Turekian (1969). Units; 10^{-9} g cm^{-2} of sea floor.

upper 2–3 km of the oceanic layer II. He then used this assumption to reassess the effects of low temperature basaltic weathering. He found a reasonably good agreement between this technique and his previous one (based on distance from centres of sea floor spreading) for the amounts of Si^{4+}, Ca^{2+}, and Mg^{2+} which are lost from the basalts during weathering, and for those of K^{+} and H_2O which are gained from sea water. However, there was disagreement between the two techniques for the balances of Fe^{2+}, Mn^{2+}, and Na^{+} and Hart (1973) concluded on the basis of the seismic data that these elements are, in fact, gained by sea water during the low temperature weathering of oceanic basalts—see Table 34.26. From this subsequent study Hart concluded that the upper 2–3 km of the oceanic layer II is a zone of chemical reaction between sea water and the oceanic crust as it moves away from the mid-ocean ridges. He further suggested that the low temperature weathering of basalt could result in the supply of manganese and iron to sea water, and that these elements could be precipitated by oxidizing reactions as metalliferous sediments around ridge crestal areas. Hart (1973) also investigated the retrograde and the greenschist metamorphism of the oceanic layer II in terms of marine trace element budgets. This involves the reaction of sea water with oceanic crust at higher temperatures. It has been suggested by Fox (1972) that, although the upper 400–1000 m of layer II is basalt, the remainder is metabasalt, mainly in the greenschist facies. Hart (1973) has divided this greenschist metamorphism into two stages. The primary stage, which takes place at temperatures of $\sim 300°C$, occurs at ridge crests. During this stage Na and Si are added to the rock (from sea water) and Ca and Fe are lost. The secondary stage, which is probably retrograde, takes place at lower temperatures and occurs progressively with increasing distance from the ridge crests. During this stage the Mg content of the rocks is increased.

Corliss (1971) made a detailed study of the deuteric* alteration of holocrystalline basalt fragments which are slowly cooled samples of submarine extrusions. He concluded that relative to the quenched flow margins, the slowly cooled interior portions of a suite of Mid-Atlantic Ridge basalts are depleted in several elements, e.g. Mn, Fe and Co which are enriched in ferro-manganese nodules and some deep-sea sediments. To explain this elemental distribution Corliss proposed that when basaltic rocks are extruded onto the sea floor the margins of the extrusions are quenched to form pillows which closely approximate in composition to the erupted liquid. However, the interiors of the flows are insulated by the pillow shells which induce slow cooling, and elements which do not readily enter the crystalline phases are fractionated into the residual silica-rich fluids. These fluids occupy accessible sites, e.g. inter-granular boundaries, in the hot solid rock and the elements are

* Alterations which occur in an igneous rock during the later stages of its solidification.

mobilized by dissolution as chloride complexes in sea water which is introduced along contraction cracks formed during cooling and solidification of the lava. Convective flow in the fracture systems may allow these metal-rich hydrothermal solutions to emerge at the sea floor, e.g. as submarine hot springs. Corliss concluded that a significant fraction of some elements, e.g. Fe, Mn and Co, found in pelagic deep sea sediments could have originated in this way, and that in particular, iron could be supplied in excess of that found in ferro-manganese oxide phases. "Excess" iron of this type had previously been identified in Pacific pelagic sediments by Goldberg and Arrhenius (1958), and by Chester and Hughes (1968). However, the most dramatic occurrence of "excess" iron is found in the metalliferous deep-sea sediments from some active centres of sea floor spreading.

The discovery of chemically specialised metal-rich sediments on the crestal portions of some active oceanic ridges has led to much speculation, and has given rise to a great deal of research over the past few years. One reason for the tremendous interest generated by these sediments is that their formation has far reaching geophysical as well as geochemical implications, and is intimately related to the history and evolution of the earth. Another reason is that the metal-rich sediments are a potential source of economically important metals.

The occurrence of metal-rich sediments in the eastern Pacific was reported by Murray and Renard (1891) and by Revelle (1944), and their special characteristics were documented by El Wakeel and Riley (1961) who described a manganese and iron-rich calcareous ooze from the East Pacific Rise. Subsequently, Boström and Peterson (1966) reported examples of metalliferous deposits rich in Fe, Mn, Cu, Cr, Ni and Pb from areas of high heat flow on the crest of this rise, and tentatively suggested that they had a hydrothermal origin. Following their initial work Boström and Peterson (1969) carried out a more detailed study of metalliferous sediments from the region of the East Pacific Rise, and were able to distinguish between those deposited on the crestal areas of the ridge and those deposited on the flanks. Their results revealed that the concentrations of Fe, Mn, As, Cr, V, Pb, Cd and Hg are highest in the crestal sediments, whereas those of Co, Ni, Cu, Mo and Zn are highest in the ridge flank sediments (see Table 34.27). Sediments from both areas are considerably depleted in Al, Ti and Si relative to "normal" Pacific pelagic clays, and on the basis of their overall elemental chemistry the authors used the ratio Al/(Al + Fe + Mn) to characterize metalliferous deep-sea sediments (see also Boström *et al.*, 1969 and Fig. 34.1). Boström and Peterson (1969) rejected both biogenous and lithogenous origins for the amorphous colloidal-sized brown metal-rich precipitates in the sediments and proposed, instead, a theory which invoked volcanic manifestations, including both magmatic and deep-seated events, as the principal source of the metals which

TABLE 34.27

Concentrations of elements in metalliferous deep-sea sediments from the East Pacific Rise (ppm on a carbonate-free basis)

Trace element	Metalliferous sediments; crest of East Pacific Rise*	Metalliferous sediments; flanks of East Pacific Rise*	Average Pacific deep sea clay†
Fe	180000	105000	65000
Mn	60000	30000	12500
Cu	730	960	570
Ni	430	675	293
Co	105	230	116
Zn	380	290	—
V	450	240	130
As	145	65	—
Cd	4	1	—
Mo	30	113	18
Cr	55	32	77
Si	61000	140000	263000
Al	5000	46300	83000
Ti	200	2350	3400

* Data from Boström and Peterson (1969).

† Data from Goldberg and Arrhenius (1958), El Wakeel and Riley (1961), Landergren (1964) and Cronan (1969).

are enriched in the sediments. They suggested that, in areas of high heat flow associated with sea floor spreading, mineralizing solutions composed of excess volatiles derived from the degassing of the upper mantle ascend along tectonically weakened zones, fissures and volcanic vents (and possibly also by migration through the sediments), and are debouched and emplaced along the crest of the rise. The energy for this process is probably derived from a local heat source associated with the rise; perhaps a recently emplaced magma. The debouched mineralizing solutions probably originate in basaltic, or ultra-basic, magmas, and are apparently relatively rich in Fe, Mn, As, Cr, Pb, Cd and possibly also in Co, Ni, Cu, Mo and Zn relative to sea water. According to Boström and Peterson (1969) the distribution of elements between the crestal and the flank deposits is related to differences in elemental residence times in the metal-rich emanation/sea water complex and does not necessarily reflect absolute elemental concentrations in the solutions. Because of differences in their residence times various elements are precipitated at different times; those with the shortest residence times being deposited adjacent to, or actually in, the upper parts of the escape fissures.

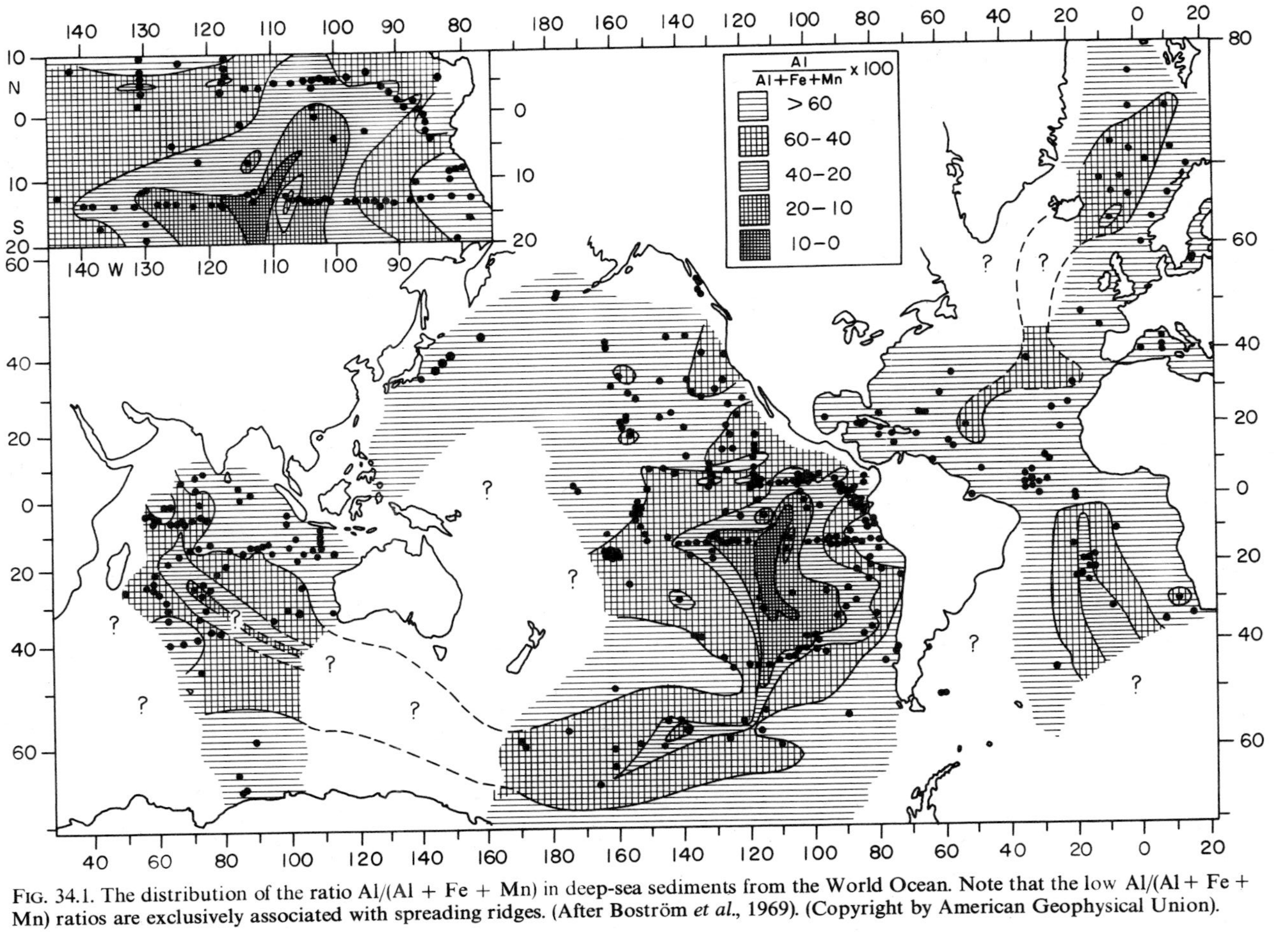

FIG. 34.1. The distribution of the ratio Al/(Al + Fe + Mn) in deep-sea sediments from the World Ocean. Note that the low Al/(Al + Fe + Mn) ratios are exclusively associated with spreading ridges. (After Boström *et al.*, 1969). (Copyright by American Geophysical Union).

On the basis of the evidence they obtained from the East Pacific Rise metalliferous sediments, Boström and Peterson (1969) suggested that the following sequence of events had occurred during the formation of the deposits. The mineralizing solutions initially debouched at the crest of the rise where they mixed with sea water. The resulting precipitates formed a sedimentary layer rich in Mn, Fe, V etc. (i.e. elements which have relatively short residence times in the mineralizing emanation/sea water complex) and poor in Al, Si, and Ti. With increasing distance from the rise crest the composition of the metal-rich complex changed, and the precipitates formed on the flanks of the ridge gave rise to sediments rich in Ni, Co, Cu etc. which became increasingly diluted with "normal" pelagic sediment components. At even greater distances from the rise crest "normal" pelagic sediments became the predominant deposit, and the precipitation of metal-rich phases eventually became zero. The basic elements of this theory are illustrated in Fig. 34.2.

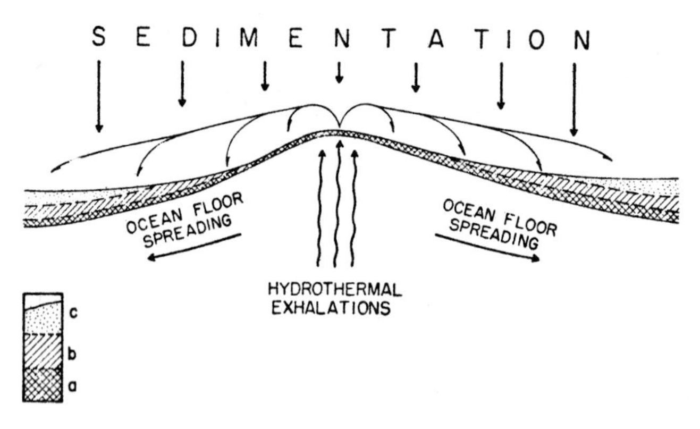

FIG. 34.2. Schematic representation of sedimentation processes on the East Pacific Rise. The formation of ferromanganoan sediments is particularly enhanced on the west of the Rise where the mineralizing solutions debouch, resulting in the formation of layer (a) which is rich in iron, manganese, vanadium etc. and poor in aluminium, silicon and titanium. With increased distance from the crest, the mineralizing solutions are changed in composition due to variations in the residence times of the various trace elements (see text); the resulting sedimentary layer (b) also becomes increasingly diluted with the products of normal pelagic sedimentation. At a greater distance from the Rise "normal" pelagic sediments (layer (c)) are formed. The upper set of arrows indicates precipitation of material from deep-water admixed with mineralizing solutions. (Boström and Peterson, 1969).

Boström and Peterson (1969) postulated their theory for the origin of metalliferous deep-sea sediments in terms of the East Pacific Rise. However, they then considered the implications of the theory within the much wider framework of the hypothesis of sea floor speading, and suggested that similar metal-rich, or ferro-manganoan, sediments may occur elsewhere in the Pacific Ocean, although they may be buried beneath a blanket of "normal" pelagic sediments. They further suggested that, in fact, the first type of sediment deposited on any newly formed crust on active oceanic ridges may be an Al-poor ferro-manganoan analogue of the East Pacific Rise deposit. This prediction was subsequently shown to be largely correct, and metalliferous deep-sea sediments have been identified at various other present day active ocean ridge sites in, for example, the Pacific Ocean—Boström *et al.* (1969); the Indian Ocean—Boström *et al.* (1969); Boström and Fischer (1971); the Atlantic Ocean—Boström *et al.* (1969); Cronan (1972). The theory outlined by Boström and Peterson (1969) also predicted that metalliferous sediments should be found at considerable depth in the sedimentary column in association with ancient centres of sea floor spreading. It was fortunate that the D.S.D.P., which was initiated in 1968, allowed the entire deep-sea sediment column to be sampled. As a result of this programme, metalliferous deep-sea deposits have been identified at the base of many drilling sites where the sedimentary column overlies basaltic basement in for example, the Pacific Ocean—von der Bosh and Rex (1970), von der Bosh *et al.* (1971); Cronan *et al.* (1972); Cronan (1973); the Atlantic Ocean—Peterson *et al.* (1970); Boström *et al.* (1969).

Both the present day and the ancient metalliferous deep-sea sediments share a number of common geological, mineralogical and chemical characteristics, some of which are listed below.

(i) Their formation appears to be associated with some form of volcanic activity, usually (past or present) active centres of sea floor spreading, and they are often found directly overlying basaltic basement.

(ii) They are relatively poor in detrital components, and usually contain abundant colloidal-sized yellow, brown or dark red spherical to sub-spherical, largely amorphous, grains of iron oxides. These oxide grains are often associated with brown to black micro-ferro-manganese nodules, fragments of volcanic glass, palagonite, smectites, zeolites, and calcareous and siliceous shell debris and phosphatic fish remains.

(iii) They have enhanced concentrations of some, or all, of the following elements, Fe, Mn, Cu, Cr, Co, Ni, Pb, As, Cd etc. and are depleted in Al, Si, and Ti, relative to "normal" pelagic clays.

The elemental analyses of a number of metalliferous sediments, both recent and ancient, are given in Table 34.28. It can be seen from this table that there are differences in the concentrations of the elements in metalliferous sediments from different regions. Compositional variations between sediments from different active oceanic ridges have been considered by Horowitz (1970), who attempted to relate them to differences in the degree of activity on the ridges themselves. From evidence based on trace element concentrations in the

TABLE 34.28

Concentrations of trace elements in some metalliferous deep-sea sediments (ppm on a carbonate-free basis)

Trace element	Crest of East Pacific Rise[a]	Flanks of East Pacific Rise[b]	Flanks of East Pacific Rise[c]	Central North Pacific[d]	Bauer Deep Central Pacific[e]	Mid-Atlantic Ridge[f]
Fe	180000	105000	175000	236000	141000	79600
Mn	60000	30000	45000	66600	46000	4100
Cu	730	960	917	1070	910	—
Ni	430	675	535	630	820	—
Co	105	230	83	86	67	—
Pb	—	—	145	—		—
Zn	380	290	358	600	330	—
V	450	240	—	—	—	—
Hg	—	—	—	—	—	414
As	145	65	—	—	—	174
Cd	4	1	—	—	—	—
Mo	30	113	—	—	—	—
Cr	55	32	—	16	13	—
Al	5000	46300	—	23900	23100	57900

[a] Average of nine near-surface deep-sea sediments; data from Boström and Peterson (1969).
[b] Average of twelve near-surface deep-sea sediments; data from Boström and Peterson (1969).
[c] Average of thirty-four ancient deep-sea sediments; data from Cronan (1943).
[d] Average of six ancient deep-sea sediments; data from Dymond *et al.* (1973).
[e] Average of seven near-surface deep-sea sediments; data from Dymond *et al.* (1975).
[f] Average of sixteen near-surface deep-sea sediments; data from Cronan (1972).

overlying sediments he concluded that the East Pacific-Antarctic Ridge is the most active, the Indian Ocean Ridge is intermediate in activity, and the Mid-Atlantic Ridge is the least active.

Following the initial investigations by Boström and Peterson (1965, 1969) various authors have attempted to define the origin of the elements which are enhanced in metalliferous deep-sea sediments. Boström and Fisher (1969) showed that mercury is enriched in the crestal sediments of the East Pacific Rise, and suggested that it originated either from the degassing of the mantle associated with crestal volcanic activity, or was adsorbed from sea water onto the colloidal precipitates which are a common constituent of the crestal sediments. However, Klein and Goldberg (1970) suggested that the enhanced mercury concentrations in the East Pacific Rise sediments could be explained by the uptake of this element by organisms in the overlying productive waters and its subsequent transport to the sea floor. Boström and Fisher (1971) showed that uranium (among other elements) is concentrated in the sediments from active ocean ridge areas in the Indian Ocean, and concluded that submarine vulcanism is an important source of this element. In contrast, Turekian and Bertine (1971) suggested that the relatively high concentrations of U and Mo in deposits on both active and inactive ridges may result, in part, from the presence of anaerobic sediments in local ephemeral basins along the ridges, and, in part, from the relatively high rate of supply of organic material to the shallow crestal regions.

It would appear from these studies that not all of the elements which are enhanced in metal-rich ridge sediments necessarily have a common source. In this context Dymond *et al.* (1973) have provided significant evidence that the various elements in a series of recent and ancient metalliferous sediments from the eastern Pacific do not have a common origin. The conclusions reached by these authors are summarized below. (i) The lanthanide elements, Sr, U and S have been incorporated into the sediments from sea water. (ii) The elements Fe, Mn, Si and Pb have probably been derived from solutions of a magmatic origin. (iii) The sources of As, V, B, Ti, Cd, Ca, Ni, and Zn could not be established. (iv) Oxygen isotopic data indicated that the sediments are in isotopic equilibrium with sea water, and that they are composed of varying proportions of two end-member phases which have different oxygen isotopic compositions, i.e. an iron-manganese hydroxyoxide and an iron-rich montmorillonite. Cronan and Garrett (1973) investigated the partition of Fe, Mn, Cu, Ni, Co and Zn among the components of a series of ancient basal metalliferous Pacific deep-sea sediments. They concluded that the bulk of the iron and the manganese occur in different sediment phases from one another, and so could have different origins; a large fraction of the iron (together with most of the Cu and Zn) is probably derived from hydrothermal sources, whereas much of the manganese (and Ni and Co) originates from sea water. However, several authors (see e.g. Bender *et al.*, 1971; Dasch *et al.*, 1971) have shown that manganese has accumulated faster in the East Pacific Rise metalliferous sediments than it has in "normal" Pacific pelagic sediments, which may indicate a local source for this element.

Clearly, there is still a good deal of uncertainty about the sources and mechanisms of incorporation of many of the elements found in metalliferous sediments, and it is unlikely that all of them are volcanically derived. There is also uncertainty about the origins of those elements which are generally agreed to have a volcanic source. In their hypothesis Boström and Peterson (1969) proposed that the East Pacific Rise metalliferous sediments were precipitated from hydrothermal solutions derived from the mantle. However, Corliss (1971) suggested that the metal-bearing solutions may have originated by the deuteric alternation of basalt, and Hart (1970, 1973) concluded that the low temperature weathering of basalt could supply manganese and iron to sea water, and that these elements could be precipitated as metalliferous sediments around ridge crests (see above). Further, Piper (1973) proposed that the Fe, Mn, Co and Cr in crestal sediments from the East Pacific Rise were precipitated from hydrothermal solutions which were formed by the reaction of sea water with newly upwelled basalt.

Boström and Peterson (1969) suggested that metalliferous deposits are the first type of sediment formed on new crust on active ocean ridges, and it was mentioned above that some D.S.D.P. cores encountered such deposits overlying basaltic basement. However, metalliferous sediments are not present at the base of the sedimentary column in all oceanic areas, e.g. they were absent from drilling sites in the Western Pacific (Winterer *et al.*, 1971). Nonetheless, there is considerable evidence that metalliferous sediments are associated primarily with past or present active centres of sea floor spreading; they are not found on inactive ridges, e.g. the Rio Grande Ridge in the Atlantic Ocean (Boström *et al.*, 1972). However, the concept that this type of open-ocean metalliferous sediment is actually restricted to active sites of sea floor spreading has been questioned by Natland (1973) who reported the occurrence of ferro-manganoan sediments above seamount-related volcanics at two D.S.D.P. sites in the northwest Pacific Ocean. Basal sediments occurring above alkali basalt at D.S.D.P. Site 183 (Aleutian Abyssal Plain) include iron-rich clays, a goethite-bearing calcareous limestone and a pyrite-bearing non-fossiliferous aragonitic limestone—see Table 34.29. This diversity of facies is similar to that found in the Red Sea sediments in deeps filled with hot metal-rich brines. These Red Sea metalliferous deposits have been described in the text edited by Degens and Ross (1969), and are thought to have originated in axial deeps by the discharge of geothermal solutions. The geothermal solutions, which are distinct from hydrothermal ones, are formed by thermal, but not chemical, transfer from a geothermal heat source. In the Red

TABLE 34.29

*Concentrations of elements in metalliferous deep-sea sediments from the northwest Pacific. (ppm on a carbonate-free basis)**

		Site 183			Site 192 (Hole 192A)	
Trace element	Aragonitic limestone	Goethite-limestone	Brown clay	Dark silty clay	Red sediment at volcanic contact	Iron oxide from basalt crack
Fe	170200	238000	73000	44000	61000	—
Mn	59900	41600	12800	936	5460	4380
Cu	870	1550	560	143	38	317
Zn	437	2	86	118	541	197
Ni	795	257	135	33	73	—
Co	167	44	35	28	69	—
Cr	33	15	15	20	12	—
Pb	181	38	34	7	4	—

* Data from Natland (1973).

Sea a convective flow is induced by the heat source and the sea water sinks, penetrating into the thick evaporite beds which underlie the permeable surface sediments. Dissolution of salt from the evaporites increases the salinity of the water and the high geothermal heat flux raises its temperature. The high density of the brine, relative to normal sea water, causes it to sink and flow northwards, to emerge ~1000 km further north at a depth of ~2000 m in the brine filled deeps in the central rift of the Red Sea. The brine becomes mineralized by extracting metals from the sediments with which it comes into contact and these metals may subsequently be precipitated as metalliferous deposits either by reaction with sea water, or by cooling (see Chapter 27). Natland (1973) concluded that the limestone of the basal sediments of D.S.D.P. Site 183 had precipitated from a near-bottom metal-rich hot saline brine. Although brines of this nature have been found in the Red Sea, which has special geological and climatical conditions, they have not been reported in the Pacific Ocean, and Natland (1973) concluded that in the open-ocean the only probable source of heat and trace metals is a volcanic one. He further suggested that the hot brines could only have existed spasmodically, and that the formation of the ferro-manganoan sediments may have been favoured by the low sedimentation rate of the other sediment components.

Natland (1973) also reported that at D.S.D.P. Site 192 (Meiji Guyot) five metres of iron and manganese-rich clays interlayered and diluted with chalk lie on extensive alkali basalt pillow lavas—see Table 34.29. He concluded from both field and textural relationships of the volcanics, and from the presence of iron oxides (with relatively high concentrations of Mn, Cu, and Zn) in crack fillings deep in the volcanics, that the metals were emplaced in the sediments diagenetically after they had been leached from the volcanics by percolating fluids during alteration and weathering. The percolating fluids are thought to have been at a low temperature by the time they reached the surface of the volcanics.

There are two very important geochemical implications which can be drawn from Natland's work. (i) At both of the D.S.D.P. sites the ferro-manganoan basal sediments occur above seamount-related volcanics, and are not associated with oceanic rise hydrothermal activity. (ii) Hot brines can probably exist in the open-ocean. However, although such brines may have been involved in the formation of the basal sediments at Site 183, the sediments at Site 192 provide evidence that they need not necessarily be invoked to explain the genesis of all ferro-manganoan basal deposits. Clearly, therefore, it is extremely dangerous in our present state of knowledge to correlate all metalliferous deep-sea sediments with hydrothermal processes associated exclusively with active ridge centres of sea floor spreading. Nonetheless, Natland's study does provide further evidence of the association of the metalliferous sediments with volcanic processes of some kind.

The detailed investigation of metalliferous deep-sea sediments is in its relative infancy. To date, a number of theories have been advanced to explain the origin of these chemically specialized deposits, and these are summarized below. (i) Precipitation from mantle-derived hydrothermal metal-rich mineralizing solutions which are debouched at active centres of sea floor spreading (see e.g. Boström and Peterson, 1969). (ii) Precipitation from metal-bearing solutions derived from the deuteric alteration of basalt (see e.g. Corliss, 1971), from the low temperature weathering of basalt (see e.g. Hart, 1970, 1973), or from the leaching of newly formed basalt (see e.g. Piper, 1973). (iii) Precipitation from open-ocean hot brines which are not necessarily associated with active centres of sea floor spreading. (iv) Precipitation from metal-rich volcanically derived solutions by bacterial activity (see e.g. von der Borsch *et al.*, 1971). (v) Deposition under reducing conditions (see e.g. Turekian and Bertine, 1971). (vi) The authigenic precipitation of iron and manganese from sea water (see e.g. Bender *et al.*, 1971). (vii) The differential transport, and open-ocean deposition, of fine-grained lithogenous material rich in trace elements (see e.g. Turekian, 1965; and subsequent sections of this chapter).

It may be concluded that there is very strong evidence for the association of metalliferous deep-sea sediments with some form of volcanic activity, particularly that occurring on active centres of sea floor spreading. However, the nature of the volcanic manifestation may vary from one area to another. Further, it is unlikely that all of the elements which are concentrated in the metalliferous sediments have a common source.

34.2.3.4. *The mechanisms by which trace elements are incorporated into deep-sea sediments.*

Some trace elements are incorporated into deep-sea sediments in association with pre-existing solid material, i.e. deposited solids which are not produced *in situ* in the sea water/sediment/interstitial water complex. Other trace elements may be incorporated into the sediments directly from sea water, or from interstitial water.

Trace elements which are associated with deposited solids may have several origins, some of which have been discussed above.

Lithogenous solids, i.e. those produced by the weathering of the earth's crust, are transported to deep-sea areas by river, ice and wind agencies, and their trace element geochemistry has been discussed in Section 34.2.3.3.1. There are two genetically different classes of trace element associated with this lithogenous material. (i) Those in "lattice-held" positions, which have a continental origin. (ii) Those in "non-lattice-held" positions, which may have been removed from the overlying waters in the river, estuarine or marine environments.

Cosmogenous solids, i.e. those having an extra-terrestrial origin, reach deep-sea areas initially via the atmosphere.

Hydrogenous solids are those produced, either partially or totally, from components dissolved in sea water by inorganic reactions. These reactions include: primary precipitation, e.g. the formation of carbonates, phosphates, sulphates, oxides etc.; co-precipitation and adsorption; ion-exchange; and, the submarine alteration of pre-existing deposited solids, which results in the formation of secondary minerals, e.g. montmorillonite and the zeolites, during which trace elements can be incorporated into both "lattice-held" and "non-lattice-held" positions.

Both quantitatively and geochemically the most important of the hydrogenous primary precipitates in deep-sea sediments are the oxides and hydroxides of manganese and iron, and ferro-manganese nodules. The nodules, which have been fully described in Chapter 28, contain relatively very high concentrations of those trace elements, e.g. Mn, Fe, Ni, Co, Cu, Zn, Pb, which are enriched in deep-sea sediments (see Table 34.7). Macro-nodules, i.e. those with diameters of centimetres of more, have their highest concentrations on the sediment surface, but micro-nodules, i.e. those with diameters of millimetres (or less), may be dispersed throughout the sediment column. They are particularly common in Pacific pelagic deep-sea clays, and their presence is probably largely responsible for the enhanced trace element concentrations in these sediments. The importance of the influence exerted by ferro-manganese nodules on the geochemistry of deep-sea clays has been stressed by many authors, and it has been suggested that much of the Ni, Co, Pb, Cu and Zn which is concentrated in pelagic clays is present within micro-nodules (see e.g. Goldberg, 1954; Goldberg and Arrhenius, 1958; El Wakeel and Riley, 1961; Landergren, 1964; Chester and Hughes, 1969).

One of the most important occurrences of iron oxides in the marine environment is in metalliferous deep-sea sediments (see Section 34.2.3.3.2). Iron oxides are also found in many other types of deep-sea sediments. Corliss (1971) concluded that metal-bearing hydrothermal solutions could supply a considerable excess of iron over that incorporated into the ferro-manganese nodule components of deep-sea sediments. "Excess" iron of this nature had previously been identified in Pacific pelagic clays by Goldberg and Arrhenius (1958), who concluded that it had originated as "free" colloidal iron which was subsequently converted to goethite in the sediment. The existence of this "excess" iron was also confirmed by Chester and Hughes (1969). The latter authors investigated the partitioning of a series of trace elements between the "lithogenous", the nodular-associated "hydrogenous", and the non-nodular-associated "hydrogenous" fractions of a North Pacific deep-sea clay which contained an abundance of micro-ferro-manganese nodules—see Table 34.30. It can be seen from this table that non-nodular associated iron constitutes ~25% of the total hydrogenous iron, and that it contains a considerable

TABLE 34.30

*The partition of trace elements among some fractions of a North Pacific clay core**

Element	Total trace element	Lithogenous fraction		Nodular-associated hydrogenous fraction		Non-nodular associated hydrogenous fraction	
	ppm	ppm	%	ppm	%	ppm	%
Fe_2O_3	7·68(%)	7·40(%)	96	0·20(%)	3	0·08(%)	1
MnO	0·66(%)	0·10(%)	15	0·52(%)	79	0·04(%)	6
Cu	314	213	67	45	14	56*	18
Ni	191	46	24	88	46	59*	30
Co	88	15	17	71	80	2	2
Pb	28	23	82	4	14	~1	4
Cr	97	92	95	3	3	2	2
V	141	119	84	8	6	1 *	10

* Data from Chester and Hughes (1969).

fraction of the hydrogenous Cu, Ni, Cr and V. Mn crusts have been discussed in Chapter 28.

It may be concluded, therefore, that iron oxides, manganese oxides and micro-ferro-manganese nodules are important agents for the concentration of some of the trace elements found in certain deep-sea clays.

Barite ($BaSO_4$) has a widespread occurrence in deep-sea sediments in which it is often associated with calcareous and siliceous tests and ferro-manganese oxides. Some of this barite is a primary precipitate of hydrogenous origin, although most of it is probably formed by some kind of biological process. Arrhenius and Bonatti (1965) have discussed the detailed distribution of barite in Pacific Ocean deep-sea sediments. They showed that the highest concentrations of barite are found in deposits from the area of the East Pacific Rise, in which it constitutes between ~5% and ~10% of the sediments (on a carbonate-free basis). The authors concluded that barium was enriched in the waters as a result of volcanic activity associated with the Rise, and was precipitated inorganically as euhedral crystals. Sediments of the Equatorial Zone have barite concentrations between ~1% and ~5% (on a carbonate free basis), and Arrhenius and Bonatti (1965) have concluded that this barium had been removed from sea water by organisms and was released as their remains were oxidized, either during descent through the water column or at the sediment surface. The barite was then incorporated into the deposits as granules. In Atlantic Ocean deep-sea sediments, barium is concentrated in sediments underlying areas of high biological productivity, e.g. off the coast of South West Africa, and on topographical highs, such as the flanks of the Mid-Ocean Ridge System. According to Turekian (1968) the barium is principally located within barite which has been precipitated biologically. Boström *et al.* (1973) have examined the distribution of barium in deep-sea sediments from the World Ocean and have demonstrated that it has both its highest concentrations, and its highest accumulation rates, on ridges which are the centres of sea floor spreading, and also in some areas underlying waters having a high primary production. The authors concluded that barium has been deposited on the deep-sea floor by both inorganic precipitation associated with volcanic activity, and by biological processes. Cronan (1974) has reviewed the occurrence of barite in ancient deep-sea sediments samples collected during the D.S.D.P., and has reported that this mineral is probably at least as abundant down the length of the sediment column as it is at the present day sediment surface.

Francolite is the only hydrogenous phosphate which has a widespread occurrence in deep-sea sediments, and it has its highest concentrations on seamounts and other topographical highs. Quantitatively, hydrogenous silicates and carbonates are not important constituents of most present day surface deep-sea sediments. However, lithified carbonates (see e.g. Thompson *et al.*, 1968; Chapter 27) and cherts (see e.g. Calvert, 1971) are important deposits which may be formed during sediment diagenesis.

Ion-exchange reactions are important for the incorporation of major elements such as Na, K, Mg and Ca into inter-sheet positions in the clay minerals, and may exert a control on the concentrations of these elements in sea water. Ion-exchange processes are reviewed in Chapter 27.

Co-precipitation, adsorption and ion-exchange reactions exert fundamental controls on the geochemistry of certain deep-sea sediments. For example, co-precipitation with, and adsorption onto, iron, manganese and ferro-manganese oxides can result in the incorporation of elements such as Cu, Ni, Co, Pb, Zn etc. into deep-sea sediments (see above). The adsorption of trace elements onto the surfaces of other sedimentary components, e.g. the clay minerals, organic material etc., is not well understood. Krauskopf (1956) made a classic study of the processes which control the concentrations of some thirteen trace elements in sea water. Although his experimental approach can be criticized on a number of grounds (see Chapter 7) his work still remains one of the most comprehensive studies of the factors controlling the distribution of trace elements in the marine environment. From the experiments dealing with adsorption, Krauskopf (1956) concluded that this process was a possible control on the concentrations of Zn, Cu, Pb, Bi, Cd, Hg, Ag and Mo, but not on those of V, Co, Ni, W and Cr. Chester (1965) demonstrated experimentally that the clay mineral illite can remove Co and Zn from sea water, and Aston *et al.* (1972a) showed that aeolian dust is capable of at least the short term uptake of Co (and by analogy, other trace elements) from the upper few metres of sea water. Other examples of the adsorption of trace elements, including radio-nuclides, from sea water have been described in Chapters 7 and 27. However, the geochemical interpretation of laboratory adsorption-desorption experiments rests heavily on the experimental techniques employed. There is no doubt that adsorbents, such as the clay minerals, can remove some trace elements from sea water when they are introduced directly into it, but this situation does not necessarily reflect that found in nature. For example, Kharkar *et al.* (1968) experimentally studied the adsorption-desorption reactions of a series of trace elements by suspended material in a simulated river–estuarine–marine system. They concluded that when a trace element is adsorbed from solution in river water by clay mineral material it is always released, at least to some extent, on contact with waters of increasing salinity. Although they agreed with the findings of Krauskopf (1965) that suspended material placed directly in sea water can absorb trace elements, they pointed out that this does not happen in nature, and that when the particles pass through waters of varying salinity river-transported solids actually add some trace elements to sea water by desorptive processes.

Chester (1972b) considered this problem and concluded that one of the critical factors which evidently controls the uptake of trace elements from sea water by suspended solids is the pathway by which the solids are transported to the oceans. He made a distinction between two types of genetically different particles brought to the oceans from the land areas. (i) Those transported by river run-off, which are introduced into sea water in a "wet" state and which pass through an estuarine environment. Subsequently, these particles may be deposited initially in the shelf regions, or be transported to deep-sea areas by surface and intermediate currents. (ii) Particles which are brought directly to deep-sea areas in a "dry" state by aeolian transport, and which fall-out into the euphotic zone where they are in contact with the marine biomass.

Quantitatively, the two most important groups of minerals formed by the submarine alteration of deposited solids are the hydrogenous smectites and the zeolites. The principal smectites, i.e. expanding lattice clays, found in deep-sea sediments are montmorillonite and nontronite, both of which may be formed by the *in situ* alteration of volcanic glass; their genesis has been discussed in Chapter 27. The zeolites are a group of hydrous sodium calcium alumino-silicates which usually contain exchangeable cations such as Na^+, K^+, Ca^{2+}, and Ba^{2+}. The origin of the zeolites, some of which are probably formed from volcanic debris, is not fully understood, and has been discussed by Cronan (1975) and in Chapter 27. Both the zeolites and the hydrogenous smectites have their highest concentrations in present day surface sediments in areas of slowly depositing deep-sea clays, e.g. in the South Pacific, where they occur in association with volcanic material and micro-ferro-manganese nodules. These clays are some of the most trace element-rich in the World Ocean—see Table 34.31. The formation of the smectites, the zeolites and the micro-nodules are all favoured by the relatively slow

TABLE 34.31
Trace elements in Atlantic, North Pacific and South Pacific deep-sea clays (ppm)

Trace element	Atlantic deep-sea clay*	North Pacific deep-sea clay†	South Pacific deep-sea clay†
Mn	4000	5465	20000
Fe	82000	52294	73500
Cu	130	531	672
Ni	79	212	380
Co	38	80	207
Ga	21	21	15
Cr	86	166	68
V	140	326	504

* Data from Wedepohl (1960), and Turekian and Imbrie (1966).
† Data from Goldberg and Arrhenius (1958).

TABLE 34.32

*Trace elements in the total sediment and "lattice-held" fractions of some montmorillonite-rich and zeolite-rich Atlantic deep-sea clays** (*values in ppm*)

Sediment	Accumulation rate;† cm/10^3 yr	Mn		Ni		Co		Cu		Zn		Cr	
		Total	"Lattice-held"	Total	"Lattice-held"	Total	"Lattice-held"	Total	"Lattice-held"	Total	"Lattice-held"	Total	"Lattice-held"
Quaternary calcareous clay	4·3	3172	351	112	79	17	18	77	26	168	100	120	153
Pliocene/Miocene montmorillonite-rich clays	3·2	1232	196	97	80	23	17	66	31	139	104	131	163
Middle Eocene and Upper Cretaceous Zeolitic clays	1·2	5397	856	148	182	70	56	159	167	180	152	80	158

* Data from Bruty *et al.* (1973) and Chester *et al.* (1975) for a core from the eastern flank of the Bermuda Rise; D.S.D.P., Leg 2, Site 9.

† Data from Peterson *et al.* (1970).

rates of accumulation of the South Pacific clays. However, the enhanced trace element concentrations in the sediments are probably a result of the presence of the micro-nodules, rather than that of either the smectites or the zeolites.

An association between the occurrence of zeolites and the enhanced trace element concentrations of the host deposits has also been shown by Bruty *et al.* (1973) for an ancient North Atlantic deep-sea sediment core (D.S.D.P. Leg 2, Site 9; Bermuda Rise) which consists of a sequence of calcareous, montmorillonite-rich and clinoptilite-rich clays—see Table 34.32. Subsequently, Chester *et al.* (1975) studied the partition of the trace elements among the components of this core and showed that those elements, i.e. Mn, Cu, Ni, Co and Zn, which are enhanced in the zeolitic clays are also enhanced in the "lattice-held" fractions of these clays (see Table 34.32). However, the authors concluded that the enhanced trace elements in the zeolitic clays had initially been incorporated into the sediments from the overlying sea water at the time of deposition in association with iron and manganese oxides. Subsequently, the trace elements had been incorporated into the lattices of the zeolites diagenetically, probably at depth in the sediment. That is, although the formation of zeolites and the incorporation of certain trace elements into the components of surface deep-sea sediments may both be related to accumulation rates, they are probably independent processes. In the same core there was no apparent concentration of trace elements into the montmorillonite-rich clays.

Biogenous components of deep-sea sediments are produced in the biosphere and include inorganic shell material (which is largely deep-sea in origin) and organic matter, which is also principally of a deep-sea origin but which may include components formed in near-shore or continental waters.

Biogenous processes involving living organisms are extremely important in the removal of some trace elements from sea water. According to Riley and Chester (1971) elements may be associated with organisms in a number of ways. These include: (i) by incorporation into hard skeletal parts; (ii) by incorporation into soft body parts; and, (iii) by association with body processes, e.g. digestive and excretory functions, which may result in the organism retaining the elements during its life or in it expelling them, e.g. as faecal pellets. In addition to the processes which involve living organisms, particulate organic compounds may remove trace elements from sea water by adsorption etc.

The principal skeletal materials in deep-sea sediments are calcium carbonate, opal and apatite. Quantitatively, calcium carbonate is by far the most important of these, and calcareous oozes cover $>50\%$ of the present day deep-sea floor. A number of workers have investigated the trace element content of the calcareous tests of marine organisms. For example, Nicholls *et al* (1959)

TABLE 34.33

Concentrations of trace elements in marine organisms (ppm dry wt.) and in some marine sediments (ppm)

	Phytoplankton Monteray Bay California[a]	Zooplankton Monteray Bay, California[a]	Zooplankton; North Pacific[a]	Microplankton; North Pacific[a]	Zooplankton; surface, North Atlantic[b]	Average; marine organisms[a,b]	Average; near-shore muds[c]	Average; deep-sea clay[d]
Fe	49–3120	54–1070	90–1720	1030–4000	567–1467	862	69000	65000
Mn	2·1–30	2·2–12	2·9–7·1	3·4–32	N.D.–23	9·3	850	6700
Cu	1·3–45	4·4–23	6·2–58	40–104	10–90	27	48	250
Ni	<0·5–13	<0·5–12	5–13	11–12	15–77	17	55	225
Co	<1	<1	<1	<1	8–20	<1	13	74
Cr	<1–21	<1	<1	<1–3·7	—	<1	100	90
V	<3	<3	<3	<3	—	<3	130	120
Ba	5–500	4–257	10–97	51–70	—	60	750	2300
Sr	53–3934	83–810	380–3000	6800–9650	57–520	862	<250	180
Pb	<1–47	<1–12	22–14	17–39	N.D.–123	20	20	80
Zn	3–703	53–279	60–750	285–4190	120–400	257	95	165
Cd	0·4–6	0·8–10	1·9–3·5	1·0–2·2	2–9	4·6	—	0·21
Hg	0·10–0·59	0·07–0·16	0·04–0·45	0·11–0·53	—	0·16	—	0·32
Al	7–2850	<8–313	9–31	72–108	—	159	80000	84000

[a] Data from Martin and Knauer (1973).
[b] Data from Martin (1970).
[c] Data from Wedepohl (1960).
[d] Data from Turekian and Wedepohl (1961), Aston *et al.* (1972a, b).

analysed a number of species of marine zooplankton and suggested that the sedimentation of pteropod shells could be a major pathway for the removal of vanadium and lead from sea water. Turekian *et al.* (1973) concluded that the calcium carbonate phase of pteropod shells is essentially pure, and that trace elements associated with the shells had been scavenged from sea water by the iron hydroxides which often coat the material of the tests. Some analyses of the calcareous tests of marine organisms and of calcareous deep-sea sediments are listed in Table 34.34a.

Results given in the literature show that some trace elements are enhanced in marine plankton by several orders of magnitude relative to sea water, and these organisms are therefore important in the cycling of the elements particularly in the upper layers of the World Ocean (see also Chapter 7). However, the collection and analysis of marine organisms presents a number of problems. According to Martin and Knauer (1973) these include: (i) the difficulty of collecting sufficient material for analysis because of the small size of the organisms; (ii) the difficulty of separating phytoplankton from zooplankton; (iii) contamination of the samples by natural inorganic solids; and (iv) contamination by rust, paint chips etc. from research vessels. There are very considerable variations in the published trace element concentrations of marine plankton, at least some of which probably result from the difficulties of collection listed above. Further, it has been pointed out that in addition to inter-species differences in trace element contents, variations in the trace element content of any single organism can be caused by factors such as the composition of the surrounding water, the principal food source and the age and recent history of the specimen (Chapter 7). For these reasons a range of values, as well as an average, is given in the compilation of trace element concentrations of marine plankton listed in Table 34.33. Despite the analytical uncertainties a number of overall trends in the trace element compositions of marine organisms can be established. (i) Relative to near-shore muds the organisms contain one or two orders of magnitude less Mn, Fe, Co, Cr, V and Ba, the same order of magnitude of Ca, Ni and Pb, and one order of magnitude more Zn. (ii) Relative to deep-sea clays the organisms contain one or two orders of magnitude less Mn, Fe, Co, Cr, V, Ba, Cu and Ni, the same order of magnitude of Pb, Hg and Zn, and one order of magnitude more Cd. It may be concluded, therefore, that relative to marine sediments the organisms are deficient in all the trace elements listed with the exception of Zn, Pb, Hg and Cd.

The photosynthetic planktonic populations of the oceans inhabit the euphotic zone. This is an extremely important marine environment because it is here that the living organisms are in contact with sea water and with suspended inorganic solids. However, the importance of marine organisms to the oceanic fluxes of trace elements is not confined to this zone because the biota are involved in their downward transport. According to Martin

(1970) marine organisms can be involved in the transport of incorporated trace elements in a number of ways. These include: the moulting of exo-skeletons; the sinking of skeletal structures after the death of the organisms; via fast-sinking of faecal pellets; via vertical migrations of organisms across mixing barriers, and, by transfer of the organisms to higher trophic levels.

Martin (1970) investigated the role played by the moulting of chitinous exo-skeletons in trace element fluxes. He compared the concentrations of Ca, Cd, Co, Cu, Fe, Mn, Ni, Pb, Sr and Zn in a series of surface zooplankton with those in a series collected at 100 m, or deeper, and found that the concentrations of Cu, Fe, Mn, Ni, Pb, Sr and Zn were higher in the samples from deeper water. He suggested that this enhancement was the result of more of the elements being adsorbed onto copepods at greater depth because the food-dependent moulting rates are lower here than they are at the surface where food is available. He further suggested that the exo-skeleton moults may be capable of elemental adsorption long after leaving the animal.

The sinking of copepod moults and other skeletal forms may result in the incorporation of trace elements into deep-sea sediments, and so in their permanent loss from sea water. A number of processes may be involved in the sinking of skeletal remains down the water column. These include: gravitational settling, which may be accelerated by phenomena such as density inversion currents; the aggregation of cells; downwelling of water; and, faecal pellet sinking. Various authors have concluded that the sinking rates of particles down the water column are considerably in excess of those predicted for gravitational settling by Stokes' Law. For example, Delany *et al* (1967) reported the presence of kaolinite concentration gradients which decreased westwards in the deep-sea sediments off West Africa. They related these gradients to the aeolian transport of kaolinite-rich material from the adjacent Sahara Desert, but pointed out that the gradients could only be maintained if the particles had fallen down the water column at accelerated sinking rates and had thus escaped lateral dispersion in the North Equatorial Current system. They suggested that one possible mechanism for this accelerated sinking was the ingestion of the dust particles by filter-feeding organisms and their subsequent excretion as aggregates in faecal pellets. Faecal pellet aggregation has been investigated by Smayda (1971) who concluded that these pellets can provide a possible means of bringing about the accelerated sinking of phytoplankton remains. Further evidence suggesting that the rate at which particles sink in the oceans is in excess of Stokes' Law predictions has been provided by Osterberg *et al.* (1963). These authors found that certain radio-nuclides with relatively short half lives, e.g. ^{95}Zr (half life, 65 days), could be detected in bottom feeders at a depth of $\sim$2800 m in the Pacific Ocean, and suggested that they may have been transported to this depth in faecal pellets. According to Smayda (1969) faecal pellets sink, at a size-dependent rate, varying between 36 and 376 m day^{-1}, i.e. for them to fall ~2800 m would take between 10 and 80 days. Faecal pellets are friable and may break up, they may also be decomposed or consumed as they descend the water column. However, under certain conditions they may play a significant role in the transport of particles, and their associated trace elements, to the sea floor. Another mechanism which has been suggested for the downward transport of trace elements in sea water is the vertical migration of marine animals across mixing barriers. For example, Pearcy and Osterberg (1967) invoked such vertical animal migration to account for the transport of ^{65}Zn across the base of the permanent halocline in the waters off the coast of Oregon. Sorokin (1972) investigated the vertical microbial transport of trace elements in the oceans and concluded that cobalt may be removed from sea water by planktonic bacteria which are subsequently utilized as food by zooplankton. During this process cobalt is excreted in faecal pellets which sink down the water column. According to Sorokin (1972) the time necessary for the complete assimilation of the dissolved cobalt in sea water by bacterial biosynthesis would be ~200 years.

The trace element composition of the total particulate material in the surface layers of sea water has been studied by Chester and Stoner (1975). This particulate material from the euphotic zone consisted of a mixture of continental weathering products (including both inorganic and organic precipitates), authigenic inorganic precipitates and organic components which included phytoplankton, zooplankton, detritus (dead organisms) and bacteria. The average elemental compositions of some total particulate samples from the euphotic zone are given in Table 34.34b. It is difficult to establish the compositions of any of the individual components which jointly comprise the total particulate material. However, some trends in the chemistry of the total particulate material can be distinguished by comparing its elemental composition to those of marine sediments and marine organisms (see Table 34.34). Using this approach Chester and Stoner (1975) drew the following conclusions. (i) Relative to near-shore muds the total particulate material is deficient in Mn, Ga, V and Ba, and is enhanced in Cu, Pb and Zn. (ii) Relative to Atlantic deep-sea clays the particulates are deficient in Mn, Co, Ga, V and Ba, they contain about the same amounts of Cu and Pb, and are enhanced in Zn. (iii) Relative to marine organisms the particulate material contains higher concentrations of all the trace elements with the exception of Zn. Chester and Stoner (1975) concluded, therefore, that the trace elements in total particulate material from the euphotic zone of the oceans have a number of sources: Mn, Co and V are probably present mainly in continentally-derived material and authigenic precipitates; Zn is probably located principally in marine organisms; and, Cu, Pb and Ba are partitioned between the inorganic components and the marine organisms.

TABLE 34.34a

Concentrations of trace elements in carbonate shell material (ppm)

	Sr	Ba	Fe	Mn	Cr	V	Ni	Co	Cu	Pb
Foraminiferal shells from deep-sea sediments[a]	1 112	10–30	1 213	335	—	15	21	—	23	138
Pteropod shells[b]	—	—	—	—	1	85	2	20	30	200
Coccolith ooze[c]	1 468	175	—	263	5	—	4	4	13	—

[a] Data from Turekian (1965b). [b] Data from Nicholls *et al.* (1959). [c] Data from Thompson and Bowen (1969).

TABLE 34.34b

The concentrations of trace elements in total particulate material from surface (0–~5 m) sea water, marine organisms and marine sediments (ppm, dry wt)

Trace element	Total surface particulate material[a]					Near-shore muds[b]	Atlantic deep-sea clays[c]	Marine organisms[d]
	N. Atlantic	S. Atlantic	Indian Ocean	China Sea	Average			
Mn	145	85	385	1501	529	850	4000	9·3
Cu	74	52	202	107	109	48	130	27
Co	11	16	14	12	13	13	38	<1
Ga	4	3	3	8	5	19	21	<1
V	38	69	60	86	63	130	130	<3
Ba	191	72	77	166	126	750	700	60
Pb	52	72	44	63	58	20	45	20
Zn	159	260	231	232	220	95	130	257
% organic carbon	24	18	14	12	17	—	—	—

[a] Data from Chester and Stoner (1973). [b] Data from Wedepohl (1960). [c] Data from Wedepohl (1960) and Turekian and Imbrie (1966). [d] Data from Table 34.33.

34.2.3.5. *The overall pattern of sedimentation.*

Some of the individual factors which control the distributions of trace elements in deep-sea sediments have been discussed in the preceding Sections. These factors include: the relative proportions of the minerals which constitute the sediments; the pathways by which the elements reach the marine environment; and, the mechanisms by which they are incorporated into the sediments themselves. However, a further, and very important, control on the trace element compositions of deep-sea sediments is the manner in which these various factors interact with each other to produce a particular type of deep-sea deposit. The extent to which these interactions occur varies from one oceanic region to another, and is dependent on the overall patterns of marine sedimentation. These patterns are, in general, governed by parameters such as the bottom topography, the geophysical conditions, the rates of supply and the proportions of lithogenous, hydrogenous and biogenous solids and dissolved material, and the rates of sediment accumulation (see Chapter 24).

It is convenient to divide the trace elements in deep-sea sediments into two broad groups: (i) those which have similar abundances in near-shore muds and deep-sea clays, examples of these elements are Cr and V; and, (ii) those which are enriched in Atlantic deep-sea clays relative to near-shore muds and which are even more enriched in Pacific deep-sea clays, examples of these elements are Mn, Ni, Co, Cu, Pb and Zn. One of the most important questions with regard to the geochemistry of deep-sea sediments is the origin of these "excess" trace elements. It was shown in Section 34.2.3.3.1 that the distributions of trace elements such as Cr and V, which are not enhanced in deep-sea sediments, are largely controlled by the input and sedimentation of lithogenous material in which they are principally located in "lattice-held" positions. In contrast, a considerable fraction of the "excess" trace elements, e.g. Mn, Ni, Co and Zn, are in "non-lattice" positions (see also Section 34.2.3.3.1). These differences in the partition characteristics of the trace elements are also reflected in their geographical distributions. For example, Turekian and Imbrie (1967) demonstrated that for Atlantic deep-sea sediments the "excess" trace elements Mn, Ni, Co and Cu had their highest concentrations (on a carbonate-free basis) in mid-ocean deposits (see Fig. 34.3a). Further, Chester and Messiha-Hanna (1970) showed that the highest "non-lattice-held" concentrations of these elements are also found in mid-ocean Atlantic deep-sea sediments (see Fig. 34.3b). It may be concluded, therefore, that the "excess" trace elements found in deep-sea clays are in "non-lattice-held" positions, and that they have been removed from the overlying waters at some time during the histories of the sedimentary components, although whether this removal took place in the marine, the estuarine, or the continental environment has given rise to much speculation. A number of theories have been advanced to explain the trace element enrichment patterns in deep-sea sediments. These theories, which are related to the overall conditions of sedimentation in the oceans, are discussed below.

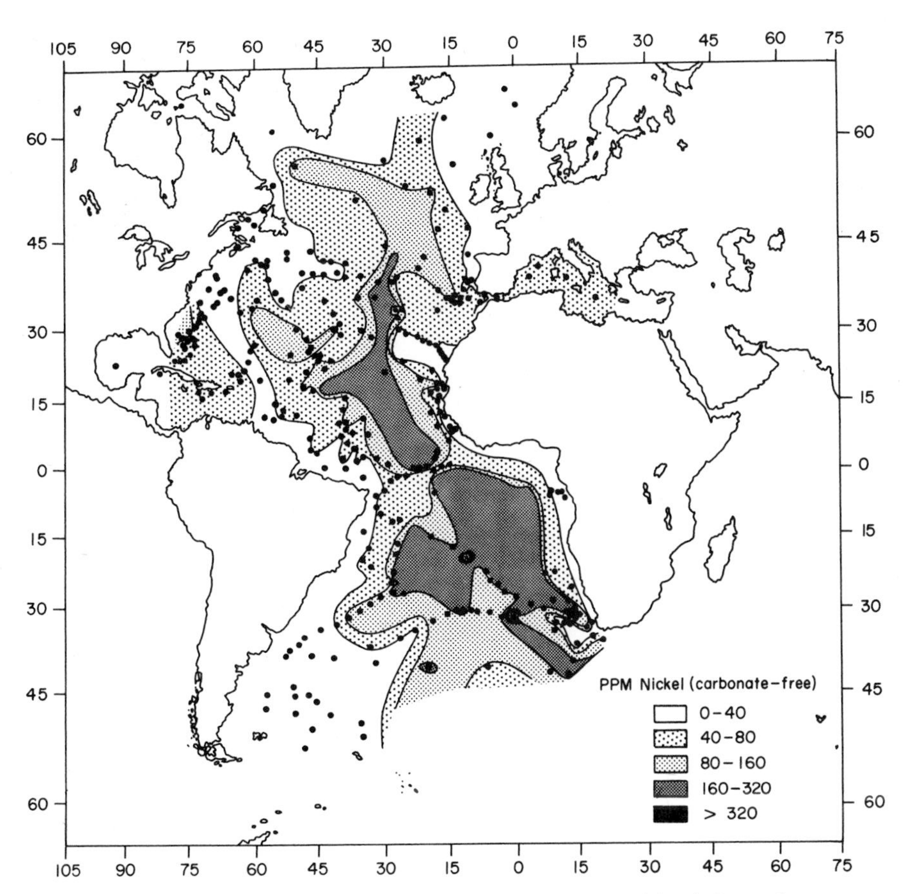

FIG. 34.3a. The distribution of nickel at the sediment surface of the Atlantic Ocean (ppm, on a carbonate-free basis, modified from Turekian and Imbrie, 1967).

Turekian (1965, 1967) has proposed that the "excess" contents of Mn, Ni, Co and Cu in deep-sea clays result to a large extent from the differential transport of non-pelagic and pelagic lithogenous components to, and within, the oceans. The predominant feature in the geographical distributions of Mn, Ni, Co and Cu in Atlantic deep-sea clays is their relatively high concentration in mid-ocean areas. The sediments deposited in these central areas are largely pelagic in character, i.e. their lithogenous material is composed of fine-grained particles which have been deposited slowly from the upper water layers. The particles are composed mainly of clay minerals (which may be coated with iron

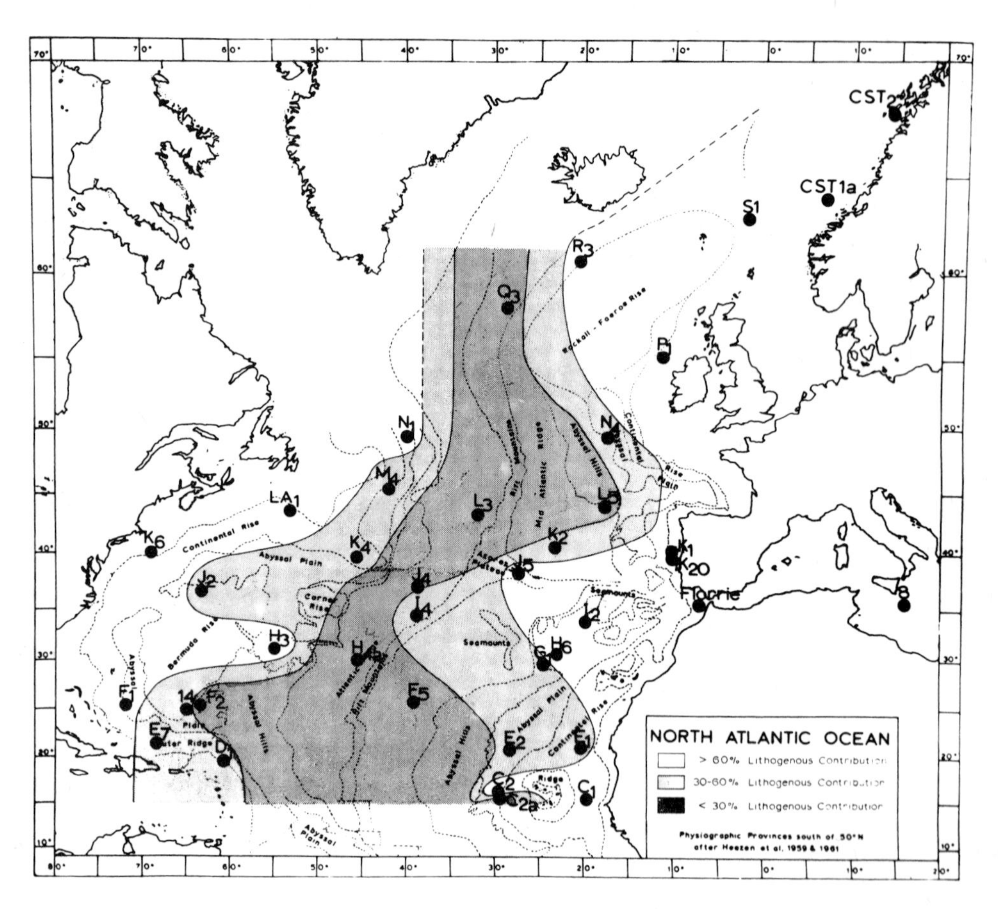

FIG. 34.3b. The distribution of "lattice-held" lithogenous nickel at the sediment surface of the North Atlantic (the "lattice-held" lithogenous contributions are expressed as a percentage of the total nickel in the sediment) (Chester and Messiha-Hanna, 1970).

and manganese oxides), together with discrete grains of manganese and iron oxides. These particles have large specific surface areas and will have acquired a relatively high content of trace elements by adsorption in the river environment. Some of these trace elements will have been desorbed on contact with sea water, but most of those held on manganese and iron oxides will probably have been retained and subsequently will have been transported to deep-sea areas in "non-lattice-held" positions on the particles. In contrast, the sediments deposited on those areas which fringe the continents, e.g. abyssal plains, are mainly non-pelagic and contain relatively large lithogenous particles which have been transported by bottom processes. These particles are enriched in the coarser-grained lithogenous minerals, e.g. quartz, and have relatively small trace element contents, with the largest fraction of the elements in "lattice-held" positions. Turekian (1965, 1967) has related the distribution of the elements to the deposition of pelagic and non-pelagic deep-sea sediments, the former having a "trace element-rich" assemblage and the later a "trace element-poor" one. He concluded that the higher contents of the "excess" trace elements in Pacific relative to Atlantic deep-sea clays arises because of the greater areal extent and remoteness from the land masses of regions of pelagic sedimentation in the former ocean. There is no doubt that (excluding specialized chemical sediments such as those deposited on the crests of active ridges) the "excess" trace elements in deep-sea clays are associated with "normal" pelagic sedimentation. However, the important point in Turekian's theory is that a high proportion of the "excess" trace elements has been acquired in the river environment, and that their removal from the sea water in deep-sea areas is only of secondary importance.

Clearly, trace elements associated with continental weathering products will be transported to deep-sea areas, and will contribute to the "excess" trace element content of pelagic clays. However, there is also no doubt that these elements are removed from sea water by various deep-sea sediment components, particularly those having a hydrogenous origin. The most important of these latter components are the ferro-manganese nodules—phases which have grown directly from material dissolved in sea water, and which exert a strong influence on the trace element geochemistry of many pelagic clays (see Section 34.2.3.4). Micro-ferro-manganese nodules are particularly abundant in Pacific pelagic clays, and those in some deposits may contain between one third and one half of the Mn, Ni, Co, Pb, Cu and Zn present in the total sediment (see e.g. Goldberg and Arrhenius, 1958; El Wakeel and Riley, 1961; Chester and Hughes, 1969; Krishnaswamy and Lal, 1972). Elderfield (1972) carried out a study of both near-shore and deep-sea ferro-manganese nodules and concluded from elemental ratios that the concentration of trace elements by the deep-sea nodules, and the enrichment of the same elements in pelagic clays, are related processes.

The association between the presence of ferro-manganese nodules and the enhanced trace element concentrations in certain deep-sea clays was investigated by Chester *et al.* (1973) who examined the partition of trace elements among the components of a series of deep-sea clays from the south west region of the North Atlantic Ocean. Chester and Messiha-Hanna (1970) had previously shown that some of these clays had enhanced concentrations of Mn, Ni and Co, and subsequently Horn *et al.* (1972) reported that ferro-manganese nodules are relatively concentrated in this region of the Atlantic. Chester *et al.* (1973) showed that the clays could be divided into two groups on the basis of their manganese concentrations: Group A, in which the sediments contain >3000 ppm Mn, and; Group B, in which they contain <3000 ppm. Mn. The average trace element contents of each group are given in Table 34.35a, from which it is apparent that, on average, the clays of Group

TABLE 34.35
*The concentrations of Mn, Ni and Co in deep-sea clays from the south west region of the North Atlantic**

A. Average concentrations in the total sediment samples of Groups A and B (see text) (in ppm)

Sample	Mn	Ni	Co	Total carbonate (wt. %)
Group A	6658	142	71	9
Group B	1518	58	27	10

B. Average concentrations in the "non-lattice-held" fractions of Groups A and B (in ppm expressed in terms of a 100% "non-lattice-held" fraction).

Sample	Mn	Ni	Co
Group A	80772	1287	879
Group B	7187	234	138

* Data from Chester *et al.* (1973).

A contain about four times as much Mn and over twice as much Ni and Co as do those on Group B. Data on the partition of the trace elements showed that the "non-lattice-held" fractions of the Group A clays contain an average of ~8% Mn, ~0·13% Ni and ~0·07% Co, concentrations which begin to approximate to those found in some ferro-manganese nodules from the same region—see Table 34.35b. The "non-lattice-held" fractions of the Group B clays, however, had very much smaller concentrations of these elements. The authors concluded that the conditions which resulted in the formation of the ferro-manganese nodules in this region are reflected in the trace element geochemistry of the Group A clays. By analogy with the mechanisms involved in nodule formation it is probable that the bulk of the "non-lattice-held" Mn, Ni and Co in the Group A clays had been removed from sea water, and had not been transported to the area from a continental or a near-shore source. The reason for the concentration of these elements in some, but not all, of the clays from this region is probably a function of factors such as the rates of sediment accumulation and the environments of deposition.

One of the characteristic features of deep-sea clays is the relatively slow rate of accumulation of their non-biogenous components. However, the rates of accumulation vary considerably from one oceanic area to another, and many authors have attempted to relate them directly to trace element concentrations in the clays themselves. One of the earliest, and still one of the most important and influential, of these attempts was made by Wedepohl (1960) who proposed the "trace element veil" theory. In essence, this theory states that the "excess" trace elements in deep-sea clays are removed from sea water homogenously, i.e. equally in all parts of the ocean, but are superimposed on deep-sea clays which accumulate at varying rates in different parts of the World Ocean. Thus, the deep-sea clays of the Pacific are thought to be more strongly enriched in "excess" trace elements than are those of the Atlantic because they are deposited at a slower average rate. Wedepohl (1960) used the average trace element composition of near-shore sediments, and those of both Atlantic and Pacific deep-sea clays, to show that, in the light of this theory, it would appear that the average deep-sea clay accumulates ~3–4 times faster in the Atlantic than it does in the Pacific. Wedepohl also concluded that volcanic emanations are an important source of supply of the "excess" trace elements found in deep-sea clays.

Turekian (1967) has criticized the "trace element veil" theory on the grounds that the "excess" trace elements have, in fact, an inhomogenous distribution in the waters of the World Ocean, and that their sources of supply and mechanisms of removal probably differ in the Atlantic and Pacific Oceans. Because of these objections Turekian (1965a, 1967) proposed his "differential transport" theory—see above.

Turekian (1965a, b) also considered the effect of accumulation rates on the trace element geochemistry of deep-sea sediments. By comparing trace element concentrations and accumulation rate data he attempted to evaluate temporal and spatial variations in trace element budgets. In order to facilitate an understanding of the processes which are involved in the incorporation of trace elements into deep-sea sediments, Turekian (1965a, b) developed several models of trace element deposition. In the formulation of these models X refers to the concentration of "clay", and Y to that of any other specified component of deep-sea sediments, and R_X and R_Y are the rates of accumulation of these components.

Model 1. This is the simplest model, in which a trace element is primarily associated with a major sediment component. If, for example, the trace element is associated exclusively with the "clay" fraction, a plot of the trace element accumulation rate (R_Y) against the clay accumulation rate (R_X) would yield a straight line through the origin, and Y would have a linear relationship with X.

Model 2. If R_Y and R_X are not associated locally in the water mass, but are both influenced by some large-scale process involving the oceans as a whole (e.g. oceanic circulation changes, slumping of sediments, bottom transport), then although there would be a correlation between R_X and R_Y, there need not necessarily be a correlation in the same sense for the concentrations X and Y.

Model 3. If part, or all, of the trace element is added to the sediment at some constant rate which is independent of the variable clay accumulation rate, then the relationship between R_X and R_Y becomes;

$$R_Y = R_Y^* + YR_X$$

where R_Y^* is the constant rate of accumulation of the trace element independent of the clay accumulation rate, and Y is the concentration of the directly associated fraction of the trace element in the pure clay fraction. A plot of R_Y against R_X would show an intercept corresponding to R_Y^* (the constant independent rate) and a slope Y corresponding to the concentration of the trace element in the clay from this source.

TABLE 34.36

*Statistics of rates of accumulation of various elements relative to that of "clay" in three Atlantic deep-sea sediment cores**

Sediment core	Fe_2O_3	Mn	Co	Ni	Cr	Pb	Cu	Sn
R 10–10								
a	27–33	5682	25·1	40·5	−35	−10·5	88	—
P (%)	10–25	5–10	10–25	10–25	20–50	2·5–5	25–50	
A 179–4								
a	−3·08	710	−7·7	−6·1	−3	2·2	7·6	
P (%)	—	25–50	10–25	25–50	>50	10–25	10–25	—
A 180–74								
a	−0·026	—	1·08	14·37	−1·62	0·77	—	7·54
P (%)	>50	—	>50	25–50	>50	>50	—	5–10

a is the intercept calculated from the regression of trace element accumulation rate on clay accumulation rate, and is measured in g cm^{-2} $(10^3\ yrs)^{-1}$ for Fe_2O_3, and in μg cm^{-2} $(10^3\ yrs)^{-1}$ for the other elements.

P (%) is the probability (in per cent) that the null hypothesis is valid in a test of the difference of values of a from a = 0. All values of $P > 5\%$ indicate that the difference between the calculated value of a and a = 0 is not significant.

* Data from Turekian (1965a).

Model 4. If the rate of clay-independent trace element accumulation and the rate of accumulation of clay (plus its associated trace elements) both vary with time and place, then the $X-Y$ (i.e. concentration) plot and the $R_X - R_Y$ (i.e. rate) plot would show a shot-gun scatter pattern.

Turekian (1965a, b) tested his models using trace element and sediment accumulation rate data from three deep-sea cores from the Atlantic Ocean; the trace elements studied were Fe, Mn, Co, Ni, Cr, Pb, Cu and Sn, and the "clay" fraction was taken as the acid-insoluble residue of the sediments. The results obtained by Turekian may be summarized as follows. (i) In all three cores the rates of trace element accumulation are related to the rates of "clay" accumulation (see Table 34.36). (ii) In core R10–10 Cr and Pb are closely associated with the "clay" fraction, i.e. there is little or no accumulation of these elements when there is zero deposition of clay. In contrast Fe, Mn, Ni and Co are controlled by at least one other sediment component or deposition process (see Table 34.36). (iii) In core A179–4, Co, Ni and Pb are closely associated with the "clay", whereas Fe, Mn, Cr and Cu are not (see Table 34.37). (iv) In core A180–74, Sn has clearly been incorporated into the sediment by a process, or component, which is not associated with the "clay" (see Table 34.37).

The models described above may each be applicable to a particular part of the ocean basins for a specific period of geological time, but considerable caution must be exercised in extending any conclusions drawn from one area,

TABLE 34.37

*Statistics on correlation of trace element concentration with clay concentration in three Atlantic deep-sea sediment cores**

Sediment core	Fe_2O_3	Mn	Co	Ni	Cr	Pb	Cu	Sn
R 10–10								
a	3·69	—	16	32	~0	1	28	—
P(%)	<0·1	—	<0·1	<0·1	>50	>50	5–10	—
A 179–4								
a	1·35	—	−4·3	8	23	1·7	21	—
P(%)	<1	—	25–50	10–25	1–2·5	25–50	<0·1	—
A 180–74								
a	−0·7	—	−0·1	2·8	3·1	2·5	—	7·6
P(%)	>50	—	>50	25–50	25–50	0·5–1	—	~1

a is the intercept calculated from the regression of trace element accumulation rate on clay accumulation rate, and is measured in g cm^{-2} $(10^3 \text{ yrs})^{-1}$ for Fe_2O_3, and in µg cm^{-2} $(10^3 \text{ yrs})^{-1}$ for the other elements.

P (%) is the probability (in per cent) that the null hypothesis is valid in a test of the difference of the values of a from a = 0. All values of P > 5% indicate that the difference between the calculated value of a and a = 0 is not significant.

* Data from Turekian (1965a).

or even from several areas, to general concepts applicable to the World Ocean sediment system. Within these limitations Turekian (1965a, b) has attempted to interpret large-scale geographical trace element distribution patterns in terms of a model which describes ocean bottom trace element economy Wedepohl (1960) concluded that the distribution of the "excess" trace elements, e.g. Mn, Ni, Co, Cu, Zn in deep-sea sediments resulted from the imposition of a constant world-wide supply of the elements onto the "clay" component which accumulates at rates which vary from one area to another—see above. Turekian (1965a, b) used Wedepohl's average values for the trace element contents of Atlantic and Pacific deep-sea clays to predict the rates of accumulation of trace elements from solution which, according to Wedepohl's theory, should be identical for both oceans. To achieve this Turekian used the following equation:

$$\bar{R}^*_Y = \frac{\bar{R}_{XA}(P - A)}{\dfrac{\bar{R}_{XA}}{\bar{R}_{XP}} - 1}$$

where $\bar{R}^*_Y$ is the average rate of accumulation of a trace element from sea water, $\bar{R}_{XA}$ and $\bar{R}_{XP}$ are the average rates of clay accumulation in the Atlantic and Pacific Oceans respectively, and A and P are respectively the concentrations of trace elements in Atlantic and Pacific Ocean deep-sea clays. Turekian tested Wedephol's "trace element veil" theory by using a series of trace element and accumulation rate data in this equation. He assumed the average clay accumulation rate for the Atlantic Ocean, i.e. $\bar{R}_{XA}$, to be 1·1 mg cm^{-2} yr^{-1}, and by taking Wedepohl's values for the average concentrations of

TABLE 34.38

Calculated values of the average rate of accumulation of trace elements from sea water (see text)†

				$\bar{R}^*_Y$ values for $\bar{R}_{XA}/\bar{R}_{XP}$ =		
Element	P (ppm)	A (ppm)	P—A	11	7	3·5
Mn	940	400	540	59	98	238
Co	110	38	72	7·9	12	29
Ni	300	140	160	18	29	70
Cu	400	130	270	30	50	119
Zn	200	130	70	7·7	13	31
Mo	46	9	37	4·1	6·8	16
Pb	110	45	65	7·1	12	29

Values for concentrations (ppm) of trace elements in Atlantic (A) and Pacific (P) deep-sea sediments from Wedepohl (1960).

$\bar{R}^*_Y$ values calculated on the basis $\bar{R}_{XA}$ = 1·1 g cm^{-2} (10^3 yrs)$^{-1}$.

† Data from Turekian (1965a).

trace elements in Atlantic and Pacific deep-sea clays he computed values of $\bar{R}_Y^*$ for each of the elements for different values of $\bar{R}_{XA}/\bar{R}_{XP}$; some of these values are given in Table 34.38. He then used the data from an Atlantic deep-sea sediment core to ascertain which values of $\bar{R}_Y^*$ were consistent with the trace element economy in this ocean (see Table 34.39). From these various calculations he concluded that the disparity between the concentrations of trace elements in deep-sea sediments of the Atlantic and the Pacific Oceans, is only compatible with a homogenous removal of trace elements from sea water imposed on a variable clay accumulation rate if the average rate of clay accumulation in the Atlantic is about fifteen times (i.e. the average $\bar{R}_{XA}/\bar{R}_{XP}$ (Min) in Table 34.39) greater than it is in the Pacific, and not ~3·4 times as suggested by Wedepohl (1960).

TABLE 34.39

*Maximum values of $\bar{R}_Y^*$ and minimum values of $\bar{R}_{XA}/\bar{R}_{XP}$ consistent with the data from core A 179–4**

Element	$\bar{R}_Y^*$ (Max.) μg/cm²/10³/yrs	$\bar{R}_{XA}/\bar{R}_{XP}$ (Min.)	River supply μg cm^{-2}(10^3 yrs)$^{-1}$
Co	7·35	11·8	2·7–27
Ni	10·7	17·4	2·7
Pb	6·7	11·7	35
Cu	19·2	16·5	157

$\bar{R}_Y^*$ (Max.) is the value at which the null hypothesis is valid at the 5% probability level for the data of core A 179–4.

* Data from Turekian (1965a).

Turekian (1965a, b) attempted to understand the processes controlling the incorporation of trace elements into deep-sea sediments by relating the vertical distributions of the elements to the accumulation rates of the non-carbonate components in sediment cores from specific locations. However, there are a number of difficulties inherent in the interpretation of trace element distribution patterns in sediments which have widely differing proportions of major components. In many deep-sea sediments the two principal components are "clay" and carbonate. In general, the carbonate component is deficient in most of the trace elements which are enhanced in deep-sea clays. For this reason trace element analyses of deep-sea sediments have often been expressed on a carbonate-free basis (see e.g. Turekian and Imbrie, 1966; Bender and Schutz, 1969; Boström and Fisher, 1969), a procedure which is designed to correct for the dilution effect of calcium carbonate and which allows the "clay" fractions of the total sediment samples to be directly compared with one another. There are, however, several drawbacks to this method of expressing deep-sea sediment analyses. Some of these have been discussed by Boström *et al.* (1972), who pointed out that for sediments which are very rich in carbonate even small errors in the determination of the $CaCO_3$ will severely affect the recalculated trace element concentrations. Equally serious objections to expressing deep-sea sediment analyses on a simple carbonate-free basis arise from the fact that no account is taken of the presence of opaline silica. In order to overcome this, Piper (1973) normalized the trace element concentrations of a series of deep-sea sediments to Al_2O_3, a technique which allows for the presence of both carbonate and opaline components in the deposits. Boström *et al* (1969) used a similar approach to identify metalliferous deep-sea sediments by characterizing them as "Al-poor" types in terms of their Al/(Al + Fe + Mn) ratios expressed on a carbonate-free basis. However, even for those deep-sea sediments which do not contain opaline silica, and for which ~90% of the carbonate-free fraction consists of SiO_2, Al_2O_3, Fe_2O_3, MnO, TiO_2 and P_2O_5, the remaining ~10% (consisting of sea salt, organic matter etc.) will result in the underestimation of the oxides. There have been various attempts to overcome this difficulty. For example, Landergren (1964) recalculated his deep-sea sediment analyses in terms of a 100% major oxide fraction, which he termed the *silicate* component Boström and Fisher (1969) also normalized sediment analyses to the major oxides after the deduction of "excess SiO_2", i.e. biogenous silica which, on the basis of an average crustal SiO_2/Al_2O_3 ratio of 4:1, they defined as; SiO_2 (excess) = SiO_2 (measured)– $4Al_2O_3$ (measured). By this method they were able to estimate the composition of the *abiogenic* component of deep-sea sediments. Boström and Fisher (1971) compared the abiogenic components of a series of Indian Ocean deep-sea sediments and were able to show that although enrichment of U and V occurs in areas of hemi-pelagic deposition, the highest concentrations of these elements are found in active ridge deep-sea sediments. They concluded that U and V, like Fe, have an important submarine volcanic source, a conclusion which it would not have been possible to draw from the raw, total sediment, trace element data. Boström *et al.* (1972) employed a more rigorously defined abiogenic component which was expressed as $\Sigma(Ti + Fe + Mn + 4Al + P_2O_5)$. This was termed the *minerogen* component, and trace element analyses were expressed on a minerogen basis (MB), which overcomes the diluting effects of both calcium carbonate and biogenous silica.

Boström and his co-workers have utilized recalculated chemical analyses and sediment accumulation rates to study the distribution and rates of accumulation of trace elements in deep-sea sediments on a World Ocean basis. For example, Boström *et al.* (1973a) were able to show that in Pacific Ocean deep-sea sediments enhanced concentrations of barium (MB) occur on active spreading ridges and under some areas of high biological productivity (see Fig. 34.4), thus indicating that this element is supplied to the deep-sea floor

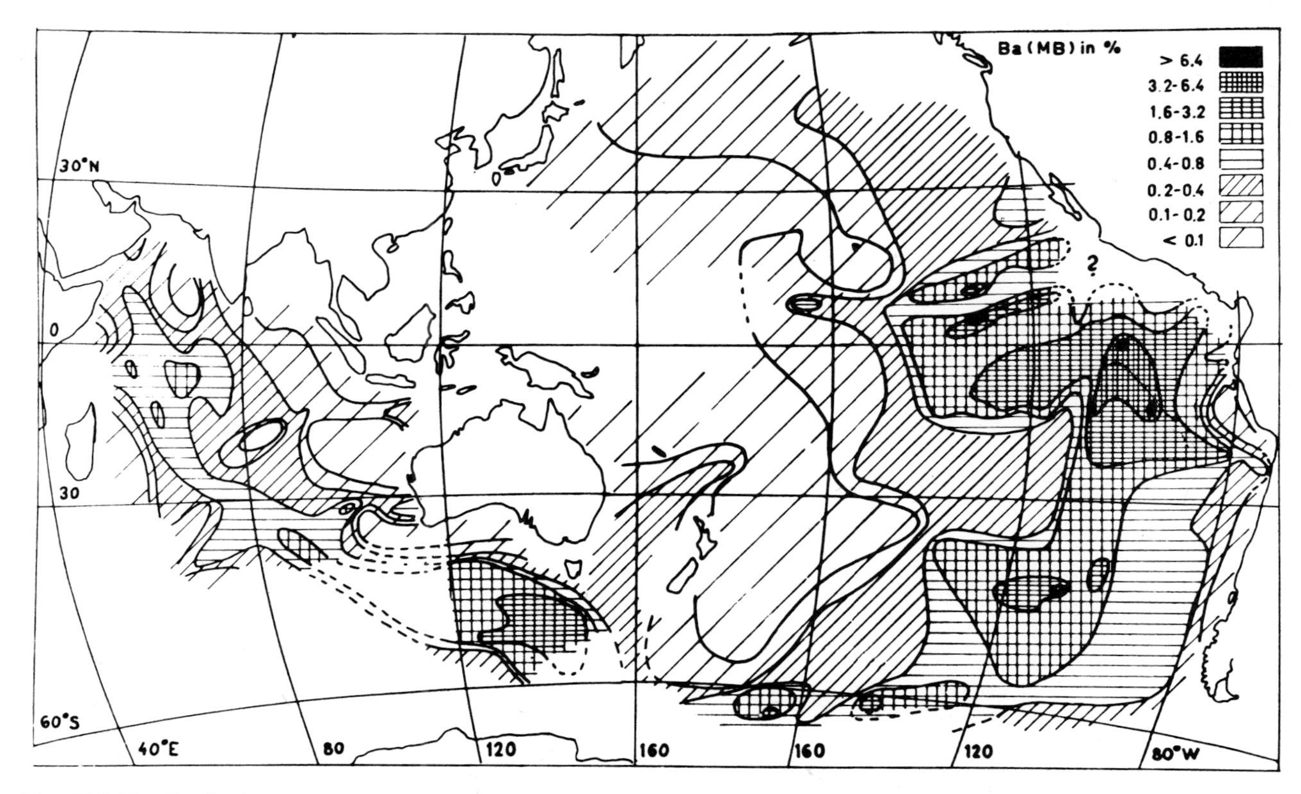

FIG. 34.4. The distribution of barium in Indo-Pacific deep-sea sediments on a minerogen basis (MB; see text) (Boström *et al.*, 1973a). (Reproduced by permission of Universitetsforlaget).

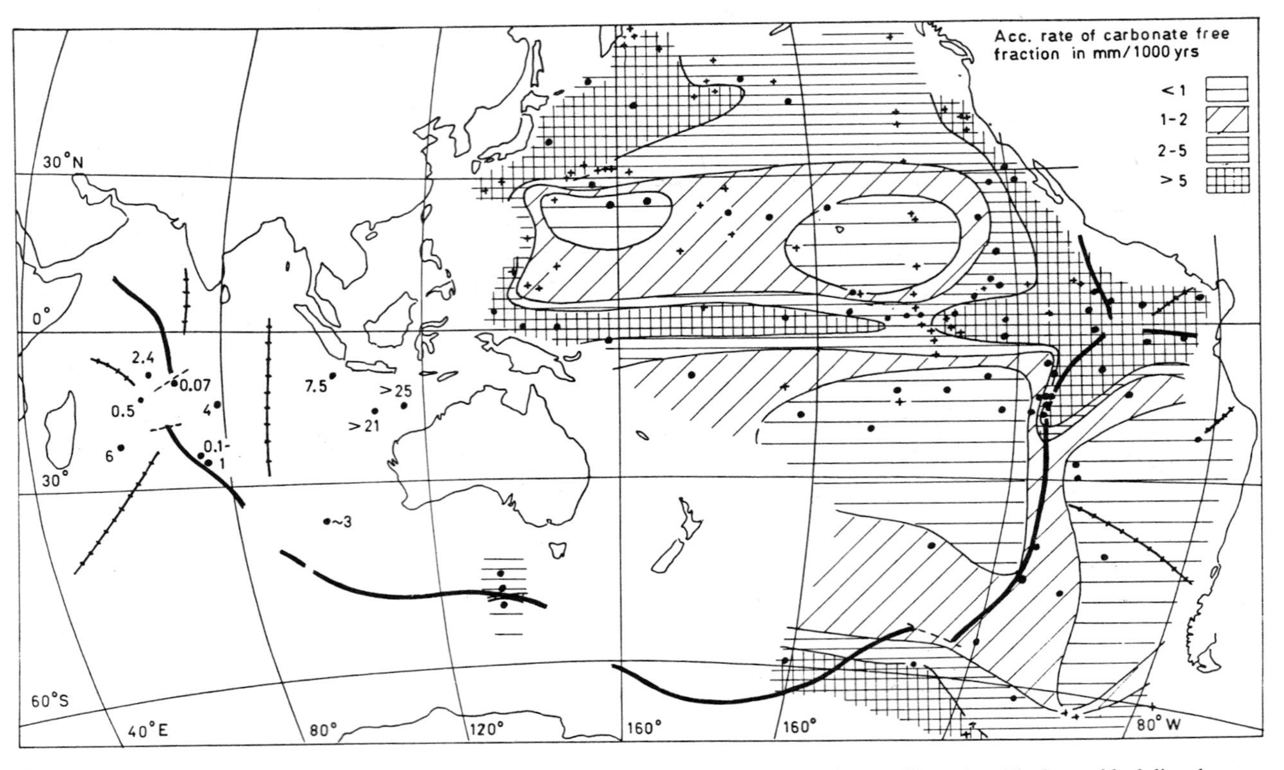

FIG. 34.5. Accumulation rates of the carbonate-free fractions of deep-sea sediments in the Indo-Pacific region. The heavy black line denotes the approximate location of the spreading Indo-Pacific Ridge System. Cross-hatched thin lines represent inactive ridges (*e.g.* the Ninety East and Nazia Ridges), or ridges where the length of the spreading axis is small compared to that of fracture zones (*e.g.* the Chile and southwest Indian Ocean ridges). Dashed lines indicate major fracture zones. Dots and crosses represent data points (Boström *et al.*, 1973b).

by both volcanic and biological processes. This kind of approach, which utilizes both the partition of trace elements among the components of deep-sea sediments and the sediment accumulation rates, yields a valuable insight into the origins of the trace elements. One of the most important examples of this has been provided by Boström *et al.* (1973b) who investigated the chemical compositions and accumulation rates of Pacific and Indian Ocean deep-sea sediments. The accumulation rates of the carbonate-free fractions of the sediments are illustrated in Fig. 34.5. and the conclusions reached by the authors are summarized below.

(i) The distribution and accumulation rates of opaline silica exhibit a pronounced latitudinal pattern (see Fig. 34.6). In mid-latitude areas <5 mg opaline silica cm^{-2} $(1000\ yr)^{-1}$ is accumulating, whereas along the equator and at high latitudes rates $\gg 150$ mg opaline silica cm^{-2} $(1000\ yr)^{-1}$ are

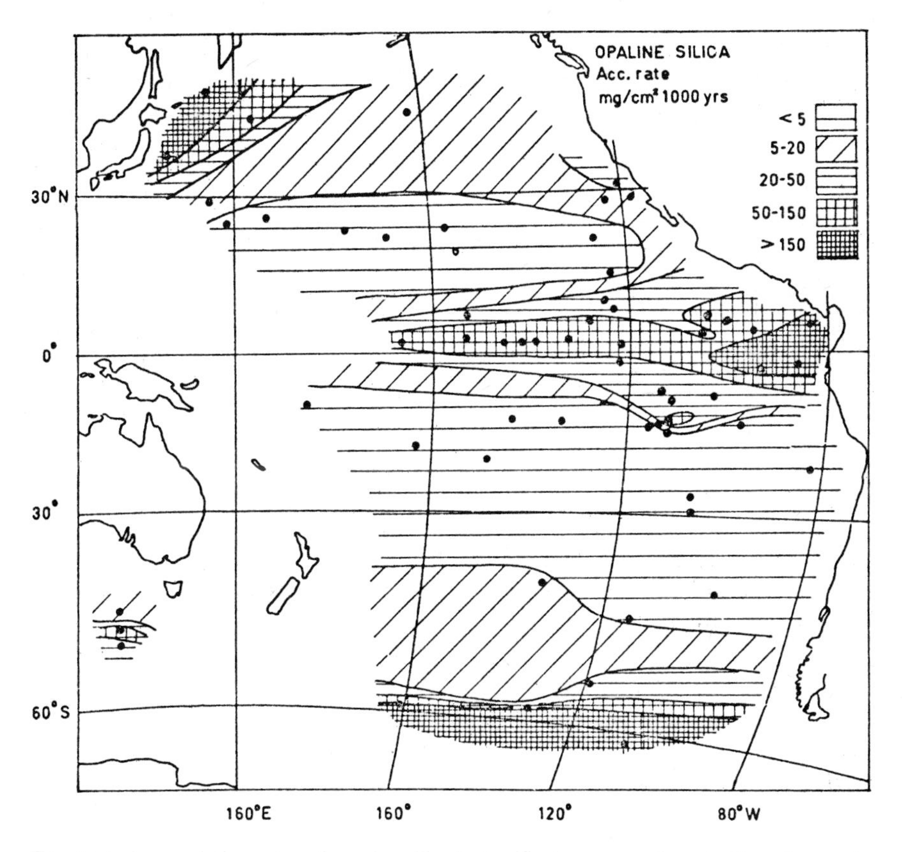

Fig. 34.6. Accumulation rates of opaline silica in Pacific deep-sea sediments (Boström *et al.*, 1973b).

common. This accumulation budget correlates with the distribution of regions of high primary production.
(ii) The distribution patterns of Al (see Fig. 34.7) and Ti (not shown) are very similar. The sediments in the central parts of the oceans have, in general,

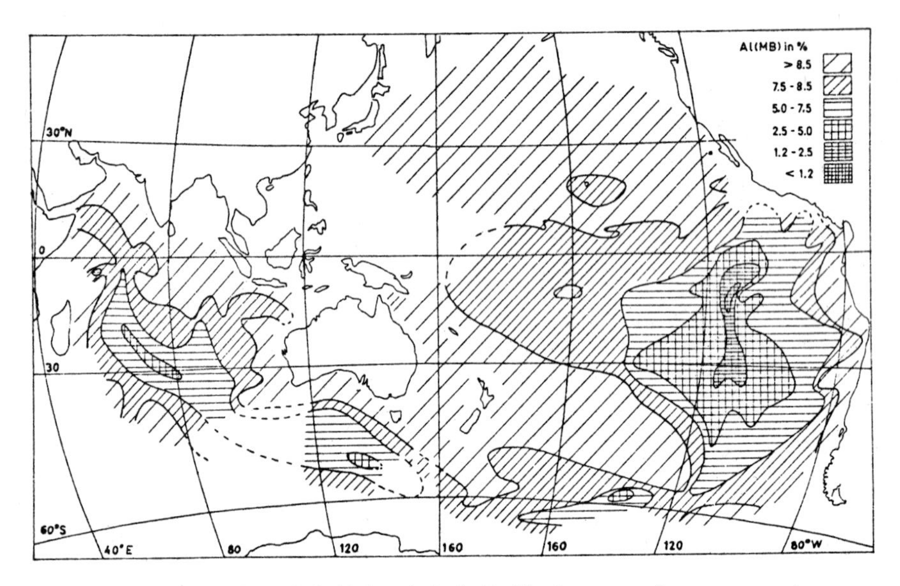

FIG. 34.7. The distribution of aluminium in Indo-Pacific deep-sea sediments on a minerogen basis (MB; see text). Note that relatively low aluminium concentrations are found on ridges with well-developed spreading (Boström *et al.*, 1973b).

relatively low concentrations of these elements, and the lowest values are found, without exception, on the active ridge centres of sea floor spreading (see Section 34.2.3.3.2). This overall distribution is confirmed by the aluminium accumulation rates (see Fig. 34.8). which show a very pronounced minimum on the East Pacific Rise.
(iii) The distribution patterns of Fe (see Fig. 34.9) and Mn (not shown) have certain similarities, but differ in one important respect. That is, close to the continents much of the iron is located in detrital material, and in these areas its distribution is similar to that of aluminium (see Fig. 34.7); however, in mid-ocean areas the patterns of iron distribution co-vary with those of manganese. The accumulation rates of both Mn and Fe exhibit pronounced maxima in the area of the East Pacific Rise, with small less well defined ones along the equator (see Fig. 34.10).

(iv) The distribution patterns and accumulation rates of Cu, Ni and Co have certain similarities to one another, and the principal features in these patterns can be illustrated with reference to that of Cu (see Figs. 34.11 and 34.12).

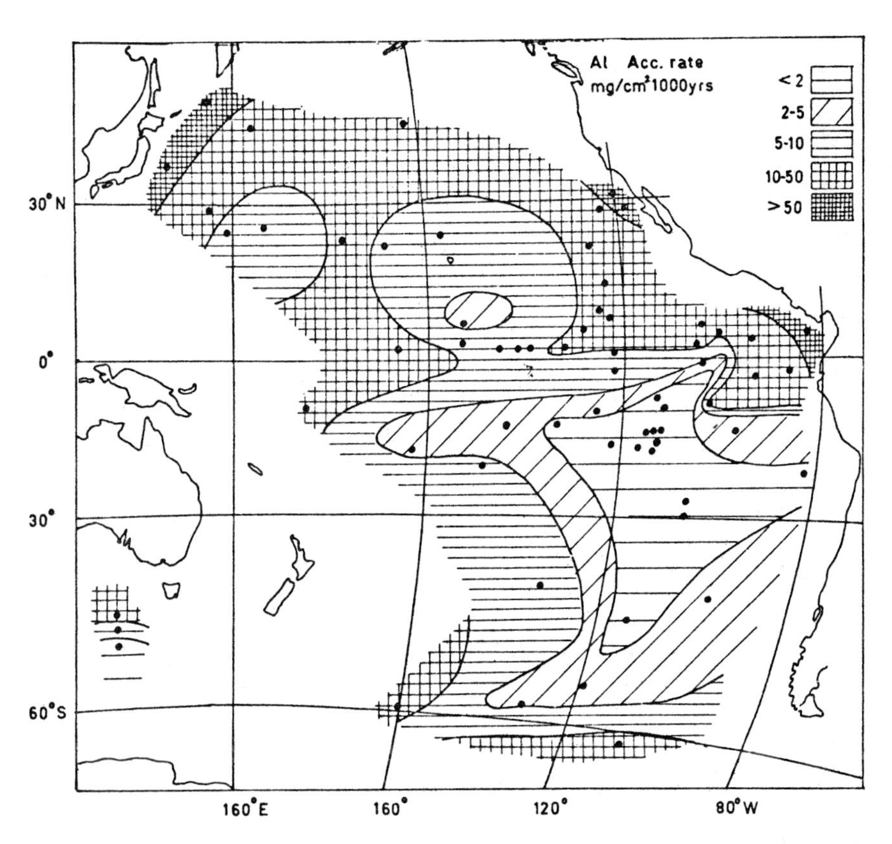

FIG. 34.8. Accumulation rates of aluminium in Pacific deep-sea sediments (Boström *et al.*, 1973b).

These three elements show a tendency to accumulate rapidly along the East Pacific Rise, but have their highest accumulation rates in the general region of the equator.

The most important generalizations which can be drawn from this phase of the study are that in the Indo-Pacific region many elements accumulate relatively rapidly in sediments close to the continents, and relatively slowly

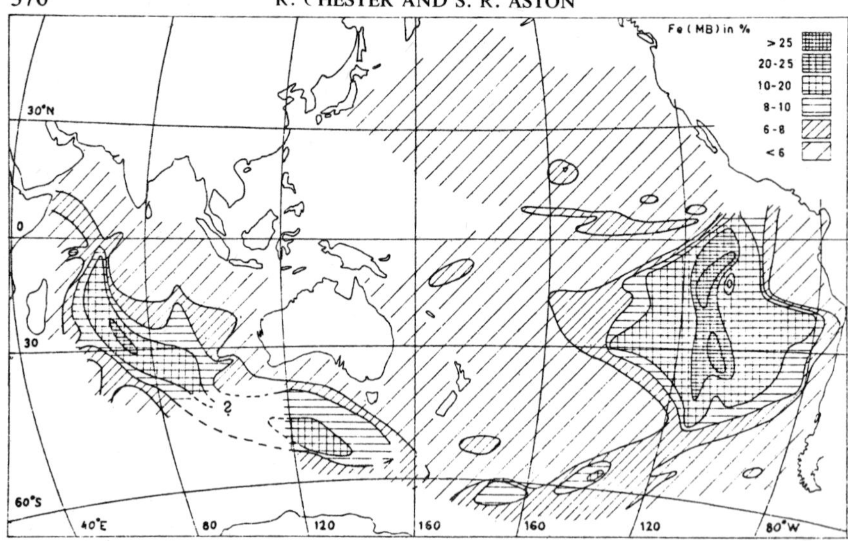

FIG. 34.9. The distribution of iron in Indo-Pacific deep-sea sediments on a minerogen basis (MB; see text) (Boström *et al.*, 1973b).

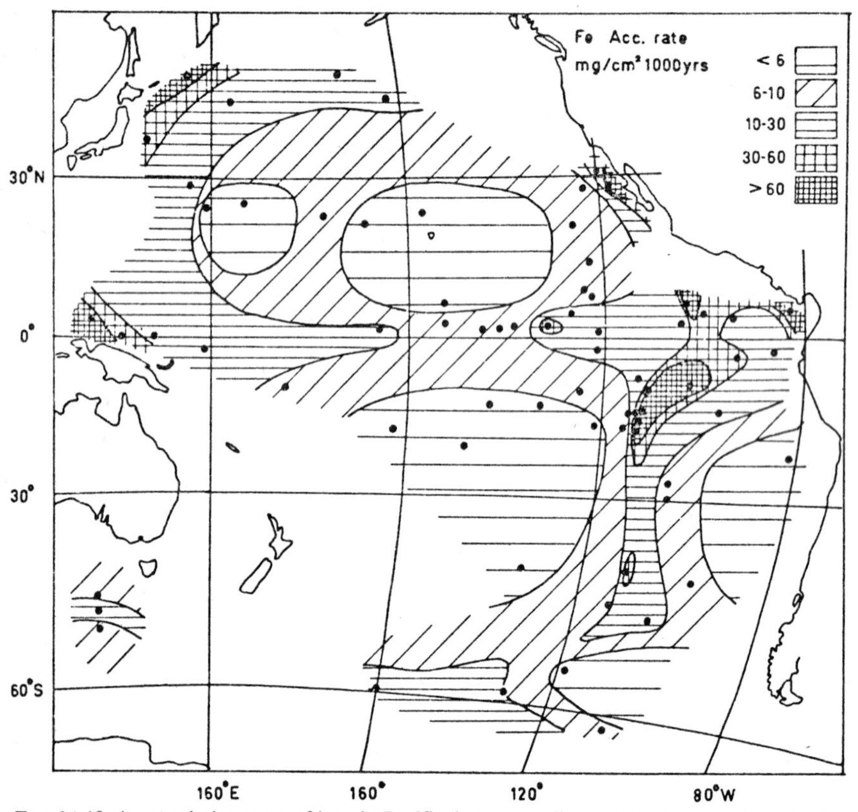

FIG. 34.10. Accumulation rates of iron in Pacific deep-sea sediments (Boström *et al.*, 1973b).

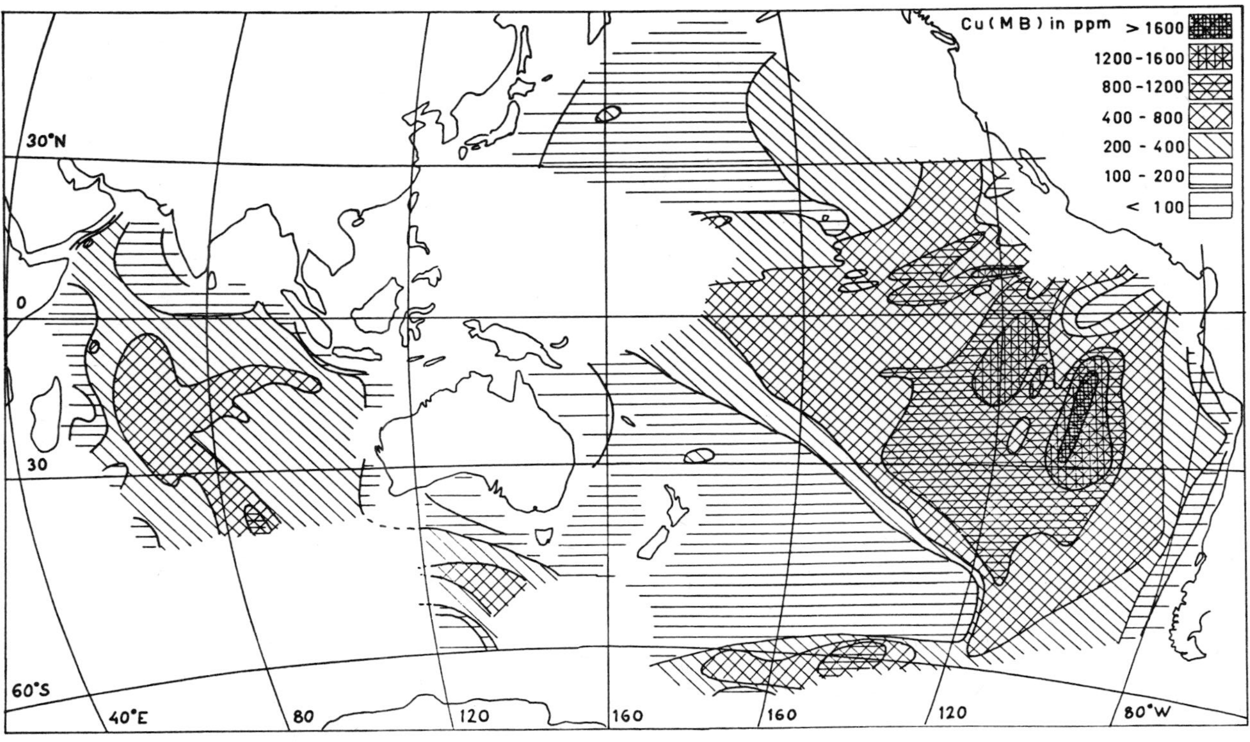

FIG. 34.11. The distribution of copper in Indo-Pacific deep-sea sediments on a minerogen basis (MB; see text) (Boström *et al.*, 1973b).

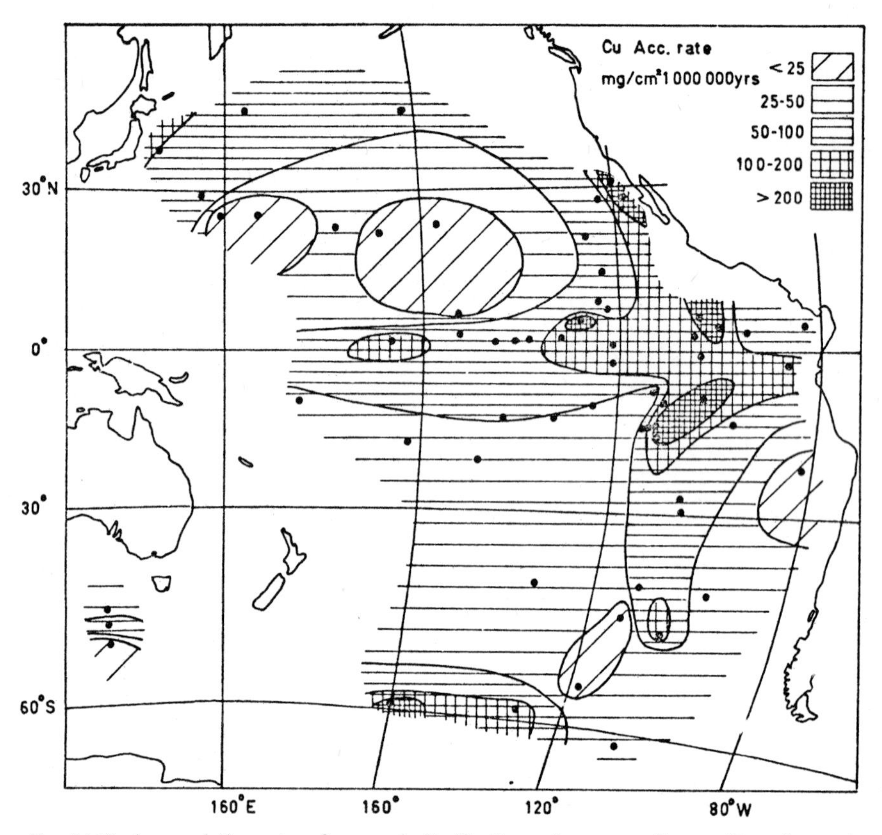

FIG. 34.12. Accumulation rates of copper in Pacific Ocean deep-sea sediments (Boström *et al.*, 1973b).

in the central parts of the oceans. However, there are specific mid-ocean areas, e.g. the East Pacific Rise and the Equatorial region, in which certain trace elements accumulate relatively rapidly.

Boström *et al.* (1973b) also investigated the provenance of Pacific Ocean deep-sea sediments. By taking the average concentrations of Fe, Ti, Al and Mn in the continental crust, the oceanic crust, and active ridge sediments, they used the relationships Fe/Ti and Al/(Al + Fe + Mn) to estimate the relative proportions of these three components in a given sediment. The results are illustrated in Fig. 34.13, from which it can be seen that, with the exception of an area around Hawaii and the southermost region, the North Pacific is totally dominated by the deposition of terrigenous sediments. In

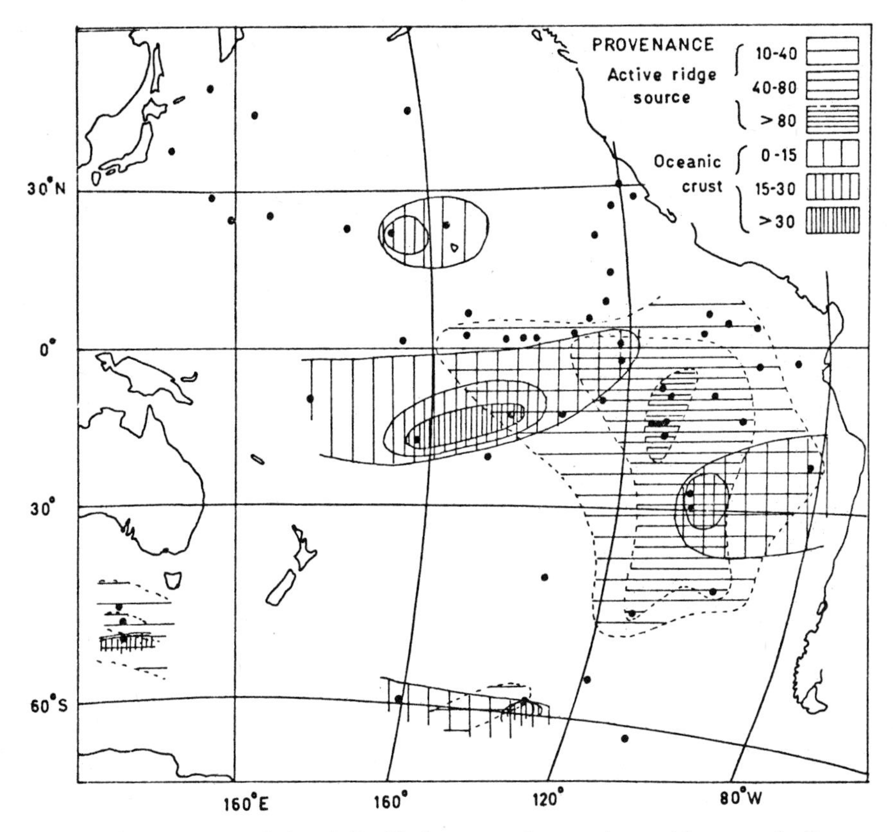

FIG. 34.13. Provenance relations in Pacific deep-sea sediments. Areas without any shading are 90–100% terrigenous—see text (Boström *et al.*, 1973b).

contrast, a large area of the South Pacific (between ~5°N and ~45°S) is influenced by basaltic (oceanic crust) and hydrothermal (active ridge) sources. Basaltic crustal material is most abundant in areas having fairly low rates of sedimentation adjacent to volcanic islands and sea mounts, and hydrothermal sources predominate in the region of the East Pacific Rise. An areal breakdown of the various sediment sources is given in Table 34.40. Table 34.41 presents two alternatives for the provenance relationships for the total open-ocean Pacific, i.e. for those areas >1000 km distant from the nearest major land mass. In Alternative I it is assumed that the sea floor has a continuous sediment cover; in Alternative II it is assumed that in most areas which have oceanic crust, and/or hydrothermal, sediment sources the sediments cover

TABLE 34.40
*Provenance of Pacific Ocean deep-sea sediments**

Range†	Average	Area 10^6 km^2	% of area investigated	Estimated % of the total area of the deep Pacific
A. (% continental influence, i.e. terrigenous source)				
95–100	~100	43·2	44·1	58
80–95	87	21·7	22·0	22
40–80	60	25·3	25·7	15
10–40	25	5·0	5·1	3
0–10	5	3·1	3·1	2
		98·3	100	100
B. (% submarine exhalative influence, i.e. active ocean ridge source)				
80–100	90	1·2	1·2	0·72
40–80	60	11·9	12·1	7·2
10–40	25	16·6	16·8	10
C. (% basaltic influence, i.e. oceanic crust source)				
0–15	7·5	14·0	14·3	8·5
15–30	22	3·5	3·6	2·1
>30	31	1·2	1·2	0·72

* Data from Boström *et al.* (1973b).
† See Fig. 34.15.

only ~10% of the sea floor. However, because even on the East Pacific Rise the sediment cover appears to be more extensive than this, the best provenance estimate probably lies somewhere between the two alternatives.

Two important conclusions may be drawn from the provenance data. (i) In the Pacific Ocean, continental sediment sources constitute ~75% of the total material brought to ocean floor. (ii) Because the Pacific Ocean has a relatively high proportion of areas of low sediment accumulation rates and of active ocean ridges, it is probable that in the World Ocean, in general, oceanic crust and active ridge sources will play an even smaller role in sediment supply than that estimated in Alternative II. The provenance relationships only refer to Al, Ti, Fe and, to a lesser extent, Mn; the latter is not a major constituent of either continental or oceanic crust, and furthermore its relationship with other elements in deep-sea sediments is complicated by remobilization processes. Despite these limitations the provenance relationships do highlight the rather insignificant role played by volcanic sources in the budget of solids brought to the World Ocean in general.

However, in some regions (e.g. those adjacent to volcanic regimes or areas of active sea floor spreading) volcanic and hydrothermal sources are extremely important for the provision of sedimentary material.

TABLE 34.41

Alternatives for the relationships between the various source components in the total mass of Pacific Ocean deep-sea sediments at distances >1000–1500km from the continents

	Alternative I. Assuming a continuous sea floor sediment cover*	Alternative II. Assuming that areas which have oceanic crust and active ridge sources have only a 10% sediment cover†
Continental source (continental crust)	76%	94%
Basaltic source (oceanic crust)	10%	4·0%
Volcanic exhalations (active ridge source)	14%	2·0%

* Data from Boström *et al.* (1975b)—see text.
† See text.

In addition to being associated with the hydrothermal activity around the area of the East Pacific Rise, the elements Fe, Mn, Cu, Ni and Co have relatively high accumulation rates in a belt at the equator—a distribution pattern which is similar to that of opaline silica. Boström *et al.* (1973b) concluded that a significant quantity of Cu, Ni and Co are transported to the sediment surface as a result of biological processes; possibly by the ingestion by filter-feeding organisms of clay particles which carry adsorbed trace elements. Because of the rapid turnovers of material in the biomass and of biologically-derived material at the equator it is probable that this biological-clay "cloud" will remain in suspension for considerable periods, and so may undergo lateral transport. In contrast, the ocean waters over much of the East Pacific Rise can be considered to be oceanic "deserts", and Boström *et al.* (1973b) concluded that much of the Cu, Ni and Co, together with the Fe and Mn which accumulate in the sediments adjacent to the Rise have been brought to the sediment surface by sub-seafloor hydrothermal processes. The problem of the source and transport of the trace elements in the Equatorial and East Pacific Rise regions was considered in more detail by Boström *et al.* (1974) who investigated the elemental compositions of a series of plankton samples from the surface waters (0·5–1·5 m) of the South Pacific. They concluded that under the Equatorial high-productivity region the elemental chemistry of the sediments shows a striking similarity to that derived from a model based on the deposition of a mixture of 75% biomass and 25% continental material (i.e. shale). In contrast, none of the tested models involving admixtures of biomass and continental material could account for the elemental composition of the sediments on the East Pacific Rise. However, the average Fe, Mn and V contents of these sediments could be simulated by an admixture of 75% biomass and 25% of an Fe-Mn-V volcanic phase; the Cu and Ni in the sediments probably have a mixed sea water–biomass–volcanic origin.

Boström *et al.* (1973a) also considered the status of macro-nodules (i.e. $\gtrsim$2 mm in diameter) in the budget of trace elements in the Pacific Ocean sedimentary cycle. From various published data the authors calculated average values for the sea floor coverage, density, growth rates and chemical compositions of macro-nodules and slabs, and estimated the accumulation rates of certain elements both in these components and in adjacent deep-sea sediments. The results are listed in Table 34.42, from which it can be seen that

TABLE 34.42

*The elemental compositions of some ferro-manganese nodules and slabs, and accumulation rates of the elements in these components and in deep-sea sediments**

		2			
Element	1	a	b	c	3
Al	3·2	0·072	0·43	0·078	4·1–18
Ti	0·67	0·015	0·090	0·016	0·19–0·88
Mn	19·3	0·43	2·60	0·47	0·25–0·99
Fe	11·7	0·26	1·58	0·29	2·5–8·7
Ni	0·66	0·015	0·089	0·016	0·011–0·038
Cu	0·39	0·0088	0·098	0·0096	0·012–0·054

1. Average composition of Pacific Ocean nodules and slabs (wt. %) from Bender (1972).
2. Accumulation rates of concretions in mg cm^{-2} (10^3 yrs)$^{-1}$; the surface unit refers to the ocean floor and not to the surface of a nodule which is four times larger than the surface of the ocean floor which it covers; density of concretions is taken as 2·45 g cm^{-3}.
   a = spherical nodules growing all round at a rate of 7 mm/10^6 yrs., and covering 33% of the ocean floor.
   b = spherical nodules growing all round at a rate of 14 mm/10^6 yrs, and covering ≈ 100% of the ocean floor.
   c = oxide slab, growing on top surface only at a rate of 10 mm/10^6 yrs, and covering ≈ 100% of the ocean floor.
3. Common accumulation rates in mg cm^{-2} (10^3 yrs)$^{-1}$ for unconsolidated sediments from areas of slow accumulation in the North Pacific.

* From Boström *et al.* (1973b)

Al, Ti and Fe always accumulate much more rapidly in the sediments than they do in the nodules, and that only under extreme conditions (column b Table 34.42). do the accumulation rates of Mn, Ni and Co in the nodules exceed those in the sediments. From these budget calculations the authors concluded that, in general, macro-nodules and slabs are of limited significance in the oceanic sedimentary budgets of Al, Ti, Fe, Cu, Ni and Co. However, it must be emphasized that these calculations do not take account of micro-nodules which may be dispersed throughout the sediment column.

Boström *et al.* (1974) considered some of the more general implications of simulating the elemental compositions of deep-sea sediments from admixtures of various "end-members", such as sea water, the marine biomass, continental crust, oceanic crust, and active ridge sources. They concluded that despite the obvious limitations of such models the elemental compositions of many deep-sea sediments can largely be described in terms of admixtures of various proportions of continental crust (shale) and the marine biomass. For other, i.e. specialised, sediments additional elemental sources such as active ridge hydrothermal exhalations are required.

One important drawback to the acceptance of these models is that little is known of the fate of elements once they have been brought to, and initially incorporated into, deep-sea sediments. A very large proportion of the organic material brought to the sediment surface is lost from slowly depositing deep-sea sediments by oxidation. During, and after, this process trace elements associated with the organic material may become available for incorporation into other sedimentary phases, e.g. ferro-manganese nodules. It is interesting to note that the biomass-associated trace elements need not all originate from the same source. For example, although the marine biomass incorporates continentally-derived and authigenic solids by filter-feeding, plankton can also remove dissolved trace elements directly from sea water itself. For such elements, therefore, the marine biomass acts as an intermediate transport mechanism but the elements themselves have a sea water, as opposed to a continental origin.

REFERENCES

Abdullah, M. I. and Royle, L. G. (1973). *Nature, Lond.* **238**, 329.

Abdullah, M. I., Dunlop, H. M. and Gardner, D. (1973). *J. Mar. Biol. Assoc. U.K.* **53**, 299.

Amin, B. S., Karkar, D. P. and Ral, D. (1966). *Deep-Sea Res.* **13**, 805.

Angino, E. E. (1966). *Geochim. Cosmochim Acta,* **30**, 939.

Angino, E. E. and Andrews, R. S. (1968). *J. Sediment. Petrology,* **38**, 634.

Arrhenius, G. O. S. (1952). *Rep. Swed. Deep-Sea Exped. 5*,

Arrhenius, G. O. S. (1963). *In* "The Sea" (M. N. Hill, ed.), Vol. 3. Wiley-Interscience, New York, 655 pp.

Arrhenius, G. O. S. and Bonatti, E. (1965). *In* "Progress in Oceanography" (M. Sears, ed.). Amer. Assoc. Advanc. Sci. Washington.
Arrhenius, G. O. S. (1967). *Deep-Sea Sedimentation Report, 14th Gen. Assem. Int. Union Geod. Geophys.*
Arrhenius, G. O. S., Bramlette, M. and Picciotto, E. (1957). *Nature, Lond.* **180**. 85.
Arrhenius, G. O. S., Mero, J. and Korkisch, J. (1964). *Science,* **144**, 170.
Aston, S. R. and Chester, R. (1973). *Estuarine Coastal Mar. Sci.* **1**, 225.
Aston, S. R., Chester, R. and Johnson, L. (1972a). *Nature, Lond.* **235**, 380.
Aston, S. R., Bruty, D., Chester, R. and Riley, J. P. (1972b). *Nature, Lond.* **237**, 125.
Aston, S. R., Chester, R. Griffiths, A. and Riley, J. P. (1972c). *Nature, Lond.* **239**, 393.
Aston, S. R., Chester, R., Johnson, L. R. and Padgham, R. C. (1973). *Mar. Geol.* **14**, 15.
Aston, S. R., Thornton, I., Webb, J. S., Purves, J. B. and Milford, B. L. (1974). *Water Air Soil Pollut.* **3**, 321.
Barker, J. L. and Anders, B. (1968). *Geochim. Cosmochim Acta,* **32**, 627.
Behairy, A. K., Chester, R., Griffiths, A. J., Johnson, L. R. and Stoner, J. H. (1975). *Mar. Geol.* **18**, M45.
Beltagy, A. I., Chester, R. and Padgham, R. C. (1972). *Mar. Geol.* **13**, 297.
Bender, M. L. (1971). *J. Geophys. Res.* **76**, 4212.
Bender, M. L. and Schutz, C. (1969). *Geochim. Cosmochim. Acta,* **33**, 292.
Bender, M. L., Broecker, W. S., Garnitz, V., Middel, U., Kay, R., Sun, S. S. and Biscaye, P. E. (1971). *Earth Planet Sci. Lett.* **12**, 425.
Bezrukov, P. L. (1960). *21st Int. Geol. Congr., Copenhagen 1960, Rep. Soviet Geologisk,* **45**.
Bien, G. S., Contois, D. E. and Thomas, W. H. (1958). *Geochim. Cosmochim. Acta,* **14**, 35.
Biscaye, P. E. (1964). *Geochem. Tech. Rep. 8*, Dept. of Geology, Yale University.
Bloxam, T. W., Aurora, S. N., Leach, L. and Rees, T. R. (1972). *Nature* (*Phys. Sci.*), *Lond.* **239**, 158.
Bogdanov, Yu. A., Kozlova, O. G. and Mukhina, V. V. (1970). *Oceanology* (*USSR*), **10**, 64.
Bonatti, E. (1965). *2nd Int. Oceanogr. Cong. Abstr. Papers,* **55**, Moscow.
Bonatti, E. (1966). *Science,* **153**, 534.
Bonatti, E. and Joensuu, O. (1967). *Abstr. Amer. Geophys. Union, Annual Meeting,* Washington, D.C., p. 17.
Bonatti, E. and Nayudu, Y. R., (1965). *Amer. J. Sci.* **263**, 17.
Bonatti, E., Fisher, D. E., Joensuu, O. and Rydell, H. S. (1971). *Geochim. Cosmochim. Acta,* **35**, 189.
Boström, K. (1967). *In* "Researches in Geochemistry" (P. H. Abelson, ed.). John Wiley, New York, 421 pp.
Boström, K., Joensuu, O. and Brohm, I. (1974). *Chem. Geol.* **14**, 255.
Boström, K. and Peterson, M. N. A. (1966). *Econ. Geol.* **61**, 1258.
Boström, K. and Peterson, M. N. A. (1969). *Mar. Geol.* **7**, 427.
Boström, K. and Fisher, D. E. (1971). *Earth Planet Sci. Lett.* **11**, 95.
Boström, K. and Fisher, D. E. (1969). *Geochim. Cosmochim. Acta,* **33**, 743.
Boström, K., Joensuu, O., Valdes, S. and Riena, M. (1972). *Mar. Geol.* **12**, 85.
Boström, K., Joensuu, O., Moore, C., Boström, B., Dalziel, M. and Horowitz, A. (1973a). *Lithos,* **6**, 159.
Boström, K., Kraemer, T. and Gartner, S. (1973b). *Chem. Geol.* **11**, 123.

Boström, K., Peterson, M. N. A., Joensuu, O. and Fisher, D. E. (1969). *J. Geophys. Res.* **74**, 3261.
Bradley, W. H. and Bramlette, M. N. (1942). *U.S. Geol. Surv., Prof. Paper,* **196**.
Bramlette, M. N. (1961). *In* "Oceanography". *Amer. Ass. Adv. Sci. Publ.* **67**.
Bramlette, M. N. and Brady, W. H. (1940). *U.S. Geol. Surv., Prof. Paper* **196A**, p. 34.
Brujevicz, S. W. (1938). *Dokl. Akad. Nauk SSSR,* **19**, 637.
Bruun, A. F., Langer, E. and Pauly, H. (1955). *Deep-Sea Res.* **2**, 230.
Bruty, D., Chester, R. and Aston, S. R. (1973). *Nature, Lond.* **245**, 73.
Bryan, W. B., Stice, G. D. and Ewart, A. (1972). *J. Geophys. Res.* **77**, 1566.
Burton, J. D. and Liss, P. S. (1968). *Nature, Lond.* **220**, 905.
Butterworth, J., Lesley, P. and Nickless, G. (1972). *Mar. Pollut. Bull.* **3**, 72.
Calvert, S. E. (1968), *Nature, Lond.* **219**, 919.
Calvert, S. E. (1971). *Nature, Lond.* **234**, 133.
Calvert, S. E. and Price, N. B. (1972). *Earth Planet Sci. Lett.* **16**, 245.
Cann, J. R. (1969). *J. Petrology,* **10**, 1.
Carroll, D. (1958). *Geochim. Cosmochim. Acta,* **14**, 1.
Cawse, P. A. and Peerson, D. H. (1972). AERE-R 7134. H.M.S.O., London.
Chester, R. (1965). *In* "Chemical Oceanography", 1st Ed. Vol. 2 (J. P. Riley and G. Skirrow, eds.). Academic Press, London and New York, 508 pp.
Chester, R. (1972a). *In* "The Changing Chemistry of the Oceans" (20th Nobel Symposium) (D. Dyrssen and D. Jagner, eds.). Almqvist and Wiksell, Stockholm, p. 291.
Chester, R. (1972b). *New Scientist*, 706.
Chester, R. and Hughes, M. J. (1969). *Deep-Sea Res.* **13**, 627.
Chester, R. and Hughes, M. J. (1967). *Chem. Geol.* **2**, 249.
Chester, R. and Hughes, M. J. (1969a). *Deep-Sea Res.* **16**, 639.
Chester, R. and Messiha-Hanna, R. G. (1970). *Geochim. Cosmochim. Acta,* **34**, 1121.
Chester, R. and Stoner, J. H. (1972). *Nature, Lond.* **240**, 552.
Chester, R. and Stoner, J. H. (1973). *Nature, Lond.* **246**, 138.
Chester, R. and Stoner, J. H. (1974). *Mar. Chemistry,* **2**, 157.
Chester, R. and Stoner, J. H. (1975). *Nature, Lond.* **255**, 50.
Chester, R. and Stoner, J. H. (1975a) *Mar. Pollut. Bull.* **6**, 92.
Chester, R., Bruty, D. and Aston, S. R. (1976). *In prep.*
Chester, R., Elderfield, H., Griffin, J. J., Johnson, L. R. and Padgham, R. C. (1972). *Mar. Geol.* **13**, 91.
Chester, R., Johnson, L. R. Messiha-Hanna, R. C. and Padgham, R. C. (1973). *Mar. Geol.* **14**, M15.
Chester, R. Stoner, J. H. and Johnson, L. R. (1974). *Nature, Lond.* **249**, 335.
Clarke, F. W. and Wheeler, W. C. (1972). *U.S. Geol. Surv., Prof. Paper,* **124**.
Coonley, L. S., Baker, E. B. and Holland, H. D. (1971). *Chem. Geol.* **7**, 51.
Cook, H. E. (1971). *Abstr. Geol. Soc. Amer.* **3**, 530.
Corliss, J. B. (1971). *J. Geophys. Res.* **76**, 8128.
Correns, C. W. (1937). *Deut. Atlant. Exped. Meteor, Wiss, Ergeb.* **3**, Teil 3.
Correns, C. W. (1941). *Akad. Wiss. Göttingen, Math. Phys., Kl.* **5**, p. 219.
Correns, C. W. (1954). *Deep-Sea Res.* **1**, 78.
Cronan, D. S. (1969). *Geochim. Cosmochim. Acta,* **33**, 1562.
Cronan, D. S. (1972). *Can. J. Earth Sci.* **9**, 319.
Cronan, D. S. (1973). *In* "Initial Reports, Deep-Sea Drilling Project," Vol. XVI, Washington.
Cronan, D. S. (1975). *In* "The Sea" (E. D. Goldberg, ed.), Vol. 5. Wiley-Interscience, New York, p. 491.

Cronan, D. S. and Garrett, D. E. (1973). *Nature (Phys. Sci.), Lond.* **242**, 88.
Cronan, D. S., Van Andel, Tj. J., Heath, G. R., Dinkelman, M. G., Bennett, R. H., Bukry, D., Charleston, S., Kanaps, A., Rodolfo, K. S. and Yeats, R. S. (1972). *Science*, **175**, 61.
Crozier, W. D. (1960). *J. Geophys. Res.* **65**, 2971.
Dasch, E. J., Dymond, J. R. and Heath, G. R. (1971). *Earth Planet Sci. Lett.* **13**, 175.
Davies T. H. and Laughton, A. S. (1972). *In* "Initial Reports, Deep-Sea Drilling Project," Vol. XII, Washington.
Deacon, M. (1971). "Scientists and the Sea." Academic Press, London and New York.
Degens, E. T. and Ross, D. A. (1969). *In* "Hot Brines and Recent Heavy Metal Deposits in the Red Sea." Springer Verlag, Berlin, 600 pp.
Del Monte, M. (1972). *J. Geol.* **38**, 213.
Delany, A. C., Delany Audrey C., Parkin, D. W., Griffin J. J., Goldberg, E. D. and Reinmann, B. E. (1967). *Geochim. Cosmochim. Acta,* **31**, 885.
Dymond, J. J., Corliss, B., Ross Heath, G., Field, C. W., Dasch, E. J. and Neeh, H. H. (1973). *Bull. Geol. Soc. Amer.* **84**, 3355.
Edwards, D. and Goodell, H. (1969). *Mar. Geol.* **7**, 207.
El Wakeel, S. K. and Riley, J. P. (1961). *Geochim. Cosmochim. Acta,* **25**, 110.
Elderfield, H. (1972). *Nature, Lond.* **237**, 110.
Elderfield, H. (1976). *Mar. Chem.* (in press).
Ericson, D. B., Ewing, M. and Heezen, B. C. (1961). *Bull. Geol. Soc. Amer.* **72**, 193.
Engel, H. E. J., Engel, C. and Haven, R. G. (1965). *Bull. Geol. Soc. Amer.* **76**, 719.
Ericson, D. B., Ewing, M. and Heezen, B. C. (1961). *Bull. Geol. Soc. Amer.* **72**, 193.
Ewing, M. and Thorndike, E. M. (1966). *Science,* **147**, 1291.
Ewing, M., Worzel, J. L. and Burk, C. A. (1969). *In* "Initial Reports, Deep-Sea Drilling Project," Vol. I, Washington.
Fleet, A. J. and Kempe, D. R. C. (1974). *In* "Initial Report, Deep-Sea Drilling Project", Vol. XXVI, Washington.
Fox, P. J. (1972). Cited in Hart (1973).
Garrels, R. M. and Mackenzie, F. T. (1971). "Evolution of Sedimentary Rocks". Norton, New York.
Gartner, S. (1970). *Science,* **169**, 1077.
Glasky, G. P. (1973). *Oceanogr. Mar. Biol. Ann. Rev.* **11**, 27.
Glass, B. P. (1972). "Antarctic Oceanology II: The Australian-New Zealand Sector" (D. E. Hayes, ed.). Amer. Geophys. Union, p. 335.
Goldberg, E. D. (1954). *J. Geol.* **62**, 249.
Goldberg, E. D. (1964). *Trans. N.Y. Acad. Sci.* **27**, 7.
Goldberg, E. D. (1967). *In* "Submarine Geology". (F. P. Shepard, ed.). Harper, New York.
Goldberg, E. D. (1968). *Earth Planet Sci. Lett.* **4**, 17.
Goldberg, E. D. and Arrhenius, G. O. S. (1968). *Geochim. Cosmochim. Acta,* **13**, 153.
Goldberg, E. D. and Koide, M. (1962). *Geochim. Cosmochim. Acta,* **26**, 417.
Goldberg, E. D. and Griffin, J. J. (1964). *J. Geophys. Res.* **69**, 4293.
Goldberg, E. D., Koide, M., Griffin, J. J. and Peterson, M. N. A. (1964). *In* "Isotopic and Cosmic Chemistry" (H. Craig, S. Mullen and G. J. Wassenburg, eds.). North-Holland, Amsterdam, p. 211.
Gordeyev, Ye. I. (1963). *Dokl. Akad. Nauk SSSR,* **149**, 181.
Graham, J. (1959). *Science,* **129**, 1428.
Gregor, B. (1968). *Nature, Lond.* **219**, 360.
Griel, J. V. and Robinson, R. J. (1952). *J. Mar. Res.* **11**, 173.

Griffin, J. J. and Goldberg, E. D. (1963). *In* "*The Sea*". Vol 3 (M. N. Hill, ed.). Wiley Interscience, New York, p. 728.
Griffin, J. J., Windom, H. and Goldberg, E. D. (1968). *Deep-Sea Res.* **15**, 433.
Grjebine, T. (1965). *Bull. Inst. Oceanogr., Monaco* **65**, 1336.
Grjebine, T., Lalou, C., Ros, J. and Capilant, M. (1964). *Amer. N.Y. Acad. Sci.* **119**, 143.
Gross, M. G. (1967). *In* "Estuaries" (G. H. Lauff, ed.). Amer. Assoc. Adv. Sci. Publ. No. 82. Washington, D.C., p. 273
Harriss, R. C. (1966). *Nature, Lond.* **212** 275.
Hart, R. A. (1970) *Earth Planet. Sci. Lett.* **9**, 269.
Hart, R. A. (1973). *Can. J. Earth Sci.* **10**, 799.
Harvey, H. W. (1937). *J. Mar. Biol. Ass. U.K.* **22**, 221.
Hathaway, J. C. and Sachs, P. L. (1965). *Amer. Mineral.* **50**, 852.
Heezen, B. C., Nesteroopf, W. D., Oberling, A. and Sabatier, G. (1965). *C. R. Acad. Sci., Paris*, **260**, 5821.
Heier, K. S. and Adams, J. A. S. (1964). *In* "Physics and Chemistry of the Earth" (L. H. Ahrens, F. Press and S. K. Runcorn, eds.), Vol. 5. Pergamon Press, London, p. 255.
Hirst, D. M. (1962). *Geochim. Cosmochim. Acta* **26**, 1147.
Holeman, J. N. (1968). *Water Resour. Res.* **4**, 737.
Horne, D. R., Ewing, M., Horn, B. M. and Delach, M. N. (1972). *Ocean Industry*, (*Jan.* 1972), p. 26.
Horne, M. K. and Adams, J. A. S. (1966). *Geochim. Cosmochim. Acta*, **30**, 279.
Horowitz, A. (1970). *Mar. Geol.* **9**, 241.
Hunter, W. and Parkin, D. W. (1960). *Proc. Roy. Soc., Lond., Ser. A*, **255**, 382.
Jacobs, M. B. and Ewing, M. (1969). *Science*, **163**, 380.
Junge, C. E. (1972). *J. Geophys. Res.* **77**, 5183.
Kharkar, D. P., Turekian, K. K. and Bertine, K. K. (1968). *Geochim. Cosmochim. Acta*, **32**, 285.
Kimo, H., Fisher, R. L. and Nasu, N. (1956). *Deep-Sea Res.* **3**, 126.
Klein, D. H. and Goldberg, E. D. (1971). *Environ. Sci. Technol.* **4**, 756.
Konovalov, B. V. and Ivanova, B. G. (1968). *Oceanology*, **9**, 482.
Krauskopf, K. B. (1956). *Geochim. Cosmochim. Acta*, **9**, 1.
Krishnaswamy, S. and Lal, D. (1972). *In* "The Changing Chemistry of the Oceans" Nobel Symposium 20 (D. Dyrssen and D. Jagner, eds.). Almqvist and Wiksell, Stockholm, p. 307.
Ku, T. L., Broecker, W. S. and Opdyke, N. (1968). *Earth Plant. Sci. Lett.* **4**, 1.
Kuenen, Ph. H. (1950). "Marine Geology". John Wiley, New York.
Laevastu, T. and Mellis, O. (1965). *Trans. Amer. Geophys. Union*, **36**, 385.
Lal, D. and Venkatavaradan, V. S. (1966). *Science*, **151**, 1381.
Lamb, H. H. (1970). *Phil. Trans. Roy. Soc., London*, **226**, 425.
Landergren, S. (1964). *Rep. Swed. Deep Sea Exped.* **10**,
Li, Y. H., Bischoff, J. and Malhieu, G. (1968). *Earth Planet. Sci. Lett.* **7**, 265.
Lisitzin, A. P. (1960). *Oceanogr. Res.* **2**, 71.
Lisitzin, A. P. (1966). *In* "Geochemistry of Silica", (M. N. Shakov, ed.). Publ. House Akad. Nauk SSSR, Moscow.
Lisitzin, A. P. (1972). *Soc. Econ. Paleontol. Mineral, Spec. Publ.* No. 17, p. 218.
Livingstone, D. A. (1963). *Data of Geochemistry, 6th Ed. U.S. Geol. Surv., Prof. Paper* 440G.
Lynn, D. C. and Bonatti, E. (1965). *Mar. Geol.* **3**, 457.

McCorkell, R., Fireman, E. L. and Langway, C. C. (1967). *Science,* **158**, 1690.
McKelvey, B. C. and Fleet, A. J. (1974). *In* Initial Reports, Deep Sea Drilling Project, Vol. XXVI, Washington.
MacKay, D. M., Halcrow, W. and Thornton, I. (1972). *Mar. Pollut. Bull.* **3**, 7.
MacKenzie, F. T. and Garrels, R. M. (1966). *Amer. J. Sci.* **264**, 507.
Manheim, F. T. (1965). *In* "Symposium on Marine Geochemistry". University of Rhode Island, Occ. Publ. No. 3, p. 214.
Manheim, F. T. (1961). *Geochim. Cosmochim. Acta,* **25**, 52.
Martin, J. H. (1970). *Limnol. Oceanogr.* **15**, 756.
Martin, J. H. and Knaver, G. A. (1973). *Geochim. Cosmochim. Acta,* **37**, 1639.
Martin, J. M. and Maybeck, M. (1976). In press.
Mellis, O. (1954). *Deep-Sea Res.* **2**, 89.
Melson, W. G. and Thompson, G. (1971). *Phil. Trans. Roy. Soc., London, A,* **268**, 423.
Merrihue, C. M. (1964). *Trans. N.Y. Acad. Sci.* **119**, 351.
Millard, H. T. and Finkelman, R. B. (1970). *J. Geophys. Res.* **75**, 2125.
Murray, J. and Irvine, R. (1894). *Trans. Roy. Soc., Edinburgh,* **37**, 712.
Murray, J. and Renard, A. F. (1891). *Sci. Rep. Challenger Exped.* **3**.
Natland, J. H. (1973). *In* "Initial Reports, Deep Sea Drilling Project," Vol. XIX Washington.
Nayudu, Y. R. (1964). *Mar. Geol.* **9**, 475.
Needham, H. D. (1962). *Deep-Sea Res.* **9**, 475.
Nicholls, G. D. (1965a). *Phil. Trans. Roy. Soc., London.* **258**, 168.
Nicholls, G. D. (1965b). *Mineral. Mag.* **34**, 373.
Nicholls, G. D., Curl, H. and Bowen, V. T. (1959). *Limnol. Oceanogr.* **4**, 472.
Ninkovich, D., Heezen, B. C., Conally, J. R. and Burdle, L. H. (1964). *Deep-Sea Res.* **2**, 605.
Olafsson, J. (1975). *Nature, Lond.* **255**, 138.
Osterberg, C., Carey, A. G. and Curl, H., *Nature, Lond.* **200**, 276.
Pearcy, W. G. and Osterberg, C. L. (1967). *Oceanol. Limnol.* **1**, 103.
Perkins, E. J., Gilchrist, J. R. S., Abbot, O. J. and Halcrow, W., (1973). *Mar. Pollut. Bull.* **4**, 59.
Peterson, M. N. A. and Goldberg, E. D. (1962). *J. Geophys. Res.* **67**, 3477.
Peterson, M. N. A., Edgar, N. T., van der Borch, C. and Rex R. W. (1970). *In* "Initial Reports, Deep Sea Drilling Project," Vol II, Washington.
Pettersson, H. H. (1945). *Medd. Oceanogr. Inst. Goteberg,* **213**, No. 5, I.
Pettersson, H. H. (1959). *Geochim. Cosmochim. Acta.* **17**, 209.
Piper, D. L. (1973). *Earth Planet. Sci. Lett.* **19**, 75.
Per hac, R. M. (1972). *J. Hydrol.* **15**, 177.
Postma, H. (1967). *In* "Estuaries" (G. H. Lauff, ed.). Amer. Assoc. Advan. Sci. Washington, D.C.
Radczewski, O. E. (1939). *In* "Recent Marine Sediments" (P. O. Trask, ed.). Amer Assoc Petrol. Geol., Tulsa, 496 pp.
Revelle, R. R. (1944). *Carnegie Inst., Washington, Publ.* 556.
Rex, R. W. and Goldberg, E. D. (1958). *Tellus,* **10**, 153.
Riedel, W. R. (1959). *In* "Silica in Sediments". Amer. Assoc. Petrol. Geol., Tulsa, p. 80.
Riley, G. A., Van Hemert, D. and Wangersky, P. J. (1965). *Limnol. Oceanogr.* **10**, 354.
Riley, G. A., Wangersky, P. J. and Van Hemert, D. (1964). *Limnol. Oceanogr.* **9**, 546.
Riley, J. P. and Chester, R. (1971). "Introduction to Marine Chemistry". Academic Press, London and New York, 465 pp.

Schaeffer, O. A., Megrue, G. H. and Thompson, S. O. (1964). *Trans. N.Y. Acad. Sci.* **119**, 347.
Schutz, D. R. and Turekian, K. K. (1965). *Geochim Cosmochim. Acta,* **29**, 259.
Siever, R. and Scott, R. A. (1963). *In* "Organic Geochemistry" (I. A. Breger, ed.). Pergamon Press, Oxford, 579 pp.
Sillén, L. G. (1961). *In* "Oceanography" (M. Sears, ed.). Amer. Assoc. Advan. Sci., Washington, D.C., 549 pp.
Sillén, L. G. (1963). *Sv. Kem. Tidskr.* **75**, 161.
Sillén, L. G. (1965). *Ark. Kemi,* **24**, 431.
Sillén, L. G. (1967). *Science,* **156**, 1189.
Skei, J., Price, N. B., Calvert, S. and Halledahl, H. (1972). *Water, Air Soil Pollut.* **1**, 452.
Smayda, T. J. (1969). *Limnol. Oceanogr.* **14**, 621.
Smayda, T. J. (1971). *Mar. Geol.,* **11**, 105.
Sorokin, Y. I. (1972). *In* "The Changing Chemistry of the Oceans." (Nobel Symposium 20) (D. Dyrssen and D. Jagner, eds.). Almqvist and Wiksell, Stockholm. p. 189.
Spencer, D. W. and Brewer, P. G. (1971). *J. Geophy. Res.* **76**, 5877.
Stetson, H. C. and Upson, J. E. (1937). *J. Sediment. Petrology,* **7**, 55.
Sverdrup, H. V., Johnson, M. W. and Fleming, R. H. (1942). "The Oceans, their Physics, Chemistry and General Biology". Prentice Hall, New York, 1087 pp.
Svirenko, I. P. (1970). *Oceanology,* **10**, 363.
Tanaka, S., Sakamoto, K., Takagi, J. and Tsuchimoto, M. (1968). *Science,* **160**, 1348.
Thompson, G. and Bowen, V. T. (1969). *J. Mar. Res.* **27**, 32.
Thompson, G., Borner, V. T., Nelson, W. G. and Cipelli, R. (1968). *J. Sediment. Petrology,* **38**, 1035.
Tilles, D. (1966). *Science,* **160**, 1968.
Turekian, K. K. (1965a). *In* "Symposium on Marine Chemistry". University Rhode Island, Occ. Pub. No. 3—1965.
Turekian, K. K. (1965b). *In* "Chemical Oceanography". (J. P. Riley, and G. Skirrow, eds.), 1st Ed. Vol. 2, Academic Press, London and New York, 508 pp.
Turekian, K. K. (1967). In "Progress in Oceanography" Vol. 4. (M. Sears, ed.). Pergamon, Press, Oxford.
Turekian, K. K. (1968). *Geochim. Cosmochim. Acta.* **32**, 603.
Turekian, K. K. (1969). *In* "Handbook of Geochemistry". Vol. 1. (K. H. Wedepohl, ed.). Springer-Verlag, Berlin, p. 296.
Turekian, K. K. and Wedepohl, K. H. (1961). *Bull. Geol. Soc. Amer.* **72**, 195.
Turekian, K. K. and Imbrie, J. (1967). *Earth Planet. Sci. Lett.* **1**, 161.
Turekian, K. K. and Scott, M. (1967). Annual Progress Report, Dept. Geology, Yale University. Yale-2912-13.
Turekian, K. K. and Bertine, K. (1971). *Nature, Lond.* **229**, 250.
Turekian, K. K., Katz, A. and Chan, L. (1973). *Limnol. Oceanogr.* **18**, 240.
Van Der Weijden, C. H., Schuiling, R. D. and Das, H. A. (1970). *Mar. Geol.* **9**, 81.
von der Borch, C. C. and Rex, R. W. (1970). *In* "Initial Reports Deep Sea Drilling Project," Vol V. Washington.
von der Borch, C. C., Nesteroff, W. D. and Galehouse, J. S. (1971). *In* "Initial Reports, Deep Sea Drilling Project," Vol VIII, Washington.
von Gumbel, C. W. (1878). *Sitzungsber. Bayer. Akad. Wiss.* **8**, 189.
Wasson, J. T., Alder, B. and Oeschger, H. (1967). *Science,* **155**, 446.
Wedepohl, K. H. (1960). *Geochim. Cosmochim. Acta,* **18**, 200.
Welby, C. W. (1958). *J. Sediment. Petrology,* **28**, 431.
Welder, F. A. (1959). *Tech. Rep. 12, Coastal Studies Inst.,* Baton Rouge, Louisiana.

Whitehouse, V. G. and McCarter, R. S. (1958–59). *Tex. Agr. Exp. Stn. Misc. Publ., Contrib. Oceanogr. Meteorol.* **4**, 193.
Williams, R. M. and Chan, K. S. (1966). *J. Fish. Res. Bd.* **23**, 575.
Winterer, E. L., Riedel, W. R., Moberly, R. M., Resig, J. M., Kroente, L. W. Gealy, E. L., Heath, G. R., Bronniman, P., Martini, E. and Worsley, T. R. (1971). *In* "Initial Reports, Deep Sea Drilling Project, Vol. XII, Washington.
Worzel, J. L. (1959). *Proc. Nat. Acad. Sci. U.S.* **45**, 349.
Wust, G. (1964). *In* "Progress in Oceanography" (M. Sears, ed.). Vol. 2. Pergamon Press, Oxford.
Yabuki, S. and Shima, M. (1972). *Rika Gaku Kenkyusho Hokoku*, **48**, 80.
Yeroshchev-Shak, V. A. (1961). *Dokl. Acad. Nauk SSSR,* **137**, 695.
Young. E. J. (1954). *Bull. Geol. Soc. Amer.* **65**, 1329.

Subject Index

D

E

H